**8**edition

D0147460

ADVENTURES *in* SOCIAL RESEARCH

To our students:
Past, present, and future—
We challenge each other and profit from it.

ADVENTURES *in* SOCIAL RESEARCH

Data Analysis Using IBM® SPSS® Statistics

Earl Babbie
Chapman University

Fred S. Halley
State University of New York, Brockport

William E. Wagner, III
California State University, Channel Islands

Jeanne Zaino
Iona College

Los Angeles | London | New Delhi
Singapore | Washington DC

Los Angeles | London | New Delhi
Singapore | Washington DC

FOR INFORMATION:

SAGE Publications, Inc.
2455 Teller Road
Thousand Oaks, California 91320
E-mail: order@sagepub.com

SAGE Publications Ltd.
1 Oliver's Yard
55 City Road
London EC1Y 1SP
United Kingdom

SAGE Publications India Pvt. Ltd.
B 1/I 1 Mohan Cooperative Industrial Area
Mathura Road, New Delhi 110 044
India

SAGE Publications Asia-Pacific Pte. Ltd.
3 Church Street
#10-04 Samsung Hub
Singapore 049483

Acquisitions Editor: David Repetto
Editorial Assistant: Lauren Johnson
Production Editor: Eric Garner
Copy Editor: Megan Granger
Typesetter: C&M Digitals (P) Ltd.
Proofreader: Barbara Johnson
Indexer: Sheila Bodell
Cover Designer: Glenn Vogel
Marketing Manager: Nicole Elliott
Permissions Editor: Karen Ehrmann

Copyright © 2013 by SAGE Publications, Inc.

Printed in the United States of America

Library of Congress Cataloging-in-Publication Data

Adventures in social research: data analysis using IBM SPSS statistics / Earl Babbie ... [et al.]. — 8th ed.

p. cm.

Includes bibliographical references and index.

ISBN 978-1-4522-0558-8 (pbk.)

1. Social sciences—Statistical methods—Computer programs. 2. SPSS for Windows. I. Babbie, Earl R.

HA32.B286 2013

300.285'555—dc23   2012018542

This book is printed on acid-free paper.

14 15 16 10 9 8 7 6 5 4 3

BRIEF CONTENTS

DETAILED CONTENTS

Part III   Bivariate Analysis   161

Part IV   Multivariate Analysis 311

Part V   The Adventure Continues 375

PREFACE

This workbook is offered to you with a number of aims in mind. To begin, we want to provide students with a practical and hands-on introduction to the logic of social science research, particularly survey research. Moreover, we want to give the students an accessible book that guides them step-by-step through the process of data analysis using current General Social Survey (GSS) data and the latest versions of IBM SPSS Statistics for either Windows or Macintosh computers. Most important, we want to involve students directly in the practice of social research, allow them to experience the excitement and wonder of this enterprise, and inspire them to pursue their own adventure in social research.

As we pursue these goals, however, there are a number of agendas in the background of this book. For example, students who complete the book will have learned a very useful, employable skill. Increasingly, job applicants are asked about their facility with various computer programs: word processing, spreadsheets, and data analysis. As of this writing, SPSS Statistics is still clearly the most popular professional program available for social science data analysis, hence our choice of it as a vehicle for teaching social research.

A Focus on Developing Professional and Intellectual Skills

What sets this book apart from others that teach SPSS Statistics or similar programs is that we cast that particular skill within the context of social research as a logical enterprise.

Thus, in addition to learning to use SPSS Statistics, students are learning the intellectual "skills" of conceptualization, measurement, and association. Even though those who know only SPSS Statistics can assist in data analysis, our intention is that our students will also be able to think for themselves, mapping out analytic paths into the understanding of social data. As they polish these intellectual skills, they should be able to progress to higher levels of research and to the administration of research enterprises.

More generally, we aim to train students who *will use* computers rather than *be used by them*. In our experience, when students first confront computers in school, they tend to fall into two groups: those who recognize computers as powerful instruments for pursuing their goals in life, or at least as the grandest of toys, and those who are intimidated by computers and seek the earliest possible refuge from them. Our intention is to reveal the former possibility to students and to coax them into that relationship with computers.

Educators are being challenged increasingly to demonstrate the practical value of instruction, no less in the social sciences than in other fields. Too often, the overreaction to this demand results in superficial vocational courses that offer no intellectual meaning or courses hastily contrived as a home for current buzzwords, whose popularity is often short-lived. We are excited to be able to offer an educational experience that is genuinely practical for students and that also represents an intellectual adventure.

Those who have taught methods or statistics courses typically find themselves with a daunting task: to ignite their often involuntary students with the fire of enthusiasm they themselves feel for the detective work of social research at its best. In this book, we seek to engage students' curiosity by setting them about the task of understanding issues that are already points of interest for them: topics such as abortion, religion, politics, poverty, gender roles, environment, sexual attitudes, mass media, gun control, child rearing, and others. For many of our readers, we imagine that mathematical analysis still smacks of trains leaving Point A and Point B at different speeds and so on. Now they are going to learn that some facility with the logic and mathematics of social research can let them focus the light of understanding on some of the dark turbulence of opinion and hysteria.

We do not tell students about opinions on abortion as much as we show them how to find out for themselves. We think that will get students to Point C ahead of either of the trains.

A Focus on Active and Collaborative Learning

As we are teaching students to learn for themselves, this book offers a good example of what educators have taken to calling "active learning." We have set up all our demonstrations so that students should be executing the same SPSS Statistics operations we are discussing at any given point. Although we may give them the "answers" to assure them that they are on the right track, we leave them on their own often enough to require that they do the work rather than simply read about it.

Finally, the culture of personal computers has been one of "collaborative learning" from its very beginning. More than people in any other field of activity, perhaps, computer users have always delighted in sharing what they know with others. There is probably no better context within which to ask for help: Those who know the answer are quick to respond, and those who do not often turn their attention to finding an answer, delighting in the challenge.

Because this book is self-contained, even introductory students can walk through the chapters and exercises on their own, without outside assistance. We imagine, however, that students will want to work together as they progress through this book. That has been our experience in student testing and in earlier courses we have taught involving computers. We suggest that you encourage cooperation among students; we are certain that they will learn more that way and will enjoy the course more. In fact, those who are initially intimidated by computers should especially be encouraged to find buddies with whom to work.

Intended for Students in Various Social Science Disciplines

This book is intended for use in any social science course that either introduces or focuses exclusively on social research methods, social statistics, data analysis, or survey research. It can be easily combined with or used as a supplement to most standard social science textbooks, including but not limited to those in fields as varied as communication science, criminal justice, health studies, political science, public policy, social work, and sociology.

As far as possible, we have designed this book to be "self-writing" and "open-ended" to ensure it is relevant to students with varying interests across numerous disciplines. Throughout the text, we encourage students to focus on issues and questions relevant to their particular areas of interest. After walking through the demonstrations that introduce the fundamentals of the data analysis process, students are given a chance to apply what they have learned. In many of the lab exercises, students are encouraged to design their own hypotheses, choose their own variables, and interpret the results. Moreover, we encourage instructors to apply the principles, techniques, and methods discussed to other data sets relevant to their fields.

Intended for Both Beginning and More Advanced Students

We have designed and structured this book to support students at a variety of levels. This includes both those students who are taking their first course in social research and more advanced students (including graduate students) who either want to hone their social research, statistical, and data analysis skills, or merely want to become acquainted or reacquainted with the latest versions of SPSS Statistics for Windows or Macintosh. More advanced students who come at this book full speed may choose either to work through the text from beginning to end or to skip around and focus on particular chapters and sections.

It is important to note, however, that because this book is "self-contained" and guides the student analyst step-by-step through the demonstrations and exercises, no previous experience with social research, statistics, computers, Windows, Macintosh, or SPSS Statistics is required. Those who have never taken a research methods, statistics, or computer-based course will find that they can easily make it through this book.

The Book and the Free Companion Website: What Is Included?

The book and *Adventures in Social Research* companion website provide the instructions and data needed to introduce students to social science data analysis. Most college and university computing services make SPSS Statistics software available to students. A student version of SPSS/PASW Statistics may be purchased through most college and university bookstores (last available up to version 18). The companion website includes a data set containing a total of 150 variables from the 2010 GSS, which can be analyzed by most versions of SPSS Statistics on Windows or Macintosh computers. As you will see, the variables cover a fairly broad terrain, although we have provided for analysis in some depth in a few instances. In addition to working their way through the demonstrations and exercises presented in the book, students will be able to find original lines of inquiry that grow out of their own interests and insights.

This book will illustrate the use of SPSS Statistics, using version 20.0 for Windows and Macintosh. While the text focuses specifically on the current version of SPSS Statistics, it can also be easily used with most of the earlier versions. Regardless of the version you are using, throughout the text we will refer to the program simply as "SPSS Statistics," which is also identical to PASW Statistics. (Versions 17 and 18 were known as PASW as part of a branding shift, before SPSS was acquired by IBM and the name reverted back to SPSS.) SPSS Statistics comes with extensive help screens. They are almost like having a coach built into your computer! Begin with the menu farthest to the right.

You can click **Help** to see the options available to you. "Topics" will usually be your most useful choice. This will give you three options: "Contents," "Index," and "Search." "Contents" and "Index" present you with two ways of zeroing in on the topic of interest to you. "Search" will search for the specific terms or keywords you indicate. You should experiment with these several options to discover what works best for you.

Organization and Content

The chapters are arranged in an order that roughly parallels the organization of most introductory social science research methods texts. Part I (Chapters 1–3) includes an overview of the essentials of social research and a description of the 2010 GSS with which you will be working. Parts II, III, and IV (Chapters 4–20) introduce readers to SPSS Statistics and data analysis, beginning with univariate analysis, then bivariate analysis, and finally multivariate analysis, respectively. Part V (Chapters 21–22) focuses on primary research and additional avenues for secondary research.

Part I includes three chapters that help prepare students for social research using the GSS. Our goal in these chapters is to give students an introduction to some of the fundamental elements of social scientific research, particularly those they will encounter later in the text.

Chapter 1 discusses the main purposes of the text and introduces students to some of the historical background that lies behind computerized social research, data analysis, and statistical software packages. It also introduces readers to the process of social research. Chapter 2 continues this focus by examining the logic of important aspects of social research, including theory, research, and measurement. Chapter 3 is designed to introduce readers to the GSS and the data sets included with this book.

Part II is designed to help students "get started" using SPSS Statistics and to give them practice in the basics of univariate analysis. Chapter 4 introduces students to SPSS Statistics by guiding them through the steps involved in launching the program, opening their data sets, and exploring the variables contained on the disk that accompanies this book.

The rest of this section (Chapters 5–9) is devoted to data analysis. In Chapter 5, we introduce frequency distributions, descriptive statistics, and saving and printing data. Chapter 6 focuses on the graphic presentation of univariate data by covering the commands for creating bar and pie charts, line graphs, and histograms. In Chapters 7 and 8, students are given a chance to practice recoding and creating composite variables. The final chapter in this section (Chapter 9) allows students to strike out on their own and apply the methods and techniques they have learned in this section to other topics.

Part III focuses primarily on bivariate analyses. In Chapters 10 through 12, we limit our discussion to the analysis of percentage tables. In Chapters 13, 14, and 15, we introduce other methods for examining the extent to which two variables are related to each other.

Chapter 13 focuses on some common measures of association for nominal and ordinal variables, including lambda and gamma. Chapter 14 focuses on correlation and regression analysis, by exploring Pearson's r, r^2, and OLS (ordinary least squares) regression. Chapter 15 introduces tests of statistical significance, such as chi-square, t tests, and ANOVA. In Chapter 16, students are once again given a chance to apply the bivariate techniques and methods discussed in Part III to other topics and issues.

Our discussion of data analysis concludes in Part IV with a discussion of multivariate analyses. Chapter 17 focuses primarily on multiple causation. Chapter 18 picks up on some of the loose threads of our bivariate analyses and pursues them further, while Chapter 19 guides students through the steps involved in creating composite measures to predict opinions. Finally, in Chapter 20, students are given a chance to apply the methods and techniques discussed in Part IV to other topics and issues.

The final section comprises two chapters that explore some further opportunities for social research. Because students often express an interest in collecting their own data, Chapter 21 focuses on primary research. We introduce students to the steps involved in designing and administering a survey, defining and entering data in SPSS Statistics, and writing a research report. This chapter is supplemented by several articles on the *Adventures in Social Research* companion website that are intended to give students additional information regarding preparing a research proposal, designing and administering a survey, constructing a sample questionnaire, and writing a research report. Chapter 22 suggests other avenues for pursuing secondary social research by focusing on the unabridged GSS, additional data sources, and other statistical software packages that students may find useful.

The text contains a codebook (Appendix A) that describes all the variables in the data files used in this book. The full GSS Codebook is included on the companion website. A sample questionnaire designed for student use (with instructions in Chapter 21) is provided in Appendix B. We have also updated and expanded the references, index, and glossary.

The *Adventures in Social Research* companion website contains not only our data set (Adventures.SAV) but also articles relevant to the discussion of primary research in Chapter 21, a comprehensive list of all the SPSS Statistics Commands introduced in the text, and recommended readings that relate to topics and issues covered in the text.

Structure of Each Chapter

Chapters include explanations of basic research principles and techniques; specific instructions regarding how to use SPSS Statistics; and demonstrations, writing boxes, a brief conclusion, a list of main points, key terms, SPSS Statistics commands introduced in the chapter, review questions, and SPSS Statistics lab exercises. Students are expected to follow along with the demonstrations in the body of each chapter. They are aided in this process by both the text, which walks them step-by-step through the process of data analysis, and screen captures, which help them understand what they should be seeing on their own monitor.

In an effort to stress the importance of describing research findings in prose, most chapters include writing boxes, which give readers an example of how a professional social scientist might describe the findings being discussed. The review questions at the end of each chapter are designed to test the students' knowledge of the material presented in the text. Because they do not require SPSS Statistics, they can be assigned as either class work or homework assignments.

In the SPSS Statistics lab exercises, students are given a chance to apply what they learned in the explanatory sections and demonstrations. These exercises generally follow a fill-in-the-blank format for presenting, analyzing, and summarizing results. Instructors may wish to assign these exercises as lab assignments to be completed either in the lab or as homework, provided students have access to SPSS Statistics, either the full or the student version, on Windows or Macintosh.

Although the book is designed to guide students through the process of computerized data analysis from beginning to end, we encourage instructors, and particularly more advanced students, to skip around and focus on chapters and sections that are of interest to them. We designed the book with the understanding that students at various levels may find different demonstrations, techniques, discussions, and methods of varying interest. Consequently, all the chapters are self-contained, and both students and instructors should feel comfortable picking and choosing among topics, issues, and material of particular interest to them.

Instructors and students who choose to take this approach may want to refer to the Table of Contents, introductions to each part, chapter conclusions, and summaries of main points to get a better sense of what sections and chapters they want to focus on.

Software Support and Service

If you or your students should run into any problems using this package, there are several sources of support that should serve your needs. Frequently, college and university computing centers have student assistants who are very helpful to new computer users. In fact, most academic computing centers employ a user services coordinator who can help faculty plan student use of the school's computers and provide aid when problems arise.

One source of SPSS Statistics assistance available via the Internet is a home page (http://www.spss.com) maintained by SPSS, an IBM company as of 2009. In addition to providing answers to frequently asked questions, it offers a variety of tips and white papers on important issues in data analysis. Specific questions may be submitted to consultants via e-mail from the home page. SPSS Statistics requests that a legitimate license or serial number be submitted with questions in order to receive a response to those questions.

For technical support, you can also call SPSS at (312) 329–2400, the number for their offices in Willis Tower (formerly the Sears Tower) in Chicago. Be forewarned that SPSS cannot give assistance with pedagogical or substantive problems and that you may have a long wait in a telephone queue for your turn to talk to a technical support person. It has been our experience that the best help comes from local resources.

Acknowledgments

In conclusion, we would like to acknowledge a number of people who have been instrumental in making this book a reality. First, Jerry Westby, of SAGE Publications/Pine Forge Press. Our thanks also go to former Pine Forge Press editor Steve Rutter and the many others at Pine Forge Press/SAGE Publications who aided us along the way, particularly Claudia Hoffman, Kristin Snow, and Vonessa Vondera for their work on the fifth edition, and Denise Santoyo, Cheryl Rivard, and Kim Suarez for their work on the sixth edition. Special thanks go also to Megan Granger and Nichole O'Grady for their work on the seventh edition, and again to Megan Granger for her work on the current edition.

We would also like to thank the many reviewers who helped us along the way: Monika Ardelt, University of Florida; Dhruba J. Bora, Wheeling Jesuit University; Tom Buchanan, The University of Tennessee at Chattanooga; James Cassell, Indiana State University; Christina DeJong, Michigan State University; Xiaogang Deng, University of Massachusetts, Boston; Stephen J. Farnsworth, Mary Washington College; Ellen Granberg, Clemson University; Jessica Greene, University of Oregon; John R. Hagen, Health Strategies, Inc.; Jade Hsuan Huang, Hawaii Pacific University; Steven L. Jones, University of Virginia; Quentin Kidd, Christopher Newport University; Wanda Kosinski, Ramapo College of New Jersey; Mary Jane Kuffner Hirt, Indiana University of Pennsylvania; Peter Allen Lee, San Jose State University; Kim Lloyd, Washington State University; William H. Lockhart, University of Virginia; Karyn D. McKinney, Penn State University, Altoona College; Maureen McLeod, The Sage Colleges; Josh Meisel, Humboldt State University; Philip Meyer, University of North Carolina; David B. Miller, Case Western Reserve University; Ojmarrh Mitchell, University of Nevada, Las Vegas; Melanie Moore, University of Northern Colorado; Thomas O'Rourke, University of Illinois at Urbana-Champaign; Graham Ousey, University of Delaware; Linda Owens, University of Illinois; Boyka Stefanova, University of Texas, San Antonio; Sue Strickler, Eastern New Mexico University; Mangala Subramaniam, Purdue University; Jennifer L. S. Teller, Kent State University; Lanny Thomson, University of Puerto Rico; John Tinker, California State University, Fresno; Madine VanderPlaat, Saint Mary's University; Roland Wagner, San Jose State University; Assata Zerai, Syracuse University; and Thomas G. Zullo, University of Pittsburgh.

We reserve our final acknowledgment for our students, to whom this book is dedicated. We recognize that we have often asked them to think and do things that they sometimes felt were beyond their abilities. We have admired their courage for trying anyway, and we have shared in their growth.

ABOUT THE AUTHORS

Earl Babbie was born in Detroit, Michigan, in 1938, but his family chose to return to Vermont 3 months later, and he grew up there and in New Hampshire. In 1956, he set off for Harvard Yard, where he spent the next 4 years learning more than he initially planned. After 3 years with the U.S. Marine Corps, mostly in Asia, he began graduate studies at the University of California, Berkeley. He received his PhD from Berkeley in 1969. He taught sociology at the University of Hawaii from 1968 through 1979, took time off from teaching and research to write full time for 8 years, and then joined the faculty at Chapman University in Southern California in 1987. Although he is the author of several research articles and monographs, he is best known for the many texts he has written, which have been widely adopted in colleges throughout the United States and the world. He also has been active in the American Sociological Association for 25 years and currently serves on the ASA's executive committee. He is also past president of the Pacific Sociological Association and California Sociological Association.

Fred S. Halley, Associate Professor Emeritus, State University of New York, Brockport, received his bachelor's degree in sociology and philosophy from Ashland College and his master's and doctorate degrees from Case Western Reserve University and the University of Missouri, respectively. Since 1970, he has worked to bring both instructional and research computer applications into the undergraduate sociology curriculum. Halley has been recognized for his leadership in the instructional computing sections of the Eastern and Midwest Sociological Societies and the American Sociological Association. At Brockport, he served as a collegewide social science computing consultant and directed Brockport's Institute for Social Science Research and the College's Data Analysis Laboratory. Off campus, Halley directed and consulted on diverse community research projects used to establish urban magnet schools, evaluate a Head Start family service center, locate an expressway, and design a public transportation system for a rural county. Now residing in Rochester, New York, he plays an active role in a faith-based mentoring program for ex-offenders, and he volunteers for Micrecycle, an organization that refurbishes computers for use by those on the other side of the computer divide in schools, day cares, youth centers, and other community organizations.

William E. Wagner, III, professor of sociology at California State University, Channel Islands, served as a member of the faculty and director of the Institute for Social and Community Research at CSU Bakersfield prior to coming to CSU Channel Islands. His MA and PhD degrees in sociology are from the University of Illinois, Chicago. He holds two separate bachelor's degrees, one in mathematics and the other in sociology/anthropology, both from St. Mary's College of Maryland. His work on topics such as urban sociology, sports, homophobia, academic status, and sexual behavior has been published in national and regional scholarly journals.

Jeanne Zaino, associate professor of political science, Iona College, earned a bachelor's degree in political science and a master's degree in survey research at the University of Connecticut, Storrs. During that time, she worked as a research assistant at the Roper Center for Public Opinion Research. She went on to earn a master's degree and PhD in political science from the University of Massachusetts, Amherst. She is currently chair of the Political Science Department at Iona College in New Rochelle, New York, where she teaches courses in American government, institutions, research methods, social statistics, public opinion, scope, and methods. She and her husband, Jeff, are the proud parents of two sons, Maxim and Logan.

Part I **Preparing for Data Analysis**

Chapter 1 Introduction: The Theory and Practice of Social Research

Chapter 2 The Logic of Measurement

Chapter 3 Description of Data Sets: The General Social Survey

In the opening chapters, we introduce you to computerized data analysis, the logic of social science research, and the data you will be using throughout this text.

Chapter 1 begins by looking at "how and why" social scientists use computers and computer programs. Then we introduce you to the two main pillars of social research: logic and observation. You will see how theory (the *logic* component) informs our investigations, making sense out of our *observations,* and sometimes offers predictions about what we will find. The other aspect of research on which we will focus in this book is the collection and analysis of data, such as those collected in a survey.

Chapter 2 delves more deeply into one central component of scientific inquiry: measurement. We look at some of the criteria for measurement quality and start examining the kinds of measurements represented by the data at hand.

In Chapter 3, we introduce you to the data you will be using throughout the text. The data come from the 2010 General Social Survey (GSS). The GSS reflects the attitudes of a representative sample of American adults on a variety of issues from religion, politics, abortion, and child rearing to the mass media, sex, the police, and immigration.

Chapter 1 **Introduction**

The Theory and Practice of Social Research

Social research is the detective work of big questions. Whereas a conventional detective tries to find out who committed a specific crime, the social researcher looks for the causes of crime in general. The logic of social scientific investigation extends beyond crime to include all aspects of social life, such as careers, marriage and family, voting, health, prejudice, environment, and poverty. In fact, anything that is likely to concern you as an individual is the subject of social science research.

Overview

The purpose of this book is to lead you through a series of investigative adventures in social research. We can't predict exactly where these adventures will lead, because you are going to be the detective. Our purpose is to show you some simple tools (and some truly amazing ones) that you can use in social investigations.

We'll also provide you with a body of data, collected in a national survey, that is so rich you will have the opportunity to undertake investigations that no one else has ever pursued.

If you have access to a computer that uses XP, Vista, or 7 (also Mac OS X 10.6 or higher), and IBM SPSS (or PASW) Statistics, this book and the website associated with it (http://www.sagepub.com/babbie8estudy/) contain everything you need for a wide range of social investigations.[1] This tool is designed specifically for exploring data. If you are already comfortable with computers, you can jump right in, and you will quickly find yourself in the midst of a fascinating computer game. Instead of fighting off alien attacks or escaping from dank dungeons, you'll be pitting your abilities and imagination against real life, but you'll be looking at a side of life that you may not now be aware of. This tool is also well designed for the creation of college term papers.

Throughout the book, we suggest ways to present the data you discover in the context of a typical term paper in the social sciences. Whereas most students are limited in their term

---

[1]Earlier versions of SPSS Statistics for Windows or Macintosh (10.0 or higher) may be used, but some of the instructions, procedures, and screens may be somewhat different from those in the book. SPSS 15.0 and higher require Windows 2000, XP, Vista, or 7. PASW Statistics 18 requires Windows 7, Vista, XP, or Mac OS X Leopard (10.5) or Snow Leopard (10.6), while SPSS 20 requires the Mac OS to be at least 10.6 (Snow Leopard) or 10.7 (Lion).

papers to reporting what other investigators have learned about society, you will be able to offer your own insights and discoveries.

Finally, the data set included here is being analyzed by professional social scientists today. Moreover, the analytical tools that we've provided for you are as powerful as those used by many professional researchers. Frankly, there's no reason you can't use these materials for original research worthy of publication in a research journal. All it takes is curiosity, imagination, practice, and a healthy obsession with knowing the answers to things. In our experience, what sets professional researchers apart from others is that they have much greater curiosity about the world around them, are able to bring powerful imagination to bear on understanding it, are willing to put in the time required of effective investigation, and are passionately driven to understand it.

Why Use a Computer?

Like physical scientists, social scientists use observations from the empirical world to develop and evaluate theory. Much of the social scientist's work involves ascertaining whether logically derived relationships expressed in social theories correspond to empirically observed relationships in social data. For instance, a theory may suggest that Catholics are more opposed to abortion than are non-Catholics, but we don't have scientific evidence until we poll Catholics and non-Catholics and evaluate their differences on abortion. To have confidence in our findings, we must poll a large number of people for their positions on abortion and their religious preferences. And for sure, we probably would feel any explanation of difference was incomplete until other factors such as each respondent's gender, age, social class, and so forth were included. We don't have to stretch our imaginations very far to realize that even simple research can soon generate a large mass of data given the number of cases and variables we need to provide credible evidence for or against a theory.

The sheer number of observations commonly made by social scientists requires a computer to make research doable. The full 2010 General Social Survey (GSS) data file we've included with this book, for example, contains more than 790 variables for more than 2,000 individuals. Initial analysis of data requires that they be sorted, categorized, and recategorized before statistics may be computed for them. With more than 1.5 million data points, a computer is clearly necessary for us to complete a meaningful analysis this semester!

SPSS Statistics

Today, the two statistical packages most widely used by social scientists are *SPSS Statistics* (originally known as the Statistical Package for the Social Sciences), temporarily branded as *PASW Statistics* (Predictive Analytics Software) for version 18, and *SAS (Statistical Analysis System).*[2] We have selected SPSS Statistics for three reasons. First, early versions of SPSS date back to 1968. The package is well known, and hardly any social scientists who earned a graduate degree in the past 40 years have not had some contact with SPSS. Second, SPSS takes you through all the basic issues of using a statistical package. This knowledge will give you a head start if you learn some other package later.

---

[2]While SPSS originally stood for "Statistical Package for the Social Sciences" (and the package is still most commonly referred to in this way), SPSS Inc. recently "updated the meaning of the letters to more accurately reflect the company and its products. Today, SPSS stands for 'Statistical Product and Service Solutions.'" The PASW acronym stands for "Predictive Analytics Software," and after SPSS Inc. became part of IBM in 2009, it was determined that the brand would then become IBM SPSS Statistics, referred to as SPSS Statistics for short.

Finally, SPSS Statistics is suitable for computers running Microsoft Windows 98, 2000, ME, XP, Vista, or 7. SPSS version 20 for Windows does require Microsoft Windows XP, Vista, or 7. There are also versions available for Apple Macintosh computers running Mac OS. SPSS Statistics version 20 for Mac does require OS X 10.6 (Snow Leopard) or above.

The SPSS Statistics Base, like a car, is sold as the basic package. Then, if the buyer wishes, it can be "souped-up" with powerful statistical accessories, all of which are beyond the scope of this book. Other upgraded "models" include SPSS Professional and the ultimate model, SPSS Premium.

SPSS also offers two packages specifically designed for students: SPSS/PASW Statistics Student Version (last offered for version 18 of PASW Statistics) and the SPSS/PASW Statistics Graduate Pack. Although both versions are available for use with Windows or Macintosh, they differ in terms of their capabilities. Unlike the SPSS Statistics Base system or the SPSS Statistics Graduate Pack, the *SPSS Statistics Student Version* is limited to 50 variables and 1,500 cases and cannot be upgraded. IBM has not authorized the student version since version 18. The student version has fewer statistical procedures but has most of the procedures that will ever be needed by an undergraduate social science major or master's-level graduate student.

The SPSS Statistics Graduate Pack contains the SPSS Statistics Base system plus two advanced statistical modules. The graduate pack is commonly available at college and university bookstores. You can learn more about IBM SPSS Statistics and the various versions available by visiting the company's online store at http://www.spss.com.[3] IBM may change the configurations or offerings of these packages or modules in the future, so be sure to check the website for the latest details.

While having a copy of SPSS Statistics on your personal computer is convenient, you may not need to purchase the software to complete the exercises in this book. Most colleges and universities offer SPSS Statistics to their students, either through all the computers connected to the school's network or in collegewide or departmental computing labs. More recently, client/server desktop application technology has been adopted at some colleges and universities. With this technology, your school can provide you with client software (e.g., Citrix) and settings to put on your personal Windows or Macintosh computer. You will then be able to run the university's version of SPSS Statistics from their server through your computer desktop application as though it were directly installed on your computer. Of course, you will need to remain connected to the Internet, and a high-speed connection is required for this software to be useful. Check with your school's IT (information technology) office to see if they have this capability.

Social Research: A Primer

This book addresses the techniques of social science data analysis. Thus, we're going to be spending most of our time using SPSS Statistics to analyze data and reach conclusions about the people who answered questions in the GSS, described in more detail in Chapter 3.

Data analysis, however, doesn't occur in a vacuum. Scientific inquiry is a matter of both observing and reasoning. Consequently, before focusing on SPSS Statistics, let's take a few minutes to consider some of the central components of social science research. We will begin here by looking at the role of theory in conjunction with the social research process. In the following chapter, we will turn our attention to another fundamental aspect of scientific inquiry—measurement. The goal is not to make you an expert in the social research process

---

[3]Throughout the book, we suggest various websites you may find useful. Keep in mind, however, that the World Wide Web is constantly changing. For this reason, some websites referred to may no longer be available; online content can change overnight. If particular websites are no longer available, you can use one of the many available search engines to locate the information you need.

but, rather, to give you the background necessary to master the techniques of data analysis presented in the remainder of the book.[4]

Theories and Concepts: Deprivation Theory

Given the variety of topics examined in social science research, no single, established set of procedures is always followed in social scientific inquiry.

Nevertheless, data analysis almost always has a bigger purpose than the simple manipulation of numbers. Our larger aim is to learn something of general value about human social behavior. This commitment lies in the realm of theory. A primary goal of social scientific research is to develop theories that help us explain, understand, and make sense of the social world.

A *theory* is a statement or set of statements describing the relationships among concepts. Theories provide explanations about the patterns we find in human social life. American social research has consistently shown, for example, that women are more religious than men. The key concepts in that observed pattern are religiosity and gender. We'll examine a theory that explains the pattern in a moment. Because concepts are the building blocks of theories, it is important that we focus briefly on what they are.

Concepts are general ideas or understandings that form the basis of social scientific research. Some of the social scientific concepts with which you are familiar might include social class, deviance, political orientations, prejudice, and alienation. The most useful concepts describe variations among people or groups. When thinking about social class, for example, we might distinguish upper class, middle class, and working class, while the concept of prejudice leads us to consider those who are more prejudiced and those who are less prejudiced.

Developing social theories is a matter of discovering concepts that are causally related to one another. We may ask questions such as "Does education reduce prejudice?" "Does gender affect how much people are paid?" or "Are minority group members more liberal than majority group members?" Because one of the subjects we are going to examine in this textbook is religiosity, we will begin with an example of a theory deriving from the sociology of religion.

The sociologists Glock, Ringer, and Babbie developed what they call the "deprivation theory of church involvement." Having asked why some church members participated more in their churches than others, the researchers' analyses led them to conclude that those who were deprived of gratification (e.g., money, prestige, power, opportunities, freedom) in the secular society would be more likely to be active in church life than would those who enjoyed the rewards of secular society. In this case, the concepts under examination are deprivation, gratification, and church involvement. Some people are more deprived of gratification than others; some people are more religiously involved than others. The research question is to find out if the degree to which people are deprived is somehow related to their degree of religious involvement.

Deprivation theory offers a plausible explanation as to how the concepts of deprivation and religious involvement are related. It gives us a possible explanation—a theory—to help us make sense of why some people are more religious or more active in church than others. In this form, however, the concepts are too general to test the theory empirically. Before a theory can be tested, another step has to be taken; namely, we need to create hypotheses. Unlike theories, well-developed hypotheses pose relationships between variables that are specific enough to permit testing.

In short, while theory is an important starting point in social science research, the empirical relationships predicted by the theory must be tested. To do that, we shift our

---

[4]If you are thinking about designing a research study or just want to learn more about the process and practice of scientific inquiry, you may find the discussion in the last two chapters and accompanying appendices a useful starting point. You may also want to browse through the Reference section on the student study website (http://www.sagepub.com/babbie8estudy/) for citations to texts that focus on the nature of social scientific inquiry, designing a research project, and other important aspects of the research process.

focus from relationships between concepts to relationships between variables, and from theories to hypotheses.

Hypotheses and Variables: Religiosity

A *hypothesis* is a statement of expectation derived from a theory that proposes a relationship between two or more variables. Specifically, a hypothesis is a tentative statement that proposes that variation in one variable "causes" or leads to variation in the other variable. We put *cause* in quotes here because more than a simple association is needed to attribute cause. Among other requirements, to be a cause, a related variable must also precede the dependent variable in time and not be related to some other variable that is also related to the dependent variable. We'll explore this further in a later chapter.

Table 1.1 illustrates the differences between theories and hypotheses. Theories specify relationships between concepts in the world of ideas, while hypotheses specify expected relationships between variables in the world of empirical experiences. *Variables* are empirical indicators of the concepts we are researching. Variables, as their name implies, have the ability to take on two or more values. For instance, people can be classified in terms of their gender (male or female) or religious involvement (involved or not involved). When we identify empirical indicators for our concepts, they become variables.

Table 1.1   Theories, Concepts, Hypotheses, and Variables

World of Ideas	Concepts:
	1. Secular deprivation
	2. Religious involvement
	Theory: The more people experience secular deprivation, the more likely they will be religiously involved.
	Variables representing dimensions of secular deprivation:
	a. Age
	b. Gender
	c. Socioeconomic status
World of Experiences	Hypotheses:
	a. As people get older, their religious participation increases.
	Independent variable: Age
	Dependent variable: Religiosity
	b. Women will have greater religious participation than will men.
	Independent variable: Gender/sex
	Dependent variable: Religiosity
	c. The lower your income, the more likely you will be to participate in religious activities.
	Independent variable: Income
	Dependent variable: Religiosity

As Table 1.2 demonstrates, each variable contains two or more *categories*. The categories of the variable "gender," for instance, are "male" and "female," while the categories of the variable "social class" may be "upper class," "middle class," and "working class."

Table 1.2   Variables and Categories

Variable	*Category*
Gender	Female
	Male
Religious involvement	Involved
	Not involved
Party identification	Democrat
	Independent
	Republican
Social class	Upper class
	Middle class
	Working class

The categories of each variable must meet two requirements: They should be both exhaustive and mutually exclusive. By *exhaustive*, we mean that the categories of each variable must be comprehensive enough that it is possible to categorize every observation. Imagine, for instance, that you are conducting a survey and one of your variables is "religious affiliation." In order to measure respondents' religion, you devise a question that asks respondents simply, "What is your religion?" Let's say you give respondents only three choices: Protestant, Catholic, and Jewish. While most Americans would identify with one of these religious traditions, the categories certainly are not exhaustive. Muslims and Hindus, among others, would not find categories descriptive of their traditions. To correct this problem, we would have to add traditions, add an "other" category, or add both so that all respondents could fit themselves into at least one category. Moreover, we'd want a "none" category for those with no religious affiliation.

Exclusive refers to the requirement that every observation fit into only one category. For instance, if we asked people for their religious affiliation and gave them the choices of Christian, Protestant, Catholic, and Jewish, the categories would not be mutually exclusive. Both Protestants and Catholics would see themselves as also being in the "Christian" category.

Looking back at the deprivation theory, then, you may recall that it was expected that poor people would be more active in the church than would rich people, given that the former would be denied many secular gratifications enjoyed by the latter. Or in a male-dominated society, it was suggested that because women are denied gratifications enjoyed by men, women would be more likely to participate actively in the church. Similarly, in a youth-oriented society, the theory would suggest that older people would be more active in church than would the young.

As Table 1.1 illustrates, one hypothesis that can be derived from the deprivation theory is that women will be more involved in church than will men. This hypothesis proposes a relationship between two variables: a dependent and an independent variable. A *dependent*

variable is the variable you are trying to explain (in this case, church involvement/religiosity), while an *independent variable* is the variable hypothesized to "cause," lead to, or explain variation in another variable (in this case, gender). Again, please note we put the word *cause* in quotes purposefully. It is a word that must be used with caution, because more than a simple association is needed to attribute causation.

Also, please note that relationships such as the one predicted in the hypotheses in Table 1.1 are *probabilistic.* The hypothesis says that women, as a group, will have a higher average level of church involvement than men will as a group. This does not mean that all women are more involved than any men. It does mean, for example, that if we asked men and women whether they attend church every week, a higher percentage of women than of men would say yes, even though some men would say yes and some women would say no. That is the nature of probabilistic relationships.

Social Research Strategies: Inductive and Deductive

After developing a hypothesis, a researcher may decide to design and conduct a scientific study to test whether there is a relationship such as the one proposed between gender and church involvement. Social scientists generally approach research in one of two ways: inductively or deductively.

In the study by Glock, Ringer, and Babbie mentioned previously, the researchers employed an *inductive research* strategy. First, they collected data regarding people's religious involvement and gender. After they completed their observations, they examined the data and constructed a theory to explain the relationships found among the variables.

An alternative and somewhat more common approach is *deductive research.* Unlike inductive research, which begins with data collection, deductive research begins with social theory. A specific hypothesis is then deduced from the theory and tested to discover whether there is evidence to support it. Most generally, the deprivation theory suggests that people who lack secular gratification will be more involved in religious activities. From that, we could derive the hypothesis that persons of lower socioeconomic status will attend church more often than will those of higher socioeconomic status. We could then collect data about people's socioeconomic status and church attendance. The data could then be examined to see if lower-status people really did attend church more than higher-status people. This would be considered deductive research because we began with the theory and tested a hypothesis with data.

Perhaps the simplest way to distinguish between inductive and deductive research approaches is by where they begin. While inductive research begins with data analysis and then moves to theory, deductive research begins with theory and then proceeds to data analysis and back to theory again. More simply, deduction can be seen as reasoning from general understandings to specific expectations, whereas induction can be seen as reasoning from specific observations to general explanations.

You can see, then, that there are many steps or stages in the social research process. When conducting deductive research, social scientists proceed from the general (theory) to the specific (data collection) and back to theory again:

1. Theory

2. *Deduce* hypotheses to test theory

3. Collect data

4. Analyze data

5. Evaluate hypotheses

In the case of inductive research, researchers move from the specific (data collection) to the general (theory):

1. Collect data

2. Analyze data

3. *Induce* a theory to account for data

While in practice the process of social research is not nearly as linear as these steps suggest, you can see that whether a researcher employs a deductive or inductive strategy, the goal is always the same: to develop theories that help us explain, make sense of, and understand human social behavior.

The possible topics for exploration are, as you can imagine, endless. Whereas some social researchers are interested in understanding religiosity, others are interested in issues such as spousal abuse, child abuse, violence in schools, unemployment, political party identification, poverty, alcoholism, drug addiction, health care, crime, starvation, overpopulation, governmental corruption, and so on. The problems and issues of concern to social scientists are as manifold and complicated as human beings themselves. Despite the diversity in questions and concerns, what connects social scientists is the belief that if used properly, the techniques and process of social science research can help us examine and begin to understand these complicated issues. Only when we understand what causes these problems, how they come about, and why they persist will we be able to solve them.

While the primary focus of this book is on one stage in the social research process, data analysis, we hope you take some time to reflect on which of the many problems in contemporary life interest you. What issues or questions are you passionate about? What social problems or issues would you like to examine, understand, and potentially address?

Theory and Research in Practice

Now that we have focused a little on the relationship between theory and the social research process, let's examine some of the theoretical work that informs two of the many subjects we are going to analyze together in this book: political orientations and attitudes toward abortion.

Example 1: Political Orientations

One of the more familiar variables in social science is political orientation, which typically ranges from liberal to conservative. Political orientation lies at the heart of much voting behavior, and it also relates to a number of nonpolitical variables, which you are going to discover for yourself shortly.

There are several *dimensions* of political orientations, and it will be useful to distinguish them here. Three commonly examined dimensions are (1) social attitudes, (2) economic attitudes, and (3) foreign policy attitudes. Let's examine each dimension briefly.

Some specific social attitudes and related behaviors might include abortion, premarital sex, and capital punishment. Let's see where liberals and conservatives generally stand on these issues:

Issue	Liberals	Conservatives
Abortion	Permissive	Restrictive
Premarital sex	Permissive	Restrictive
Capital punishment	Opposed	In favor of

In terms of economic issues, liberals are generally more supportive than conservatives of government programs such as unemployment insurance, welfare, and Medicare and of government economic regulation such as progressive taxation (the rich taxed at higher rates), minimum-wage laws, and regulation of industry. By the same token, liberals are likely to be more supportive of labor unions than are conservatives.

Example 2: Attitudes Toward Abortion

Abortion is a social issue that has figured importantly in religious and political debates for years. The GSS contains several variables dealing with attitudes toward abortion. Each asks whether a woman should be allowed to get an abortion for a variety of reasons. The following list shows these reasons, along with the *abbreviated variable names* you'll be using for them in your analyses later on.

Abbreviated Variable Name	
ABDEFECT	Because there is a strong chance of a serious defect
ABNOMORE	Because a family wants no more children
ABHLTH	Because the woman's health would be seriously endangered
ABPOOR	Because a family is too poor to afford more children
ABRAPE	Because the pregnancy resulted from rape
ABSINGLE	Because the woman is unmarried
ABANY	Because the woman wants it, for any reason

Before we begin examining answers to the abortion attitude questions, it is worth taking a moment to reflect on their logical implications. Which of these items do you suppose would receive the least support? That is, which will have the smallest percentage of respondents agreeing with it? Think about that before continuing.

Logically, we should expect the smallest percentage to support ABANY, because it contains all the others. For example, those who would support abortion in the case of rape might not support it for other reasons, such as the family's poverty. Those who support ABANY, however, would have to agree with both of those more specific items.

Three of the items tap into reasons that would seem to excuse the pregnant woman from responsibility:

Abbreviated Variable Name	
ABDEFECT	Because there is a strong chance of a serious defect
ABHLTH	Because the woman's health would be seriously endangered
ABRAPE	Because the pregnancy resulted from rape

We might expect the highest percentages to agree with these items. We'll come back to this issue later to find out whether our expectations are correct.

When we analyze this topic with data, we will discover useful ways of measuring overall attitudes toward abortion. Once we've done that, we'll be in a position to find out why some people are generally supportive and why others are generally opposed.

Conclusion

This book has two educational aims. First, we want to share with you the excitement of social scientific research. You are going to learn that a table of numerical data, which may seem pretty boring on the face of it, can hold within it the answers to many questions about why people think and act the way they do. Finding those answers requires that you learn some skills of logical inquiry. Second, we will show you how to use a computer program that is popular among social scientists. SPSS Statistics is the tool you will use to unlock the mysteries of society, just as a biologist might use a microscope or an astronomer a telescope.

Before getting started using SPSS Statistics, however, it is important that you have at least an initial appreciation for social research. In this chapter, we focused in particular on the relationship between theory and the social research process. This examination will continue throughout the book. While most of our attention will focus on the skills of analyzing data, we will always want to make logical sense out of what we learn from our manipulations of the numbers. Measurement is a fundamental topic that bridges theory and research. We turn our attention to that topic next.

Main Points

- The main purpose of this text is to introduce you to the logic and practice of social scientific research by showing you some simple tools you can use to analyze real-life data.
- Social and behavioral scientists' use of computing machinery has evolved over many years because of the need to analyze large amounts of data.
- SPSS Statistics is a widely used state-of-the-art statistical software program that will take you through all the basics of using any sophisticated statistical package.
- A theory is a general statement or set of statements that describes and explains how concepts are related to one another.
- A hypothesis is a tentative statement of expectation derived from a theory.
- A hypothesis proposes a relationship between two or more variables (the independent and dependent variables) that can be tested by researchers employing scientific methods.
- The categories of variables must be both exhaustive and mutually exclusive.
- When a social scientist proceeds from theory to hypotheses development, data collection, and data analysis, the process is called deductive research.
- When a social scientist moves from data collection to data analysis and then induces a general theory based on those observations, the process is called inductive research.
- Theoretical work informs all the subjects we are going to analyze in this book and, indeed, all questions and issues of relevance to social scientists.

Key Terms

Theory	Dependent variable
Exhaustive	Dimensions
Inductive research	Variables
Concepts	Independent variable
Exclusive	Abbreviated variable names
Deductive research	Categories

Hypothesis   Probabilistic
PASW Statistics   SAS (Statistical Analysis System)
SPSS Statistics   SPSS Statistics Student Version

Review Questions

1. What are the two statistical packages most widely used by social scientists today?

2. Which of the versions of SPSS Statistics described is the least powerful in terms of the number of cases and variables it can handle? (Hint: It was last made available for SPSS/PASW version 18.)

3. Which version of SPSS Statistics is the most powerful in this regard?

4. What version (or versions) of SPSS Statistics are you using?

5. Name two tasks for which a statistical package such as SPSS Statistics can be used.

6. What is the primary goal of social scientific research?

7. Name two social scientific concepts.

8. What is the relationship between theories and hypotheses?

9. Complete the following statement: Theories are to concepts as hypotheses are to _____
_____.

10. Does a hypothesis propose a relationship between dimensions or variables?

11. The categories of each variable should meet what two requirements?

12. What, if anything, is the problem with the following categories of the variable "political views": liberal and moderate? If there is a problem, how might you correct it?

13. What, if anything, is the problem with the following categories of the variable "political perspective": liberal, Democrat, Republican, and conservative? If there is a problem, how might you correct it?

14. Construct a hypothesis based on the deprivation theory of church involvement using level of education as your independent variable.

15. List the categories of the variables you used to construct your hypothesis in response to the previous question (#14).

16. Construct potential hypotheses to relate the following concepts, and identify the independent and dependent variable in each hypothesis. In addition, list the categories of each variable.
    a. Age and health
    b. Race and attitude toward affirmative action
    c. Gender and income

17. Which of the following is not a dependent variable: grade point average, church attendance, age, number of children?

18. Which of the following is not a variable: occupation, amount of television viewing, female, education level?

19. Consider the following hypothesis: People who earn more than $50,000 a year are more likely to vote Republican than people who earn less than $50,000 a year. Does this mean that all people who earn more than $50,000 a year vote Republican? Why or why not?

20. Is the following statement T (true) or F (false)? A researcher who begins by collecting data and then develops a theory to explain his or her findings is engaged in deductive research.

21. A researcher formulates a hypothesis based on the "magic bullet theory" and then selects independent and dependent variables to test this hypothesis. What process is the researcher engaged in?

22. A researcher collects data on the spread of AIDS in the United States and then, based on his or her findings, develops a theory to explain why the rate of exposure and infection to the disease is higher among certain racial and ethnic groups than among others. In what process is the researcher engaged?

Chapter 2   The Logic of Measurement

Measurement is one of the most fundamental elements of science. In the case of social research, the task typically is one of characterizing individuals in terms of the issues under study. Thus, a study of voting will characterize respondents in terms of the candidate for whom they plan to vote. A study of abortion attitudes will describe people in terms of their attitudes on that topic.

Validity Problems

Validity is a term used casually in everyday language, but it has a precise meaning in social research. It describes an indicator of a concept. Most simply, an indicator is said to be valid if it really measures the concept it is intended to measure; it is invalid if it doesn't.

As a simple example, let's consider political orientations, ranging from very liberal to very conservative. For an example of a clearly valid measure of this concept, here's the way the General Social Survey (GSS) asked about it.

POLVIEWS: We hear a lot of talk these days about liberals and conservatives. I'm going to show you a seven-point scale on which political views that people might hold are arranged from extremely liberal to extremely conservative. Where would you place yourself on this scale?

1. Extremely liberal

2. Liberal

3. Slightly liberal

4. Moderate, middle-of-the-road

5. Slightly conservative

6. Conservative

7. Extremely conservative

At the opposite extreme, a simple question about respondent gender would not be a valid measure of political orientations. Political orientations and gender are different concepts. But now let's consider another questionnaire item that does not come from the GSS.

This item lies somewhere in between these two extremes of validity with regard to measuring political orientations.

Which of these two political parties do you most identify with?

1. Democratic Party

2. Republican Party

3. Neither

This second item is another reasonable measure of political orientation. Moreover, it is related to the first, because Democrats are, on the whole, more liberal than Republicans. On the other hand, there are conservative Democrats and liberal Republicans. If our purpose is to tap into the liberal–conservative dimension, the initial item that asks directly about political orientation is obviously a more valid indicator of the concept than is the item about political party.

This particular example offers us a clear choice as to the most valid indicator of the concept at hand, but matters are not always that clear-cut. If we were measuring levels of prejudice, for example, we could not simply ask, "How prejudiced are you?" because no one is likely to admit to being prejudiced. As we search for workable indicators of a concept such as prejudice, the matter of validity becomes something to which we must pay careful attention.

Validity is a concern not only when you collect and analyze your own data (what is known as *primary research*) but also when you reanalyze data previously collected by someone else, as we do in this book. The process of reanalyzing someone else's data is referred to as *secondary analysis.* Even if you can think of a survey question that would have captured your concept perfectly, the original researchers might not have asked it. Hence, you often need to use ingenuity in constructing measures that nevertheless tap the quality in which you are interested. In the case of political orientations, for example, you might combine the responses to several questions: asking for attitudes about civil liberties, past voting behavior, political party identification, and so forth. We'll return to the use of multiple indicators shortly.

In large part, the question of validity is settled on prima facie grounds: We judge an indicator to be relatively valid or invalid on the face of it. It was on this basis that you had no trouble seeing that asking directly about political orientation was a valid indicator of that concept, whereas asking a person's gender was definitely not a valid measure of political orientation. Later in the book, we'll explore some simple methodological techniques that are also used to test the validity of measures.

Reliability Problems

Reliability is a different but equally important quality of measurements. *Reliability* refers to the quality of a measuring instrument that would cause it to report the same value in successive observations of a given case (provided the phenomenon being measured has not changed). For instance, if you step on a bathroom scale five times in a row and each time it gives you a different weight, the scale has a reliability problem. Conversely, if you step on a bathroom scale five times in a row and the scale gives you the same weight each time (even if the weight is wrong), the scale is reliable.

Similarly, if your statistics instructor administered the same test to you three times and each time you got a different score even though your knowledge of statistics had not changed in the interim, the test would have a reliability problem. Conversely, if your instructor administered the same test three times and your score was the same each time, the test would be reliable.

In the context of survey research, reliability also refers to the question of whether we can trust the answers that people give us even when their misstatements are honest ones.

For instance, in medical research, some patients report a particular organ has been removed in one survey, only to indicate in subsequent surveys that they still have that organ.[1] Similarly, students of voting behavior regularly encounter individuals who claim they did vote in the past presidential election and then, in subsequent surveys, claim they either did not vote or do not remember whether they voted. As noted previously, these statements are often honest ones. It is difficult enough for most of us to recall what we did a few months ago, let alone several years ago.

Conceptually, the test of reliability is whether respondents would give the same answers repeatedly if the measurement could be made in such a way that (a) their situations had not changed (e.g., they hadn't had additional surgery to remove organs) and (b) they couldn't remember the answer they gave before.

As we suggested, testing the reliability of an item empirically requires multiple measures (i.e., your instructor must administer the statistics test at least two or more times in order to determine its reliability). However, we can sometimes assess the reliability of a single item on its practicality. Years ago, one of us was asked to assist on a survey of teenage drivers in California. Over researcher objections, the client insisted on asking the question, "How many miles have you driven?" and providing a space for the teenager to write in his or her response.

Perhaps you can recognize the problem in this question by attempting to answer it yourself. Unless you have never driven an automobile, we doubt that you can report how many miles you have driven with any accuracy. In the survey mentioned, some teenagers reported driving hundreds of thousands of miles. By the way, if we administered the same survey to these respondents again, we would most likely have gotten widely different answers to this question, even if they had not driven a significant number of miles in the interim.

A better technique in that situation, however, would be to provide respondents with a set of categories reflecting realistically the number of miles respondents are likely to have driven: fewer than 1,000 miles; 1,000 to 4,999 miles; 5,000 to 9,999 miles; and so on. Such a set of categories gives respondents a framework within which to place their own situations. Even though they still may not know exactly how much they have driven, there will be a fair likelihood that the categories they choose will actually contain their correct answers. The success of this technique, of course, depends on our having a good idea in advance of what constitutes reasonable categories, determined by previous research, perhaps. As an alternative, we might ask respondents to volunteer the number of miles they have driven but limit the time period to something they are likely to remember. Thus, we might ask how many miles they drove the preceding week or month.

Distinguishing Between Validity and Reliability

Perhaps the difference between validity and reliability can be seen most clearly in reference to a simple bathroom scale. As we noted earlier, if you step on a scale repeatedly (scales don't remember) and it gives you a different weight each time, then the scale has a reliability problem. On the other hand, if the scale tells you that you weigh 125 pounds every time you step on it, it's pretty reliable, but if you actually weigh 225 pounds, the scale has a problem in the validity department; it doesn't indicate your weight accurately.

Both validity and reliability are important in the analysis of data. If you are interested in learning why some people have deeply held religious beliefs and others do not, then asking people how often they attend church will be problematic.

This question doesn't really provide a valid measure of the concept that interests you, and anything you learn will explain the causes of church attendance, not religious belief.

---

[1]The authors are grateful to Professor Randall MacIntosh, California State University, Sacramento, for this suggestion.

And suppose you asked people how many times they had attended church in the past year; any answers you received would probably not be reliable, so anything you might think you learned about the causes of church attendance might be only a function of the errors people made in answering the question. (It would be better to give them categories from which to choose.) You would have no assurance that another study would yield the same result.

Multiple Indicators

Often, the solution to the problems discussed above lies in the creation of *composite measures* using *multiple indicators*. As a simple example, to measure the degree to which a sample of Christian church members holds the beliefs associated with Christianity, you might ask them questions about several issues, each dealing with a particular belief, such as the following:

- Belief in God
- Belief that Jesus was divine
- Belief in the existence of the devil
- Belief in an afterlife: heaven and hell
- Belief in the literal truth of the Bible

The several answers given to these questions could be used to create an overall measure of religious belief among the respondents. In the simplest procedure, you could give respondents 1 point for each belief to which they agreed, allowing you to score them from 0 to 5 on the index. Notice that this is the same logic by which you may earn 1 point for each correct answer on an exam, with the total score being taken as an indication of how well you know the material.

Some social science concepts are implicitly multidimensional. Consider the concept of *social class*, for example. Typically, this term is used in reference to a combination of education, income, occupation, and sometimes dimensions such as social class identification and prestige. This would be measured for data analysis through the use of multiple indicators.

When it becomes appropriate in the analyses we are going to undertake together, we'll show you how to create and use some simple composite measures.

Levels of Measurement

As we convert the concepts in our minds into empirical measurements in the form of variables, we sometimes have options as to their level of statistical sophistication. Specifically, there are a number of different possibilities regarding the relationships among the categories constituting a variable. In social research, we commonly speak of four *levels of measurement*: nominal, ordinal, ratio, and interval.

Nominal Variables

Some variables simply distinguish different kinds of people. Gender is a good example of this; it simply distinguishes men from women. Political party distinguishes Democrats from Republicans and from other parties. Religious affiliation distinguishes Protestants, Catholics, Jews, and so forth. We refer to these measurements as nominal in that term's sense of naming. *Nominal variables* simply name the different attributes constituting them.

The attributes constituting a nominal variable (e.g., gender, composed of male and female) are simply different. Republicans and Democrats are simply different from each other, as are Protestants and Catholics. In other cases, however, we can say more about the attributes making up variables.

Ordinal Variables

Many social scientific variables go a step beyond simply naming the different attributes constituting a variable. *Ordinal variables* arrange those attributes in some order: from low to high, from more to less, and so on. Whereas the nominal variable "religious affiliation" classifies people into different religious groups, "religiosity" might order them in groups such as very religious, somewhat religious, and not at all religious.

And where the nominal variable "political party identification" simply distinguishes different groups (e.g., Democrats and Republicans), an ordinal measure of political philosophy might rank order the very liberal, the somewhat liberal, the middle-of-the-road, the somewhat conservative, and the very conservative. Ordinal variables share the nominal variable quality of distinguishing differences among people, and they add the quality of *rank ordering* those differences.

At the same time, it is not meaningful to talk about the distances separating the attributes that make up an ordinal variable. For example, we have no basis for talking about the amount of liberalism separating the very liberal from the somewhat liberal or the somewhat liberal from the middle-of-the-road. We can say that the first group in each comparison is more liberal than the second, but we can't say by how much.

Ratio Variables

Some variables allow us to speak more precisely about the distances between the attributes constituting a variable. Consider age for a moment. The distance between 10 years old and 20 years old is exactly the same as that between 60 years old and 70 years old. Thus, it makes sense to talk about the distance between two ages (i.e., they are 10 years apart).

Moreover, *ratio variables* such as age have the additional quality of containing a genuine zero point—in this case, no years old. This quality is what allows us to examine ratios among the categories constituting such variables. Thus, we can say that a 20-year-old is twice as old as a 10-year-old. By comparison, notice that we would have no grounds for saying one person is twice as religious as another.

Ratio variables, then, share all the qualities associated with nominal and ordinal variables, but they have additional qualities not applicable to the lower-level measures. Other examples of ratio measures include income, years of schooling, and hours worked per week.

Interval Variables

Rarer in social research are variables that have the quality of standard intervals of measurement but lack a genuine zero point, called *interval variables*. One example is intelligence quotient (IQ). Although IQ is calculated in such a way as to allow for a score of zero, that would not indicate a complete lack of intelligence, because the person would at least have been able to take the test.

Moving outside the social sciences, consider temperature. The Celsius and Fahrenheit measures of temperature both have 0° marks, but neither represents a total lack of heat, given that temperatures below zero are possible. The Kelvin scale, by contrast, is based on an absolute zero, which does represent a total lack of heat (measured in terms of molecular motion).

For most statistics used by social scientists, interval and ratio scales may be considered the same. When we start using SPSS Statistics, we'll see that its creators have lumped interval and ratio variables into a single category called *scale*. Although these variables may be combined for practical purposes, the distinction between them helps us understand why a negative income might be interpreted as debt and why a negative age is impossible!

Measurement and Information

Knowing a variable's level of measurement is important for selecting an appropriate statistic. Variables of different levels of measurement contain different amounts of information. The only information we have about nominal variables is the number of cases that share a common attribute. With ordinal variables, in addition to knowing how many cases fall in a category, we know a greater-than/less-than relationship between the cases. Variables measured at the interval level have points equidistant from one another. With equidistant points, we know how much greater than or less than cases are from one another. Finally, with ratio variables, we have all the characteristics of nominal, ordinal, and interval variables, plus the knowledge that zero is not arbitrary but means an absence of the phenomenon.

The statistics that SPSS Statistics has been programmed to compute are designed to make maximum use of the information preserved in a level of measurement. Using the mode on a sample of grade point averages ignores information used by the mean. Conversely, using the mean for a sample of religious preferences assumes information (equidistant points) not contained in a nominal measure. Responsible use of statistics requires selecting a statistic that matches the data's level of measurement. We'll talk about this more later. Right now, we want you to know that being able to identify a variable's level of measurement is essential for selecting the right statistical tool. We don't want to see you using a screwdriver when you need a hammer.

Table 2.1 displays the three primary levels of measurement that we discuss in this book: nominal, ordinal, and interval/ratio (scale). We purposefully designed the table as a series of steps in order to remind you that there is a hierarchy implied in the levels of measurement idea. Variables at the nominal level (bottom step) contain the least amount of information, followed by variables at the ordinal level (middle step) and, finally, variables at the interval/ratio level (highest step), which contain the most information. You should also note that as you move "up" from the nominal to the ordinal and finally to the interval/ratio level, each has the qualities of the level(s) below it, plus a new trait.

Table 2.1   Levels of Measurement

		INTERVAL/RATIO Distance between categories is meaningful *Ex. Income (measured in thousands of dollars); Age (measured in years)*
	ORDINAL Categories can be rank ordered *Ex. Social class (lower, working, middle, upper); Attitudes toward gun control (strongly oppose, oppose, favor, strongly favor)*	
NOMINAL Categories differ in name *Ex. Gender (male, female); Party identification (Democrat, Republican)*		

Measurement Options

Sometimes you will have options regarding the levels of measurement to be created in variables. For instance, although age can qualify as a ratio variable, it can be measured as ordinal (e.g., young, middle-aged, old) or even as nominal (baby boomer, not baby boomer).

The significance of these levels of measurement will become more apparent when we begin to analyze the variables in our data set. As you'll discover, some analytic techniques are appropriate to nominal variables, some to ordinal variables, and some to ratio variables. On the one hand, you will need to know a variable's level of measurement in order to determine which analytic techniques are appropriate. On the other hand, where you have options for measurement, your choice of measurement level may be determined by the techniques you want to employ.

Classifying Variables as Discrete or Continuous

In addition to distinguishing variables by their level of measurement, researchers sometimes classify variables as discrete or continuous. Just as distinguishing between levels of measurement helps us choose appropriate statistics, so, too, does knowing whether variables are discrete or continuous.

Unlike levels of measurement, which tell us the amount of information provided by a measure, the discrete versus continuous distinction gives us information about the underlying characteristics of a variable. In particular, it refers to the phenomenon's divisibility, or whether the values of a variable can be subdivided into ever smaller units, as described in Table 2.2.

Discrete variables are variables whose values are completely separate from one another, such as RACE or SEX. These are variables with a limited number of distinct (i.e., discrete) values or categories that cannot be reduced or subdivided into smaller units or numbers. Discrete variables can be nominal (sex), ordinal (class rank), or interval/ratio (number of siblings). All these variables are discrete because their values cannot be subdivided or reduced. A respondent may, for instance, have 1 sibling, but she cannot have 0.5 or 0.25 siblings. People come in discrete units of 1; they cannot be subdivided into smaller units.

Continuous variables, on the other hand, are variables whose values can be infinitely subdivided, such as AGE or EDUC (education measured in years). These variables are continuous both because they have a time dimension and because time can be infinitely subdivided (i.e., years, months, weeks, days, hours, minutes, seconds, etc.). The level of measurement for continuous variables can be either interval/ratio (age measured in years) or ordinal (age measured as infant, toddler, adolescent, preteen, teenager, etc.).

Table 2.2 Tips for Distinguishing Between Discrete and Continuous Variables

Discrete variable	Example: Number of siblings
	■ Unit CANNOT be reduced to ever smaller units
	■ There is NOT an infinite number of other possible categories between any two categories of this variable (i.e., 1 and 2 siblings, or 2 and 3 siblings, etc.)
	■ Variables can be nominal, ordinal, or interval/ratio (scale)
Continuous variable	Example: Age measured in years
	■ Unit CAN be further reduced to smaller units (i.e., months, weeks, days, hours, minutes, seconds, etc.)
	■ There ARE an infinite number of other possible categories between any two categories of this variable (i.e., 19 and 20 years old, 20 and 21 years old, etc.)
	■ Variables can be interval or ratio (scale)

Conclusion

Measurement is a fundamental aspect of social science research. It may be seen as the transition from concepts to variables—from sometimes ambiguous mental images to precise, empirical measures. Whereas we often speak casually about such concepts as prejudice, social class, and liberalism in everyday conversation, social scientists must be more precise in their use of these terms.

Chapters 1 and 2 give you a brief overview of two important issues in social scientific inquiry that are directly relevant to our primary focus—computerized statistical analysis. The chapters that follow build on this discussion of theory and measurement and show you the concrete techniques you need to engage in data analysis.

Main Points

- Measurement is a vital component of social scientific research.
- In designing and evaluating measurements, social scientists must pay particular attention to the problems of validity and reliability.
- A common remedy for problems of validity and reliability is the construction of composite measures using multiple indicators.
- Level of measurement signifies the different amounts and types of information obtained about a variable and is essential for selecting appropriate statistical tools.
- The four levels of measurement are nominal, ordinal, ratio, and interval.
- Variables of different levels of measurement contain different amounts of information.
- There is an implied hierarchy in the levels of measurement idea.
- In addition to classifying variables by their level of measurement, researchers also distinguish between discrete and continuous variables.
- Discrete variables cannot be infinitely subdivided into ever smaller units, whereas continuous variables can.

Key Terms

Validity
Ordinal variables
Primary research
Ratio variables
Secondary analysis
Interval variables
Reliability

Composite measures
Discrete variables
Multiple indicators
Continuous variables
Levels of measurement
Nominal variables
Scale

Review Questions

1. A researcher sets out to measure drug use on U.S. college campuses by asking a representative sample of undergraduates whether they are currently receiving federal grants or loans. What is the problem with this measure?

2. A researcher asks a representative sample of baby boomers how much alcohol they consumed during their college years and leaves a space for them to write in their response in terms of the actual number of drinks. Four weeks later, the researcher administers the same questionnaire to the same respondents. This time, however, more than half the respondents report consuming much less alcohol during their college years than they reported just a month earlier. What is the problem with this measure?

3. Multiple indicators are useful in dealing with what types of problems?

4. Name one reason why it is important to know or be able to identify a variable's level of measurement.

5. Ordinal variables have all the qualities of variables at which other level of measurement?

6. Ratio variables have all the qualities of variables at which other levels of measurement?

7. The creators of SPSS Statistics have combined ratio and interval variables into one category that they refer to as _____.

8. A variable whose values can be infinitely subdivided is called a _____ variable.

Identify the level of measurement of each of the following variables (Questions 9–11):

9. A researcher measures respondents' attitudes toward premarital sex by asking the following question: "If a man and woman have sexual relations before marriage, do you think it is always wrong, almost always wrong, wrong only sometimes, or not wrong at all?"

10. A researcher measures the amount of television viewing by asking the following question: "On the average day, how many hours do you personally watch television?" Respondents are then asked to fill in the actual number of hours in the space provided.

11. A researcher measures marital status by asking respondents whether they are currently married, widowed, divorced, separated, or never married.

12. Classify the variables in Questions 9 through 11 as either discrete or continuous.

Indicate whether the following statements (Questions 13–18) are T (true) or F (false):

13. Certain variables can be measured at both the nominal and ordinal levels.

14. You are invited to screen a new movie and then asked to rate it as either excellent, good, fair, or poor. The level of measurement is nominal because the ratings differ in name.

15. A researcher asks respondents to indicate the last four digits of their social security numbers. The level of measurement for this variable is interval/ratio because the distance between categories is meaningful.

16. A researcher asks respondents how many siblings they have. This variable can be categorized as continuous.

17. A researcher asks respondents how long they have lived at their current residences. This variable can be categorized as continuous.

18. Discrete variables can be either nominal or ordinal but not interval/ratio.

Complete the activities below (Questions 19–22):

19. Construct measures of "annual income" at two levels of measurement.

20. Classify the variables you constructed in response to Question 19 as either discrete or continuous.

21. Construct measures of "individual age" at two levels of measurement.

22. Classify the variables you constructed in response to Question 21 as either discrete or continuous.

Chapter 3 **Description of Data Sets**

The General Social Survey

The data we provide for your use here are real. They come from the responses of 2,044 adult Americans selected as a representative sample of the nation in 2010. These data are a major resource for professional social scientists and are the basis of many published books and articles.

The *General Social Survey (GSS)* is conducted regularly by the *National Opinion Research Center (NORC)* in Chicago, with financial support from the National Science Foundation and private sources. The purpose of the GSS program is to provide the nation's social scientists with accurate data for analysis.

This survey was the brainchild of Jim Davis, considered one of the most visionary social scientists of our lifetime. The GSS, which began in 1972, was conducted annually until 1994, when it became biennial.[1] Over the past 38 years, the GSS has asked more than 48,500 respondents more than 4,500 questions on topics ranging from attitudes toward abortion to star signs in the zodiac. In 2010 alone, the GSS asked questions on issues such as Internet use, religious transformations, spirituality, genetic testing, heredity, stress, violence in the workplace, immigration, altruism, alcohol consumption, sexual behavior, social networks, and group membership.

While this chapter provides you with a brief overview of some of the central components of the GSS, you can access further information about the GSS and other data by visiting one of the following websites:

A. NORC is a nonprofit corporation, affiliated with the University of Chicago, that conducts the GSS. Visit the NORC home page at www.norc.uchicago.edu. Once there, click on *Projects* → *General Social Survey* OR go to http://www.norc.org/Research/Projects/Pages/general-social-survey.aspx.

B. The Inter-University Consortium for Political and Social Research (ICPSR) is associated with the University of Michigan. Part of its mission is to maintain and provide access to a vast archive of social science data for research and instruction. ICPSR's home page can be found at www.icpsr.umich.edu.

---

[1]Funding shortages precluded GSS studies in 1979, 1981, and 1992.

C. The Roper Center for Public Opinion Research is an archive of public opinion research, affiliated with the University of Connecticut, that provides access to the GSS. The Roper Center home page is at www.ropercenter.uconn.edu. Click on *Quick Links* → *General Social Surveys* OR visit http://www.ropercenter.uconn.edu/data _access/data/datasets/general_social_survey.html.

D. You can also access this information on the World Wide Web by using one of your favorite search engines. Simply search using one of the following terms: *General Social Survey (GSS), GSS 2008, National Opinion Research Center (NORC), Inter-University Consortium for Political and Social Research (ICPSR),* or *Roper Center for Public Opinion Research*. If you do such a search, you will find that Canada also has a GSS sponsored by Statistics Canada, the equivalent of our Census Bureau.

Sampling

The data provided by the GSS are a representative sample of American adults. This means that anything we learn about the 2,044 people sampled can be taken as an accurate reflection of what all (non-institutionalized, English-speaking) American adults (18 years of age or older) would have said if we could have interviewed them all.

Creating a truly representative sample requires developing a selection technique that ensures every person in the population has an equal probability of being included. Since the previous GSS, the U.S. Postal Service and the U.S. Census Bureau have made lists of household units available for 72% of the U.S. population. With the aid of computers, random samples of households are drawn from the lists for inclusion in the GSS sample. For the 28% of the population not on household lists, random samples of census tracts and enumeration districts are drawn. Once tracts and districts are selected, housing units are randomized and selected for inclusion in the GSS sample.

This complex and sophisticated sampling process makes it possible for the responses of 2,044 individuals to provide an accurate reflection of the feelings of all adult Americans. The U.S. Census Bureau uses similar techniques for the purpose of government planning and by polling firms that predict voting behavior with relative accuracy.

For a detailed description of the sample design the GSS uses, see "Appendix A: Sampling Design and Weighting" in the *General Social Surveys 1972–2010 Cumulative Codebook*.

We have created a "Sample" data file by reducing the size of the GSS sample at the *Adventures in Social Research* student study website (http://www.sagepub.com/babbie8estudy/) to 1,500 cases (through random subselection) so that those who are using the student version (PASW/SPSS version 18 and earlier for both Windows and Macintosh) can access the GSS 2010 data. As we noted in Chapter 1, whereas the professional and graduate versions of SPSS Statistics for Windows and Macintosh are, for all practical purposes, limited only by the size of the computer on which they are installed, the student version on both Macintosh and Windows is limited to 1,500 cases and 50 variables.

We have also selected 150 of the variables from the GSS to include in the Adventures.SAV data file. All 2,044 cases (respondents) across those 150 variables were retained. This data file will be used in the demonstrations throughout the book and should also be used to complete all the laboratory exercises at the end of each chapter, starting with Chapter 4.

The *Adventures in Social Research* student study site (http://www.sagepub.com/babbie8estudy/) includes three data files, named **Adventures.SAV** (used for *Adventures in Social Research* and including 2,044 cases across 150 variables), **Sample.SAV** (a sample of 1,500 cases and 45 variables from the Adventures.SAV file), and **GSS2010.SAV** (the full GSS 2010 data set, including 2,044 cases across 790 variables).

The data you have at hand, then, can be taken as an accurate reflection of the characteristics, attitudes, and behaviors of Americans 18 years of age and older in 2010. This statement needs to be qualified slightly, however. When you analyze the data and learn that 44%

(variable: ABANY, GSS, 2010) of the sample said that they supported a woman's unrestricted right to have an abortion for any reason, you are safe in assuming that about 44% of the entire U.S. adult population feels that way. Because the data are based on a sample rather than on asking everyone, however, we need to anticipate some degree of sampling error. You can think of *sampling error* as the extent to which the responses of those sampled differ from the responses of the larger population (English-speaking persons 18 years of age and older living in non-institutionalized arrangements within the United States in 2010). As a general rule, the greater the sampling error, the less representative the sample. It would not be strange, based on the example given, to discover that 42% to 46% of the total adult population (rather than exactly 44%) support a woman's unrestricted right to have an abortion for any reason. It is inconceivable, however, that as few as 10% or as many as 90% hold the opinion in question.

As a rough guideline, you can assume that the sampling error in this data set is plus or minus only a few percentage points. Later in the book, we'll talk about how to calculate sampling error for specific pieces of data.

Even granting the possibility of sampling error, however, our best estimate of what's true among the total U.S. population is what we learned from the probability sample. Thus, if you had to bet on the percentage of the total U.S. population who supported a woman's unrestricted right to an abortion, you should put your money on 44%. You would be better off, however, betting, say, between 42% and 46%.

Data Collection

The GSS data were collected in face-to-face household interviews. Once the sample households were selected, professional interviewers were dispatched to call on each one. The interviewers asked each of the questions and wrote down the respondents' answers. Each interview took about 90 minutes.

To maximize the amount of information that can be collected in this massive interviewing project, the GSS uses a split-ballot design, such that NORC asked some questions in a random subsample of the households and other questions in the other households. Some questions were asked of all respondents. When we begin analyzing the GSS data, you will notice that some data items have a substantial number of respondents marked *missing data*. For the most part, this refers to respondents who were not asked that particular question.

Although some of the questions were posed to only a subsample of households, you can still take the responses as representative of the U.S. adult population, except that the degree of sampling error, mentioned above, is larger.

For more information about how the GSS data were collected, see "Appendix B: Field Work and Interviewer Specifications" and "Appendix C: General Coding Instructions" in the *General Social Surveys 1972–2010 Cumulative Codebook*.

The Codebook: Appendix A

The questionnaire items included in the files you will be using throughout this text (Adventures.SAV) are listed in Appendix A of this book. We attempted to choose variables that are not only interesting but relevant to students from a variety of social science disciplines, including communications, criminal justice, education, health studies, political science, public administration, social work, and sociology.

Before proceeding to the next chapter, you may want to take a few minutes to review the variables in Appendix A. Before long, you'll be getting much more familiar with them. As you analyze survey data, it is important to know exactly how questions were asked if you are to understand the meaning of the answers given in response. Appendix A includes the following information for each of the variables contained in your Adventures.SAV file:

- Abbreviated variable names (used by SPSS Statistics to access variables)
- Question wording (how the interviewer asked the question)
- *Values* (sometimes called *numeric values* or *numeric codes*, used to code responses)
- *Value labels* (used to identify categories represented by values. Please note that in Appendix A we have excluded the following value labels: *NA (no answer)*, *DK (don't know)*, and *IAP (inapplicable)*. These labels refer to cases when the respondent offered "no answer," said he or she "did not know," or the question was not asked and, thus, is "inapplicable.")
- Level of measurement for each variable

Dataset 1: Adventures.SAV

Adventures.SAV will be used and referred to in the demonstrations in the body of the following chapters. These examples are basically demonstrations that you can "follow along with" on your own computer. These data are also used for the exercises at the end of each chapter, starting with Chapter 4. This file contains all 2,044 cases and 150 variables drawn from the 2010 GSS. The items are arranged according to the following subject categories: abortion, children, family, religion, social–political opinions, sexual attitudes, sex roles, police, health, mass media, national government spending priorities, teen sex, affirmative action, equalization, terrorism preparedness, and voting.

Dataset 2: Sample.SAV

Sample.SAV is made available for those using an older student version of SPSS. This file contains 1,500 cases and 45 variables drawn from the 2010 GSS. The file is provided so that you can experiment with the data using your student software, but it cannot be used for many of the examples and exercises in the book because it does not include all the variables necessary to do such.

Dataset 3: GSS2010.SAV

GSS2010.SAV is the full 2010 GSS file. It is like the Adventures.SAV file in that it includes all 2,044 cases, but it also contains all 790 original variables. The *GSS Cumulative Codebook*, also provided on the *Adventures in Social Research* website, provides exact details for each of the variables in this dataset.

Conclusion

After reading this chapter and Appendix A, you should be familiar with the GSS and the variables with which you will be working. The data you will be using are real and can be taken as an accurate reflection of the attitudes, opinions, beliefs, characteristics, and behaviors of adult Americans in 2010.

In the next chapter, we are going to get started using SPSS Statistics. Once you have learned how to launch an SPSS Statistics session and access the files on the *Adventures in Social Research* website (http://www.sagepub.com/babbie8estudy/), you will be ready to begin exploring your data. With the help of SPSS Statistics and some simple tools, we think you will find that the possibilities for discovery can be both rich and rewarding.

Main Points

- The General Social Survey (GSS) is a national survey of adult Americans that has been conducted annually or every other year since 1972.
- The 2010 data are based on a sample of 2,044 adult Americans and can be taken as an accurate reflection of non-institutionalized, English-speaking adult Americans in 2010.
- The data were collected in face-to-face household interviews averaging about 90 minutes each.
- The website associated with this book (http://www.sagepub.com/babbie8estudy/) contains three data files: Adventures.SAV, Sample.SAV, and GSS2010.SAV.
- Adventures.SAV is the file you will be using as you work your way through the chapters and to complete the exercises at the end of each chapter.
- The Adventures.SAV data file contains data from all 2,044 respondents and 150 variables from the full GSS 2010 dataset.

Key Terms

General Social Survey (GSS)	NA (no answer)
GSS2010.SAV	Adventures.SAV
National Opinion Research Center (NORC)	IAP (inapplicable)
Values	Sampling error
Value labels	DK (don't know)
Sample.SAV	Missing data

Review Questions

1. What is the General Social Survey (GSS)?

2. When were the data discussed in this chapter collected?

3. If the data show that 66% of respondents favor capital punishment, what will be your best estimate in terms of the percentage(s) of adult Americans who feel this way?

4. Is it possible, based on this example, that as little as 10% to 15% of adult Americans favor capital punishment? Why or why not?

5. When social scientists refer to "sampling error," to what are they referring?

6. Would you agree that as a general rule, the smaller the sampling error, the less likely that the data are representative of the population?

Indicate whether the following statements (Questions 7–11) are T (true) or F (false):

7. Respondents to the 2010 GSS were interviewed over the telephone.

8. Each interview takes about $1\frac{1}{2}$ hours.

9. When an interviewer forgets to ask a respondent a particular question, this is called "missing data."

10. The label "DK" is used when a respondent refuses to answer a question.

11. The label "IAP" is used when a respondent is not asked a particular question.

In order to answer Questions 12 through 14, consult Appendix A:

12. List three items from the Adventures.SAV file that deal with sexual attitudes.

13. List three items from the Adventures.SAV file that deal with opinions about who should be able to teach.

14. The Adventures.SAV file contains several variables addressing a variety of issues. Once you have looked through the codebook in Appendix A, answer the following questions: Which of the issues most interest you? What would you like to learn about these issues? Why?

Part II   Univariate Analysis

Chapter 4 provides you some initial familiarization with the version of SPSS Statistics you will be using throughout the text. This and each of the following chapters contain a series of demonstrations you can follow along with on your computer. When the images on your computer screen match ours in the book, you will know that you successfully used SPSS Statistics commands. In Chapter 4, for instance, we begin with Demonstration 4.1, "Starting an SPSS Statistics Session," and end with Demonstration 4.7, "Ending Your SPSS Statistics Session." By the time you work your way through these demonstrations, you should be fairly comfortable accessing SPSS Statistics, moving through the Data Editor, opening a data file, and ending your SPSS Statistics session. You can further check your abilities by completing the lab exercise at the end of the chapter.

In the body of each of the following chapters, we are going to pay special attention to three concepts: religiosity, political orientations, and attitudes toward abortion. In the exercises at the ends of these chapters, you will have a chance to explore other issues such as sex roles, law enforcement, health, mass media, national government spending priorities, teen sex, affirmative action, and terrorism preparedness. We've chosen these topics on the basis of general interest and the possibilities they hold for analysis.

Chapter 5 begins our discussion of univariate analysis, the analysis of one variable at a time. In this chapter, we will examine several different ways we might measure the religiosity of the respondents to the General Social Survey (GSS), distinguishing the religious from the nonreligious and noting variations in between. In so doing, we are going to learn how to instruct SPSS to create frequency distributions, produce descriptive statistics, and save and print our output.

In Chapter 6, we continue our focus on univariate analysis by looking at differences in political orientations. We will show you how to present your data in graphic form by reviewing commands for creating bar charts, pie charts, line graphs, and histograms.

Chapters 7 and 8 provide you with two additional skills that many social researchers find useful: recoding and creating composite measures. In Chapter 7, we teach you how to combine the categories from one variable to create a new variable. This process, known as recoding, is most useful when you are working with variables with a large number of categories. In addition to recoding, social researchers often combine several responses into more sophisticated measures of the concepts under study. In Chapter 8, you'll learn a few basic techniques for doing this.

Finally, Chapter 9 suggests a number of other topics you might be interested in exploring: desired family size, child training attitudes, sexual behavior, and prejudice. We'll give you some guidance in applying the techniques we've focused on in the previous chapters to examine these topics. The major goals of this chapter are not only to review the SPSS techniques discussed in the previous chapters but to give you the opportunity to strike out on your own and experience some of the open-endedness of social research. The exercises at the end of this chapter are designed to assist you in the process of writing up your research results.

Chapter 4   Using SPSS Statistics

Some Basics

Like most data analysis programs, SPSS Statistics is capable of computing many different statistical procedures with different kinds of data. This makes SPSS Statistics a very powerful and useful tool, but because of its generalization, we need to specify what we want it to do for us.

In many ways, SPSS Statistics is a vehicle for discovering differences and relationships in data, the same way a car is a vehicle for discovering places we have not yet visited. The car does not know where we want to go or what we wish to see. We, rather than the car, plan the trip and set the direction. Similarly, when we use SPSS Statistics, we choose the data we wish to explore and select the statistical procedures we wish to use. Sitting at our computer keyboards, we are in SPSS's driver's seat.

Demonstration 4.1: Starting an SPSS Statistics Session

Starting an SPSS Statistics session is simple. Once you have started Windows (7, Vista, XP) or the Macintosh OS (10.6x: Snow Leopard or 10.7x: Lion), you will see a "desktop" that may contain several small graphics (called icons) representing programs. The desktop also contains access to pull-down menus and can have a dock, on which program or file icons are displayed. At this point, you have several options, and two of the most basic are summarized in SPSS Statistics Command 4.1.

SPSS Statistics Command 4.1: Starting an SPSS Statistics Session

Once you are in Windows (or Mac OS), you have two major options:

Option 1
Double-click **SPSS Statistics icon**

Option 2
Click **Start** → **Programs** or **All Programs** → **SPSS Statistics** or **PASW Statistics**

For Macintosh OS users, find the SPSS Statistics icon in the Applications folder or on the Dock. (If it's not there already, you may wish to place it on the dock.)

Whenever we introduce a new command, we will provide you with a summary of the command in a box such as the one above (SPSS Statistics Command 4.1). The box contains a brief description of the command, as well as a summary of how to accomplish the procedure. You will also find a list of new SPSS Statistics commands at the end of the chapter. In addition, the *Adventures in Social Research* student study website (http://www.sagepub.com/babbie8estudy/) contains more resources related to SPSS Statistics commands introduced in the text.

If neither of these alternatives works, you can panic if you like. Better yet, verify that SPSS Statistics has been installed on your computer or ask your instructor for assistance.

After launching SPSS Statistics, you may see an SPSS Statistics *dialog box* asking, "What would you like to do?"[1] For now, click on the **Cancel** button or the **X** (Close button) in the upper right-hand corner, and this box will disappear. SPSS Statistics will finish loading, and you should see something like what is shown in the following screen.

Once SPSS Statistics opens, you will notice that most of the screen is taken up by the SPSS Statistics *Data Editor*. You can find that name in the *title bar* on the top left-hand side of your screen. In the example, note that it is labeled "IBM SPSS Statistics Data Editor."

The Data Editor displays the contents of the active file. Because we haven't loaded any data yet, the screen is currently empty (thus, the name "Untitled1").

If you look to the bottom left-hand corner of the screen, you will notice two tabs representing the two primary components of the Data Editor: *Data View* and *Variable View*. Both screens contain important information about the data with which you are working. The Data View screen, the one you are looking at now, is designed to hold raw data for analysis. The Variable View screen contains information about that data.

---

[1]If you do not see this dialog box, don't worry. It just means that someone using SPSS Statistics before you requested that this box NOT be displayed every time the program is opened.

We are going to examine each aspect of the Data Editor in turn, beginning with the Data View screen.

Demonstration 4.2: Exploring the Data View Portion of the Data Editor

Menu Bar

The Data View screen is designed to hold data for analysis. If you wished, you could enter data directly into the screen now and analyze it. Instead, however, we are going to load a portion of the General Social Survey (GSS) data set into the screen in a moment. Directly beneath the title bar is a set of menus called a *menu bar*, running from "File" on the left to "Help" on the right. You are going to become very familiar with these menus because they are the control system or the primary means through which you will operate SPSS Statistics.

As a preview, click **File** on the menu bar, and a drop-down menu will appear. Notice how some commands in the drop-down menu appear black, whereas others are faint gray. Whenever you see a list such as this, you can execute the black commands (by clicking on them), but the gray ones are not currently available to you. Right now, for example, you can **Open** a data set, but you can't click **Save** because there's nothing to be saved at this time.

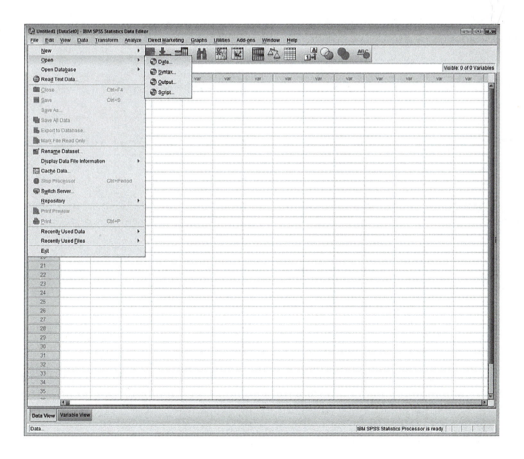

You will also notice that some of the options in the drop-down menu (i.e., New, Open, Open Database, Display Data File Information, Repository, Recently Used Data, and Recently Used Files) are followed by a right-pointing arrow. The arrow signifies that several additional commands will appear on a submenu when the category is picked. If you hold your cursor over **Open** for a moment, you will notice that a submenu of options appears on the screen. Now click on **File** again, and the list disappears. Do that now to get rid of the File menu, and we'll come back to it shortly.

Getting Help

At the far right-hand side of the menu bar, you will see the SPSS Statistics Help menu, something that you may want to take note of in case you need to use it later. As you are now aware, SPSS Statistics is a powerful state-of-the-art statistical package that allows users to accomplish numerous tasks and procedures. While this textbook introduces you to a variety of SPSS Statistics commands, options, and procedures, we cannot hope to cover all the program's capabilities. If you find that you want to use SPSS Statistics to perform a procedure not covered in the text or if you have a question or problem that hasn't been addressed in the book, you may want to consult the Help feature. All you have to do is click on **Help**, and a drop-down menu containing several options will appear.

The first option available in the Help menu is "Topics." Click on **Topics** and you will see that this box gives you several options for getting assistance. The Index tab, for instance, is organized in alphabetical order and provides a searchable index that makes it easy to find specific topics; the Search tab allows you to type in a word or phrase to be found. To close this box, simply click on the **X** (Close button) in the upper right-hand corner.

Another Help feature you may find useful is the online tutorial that gives a comprehensive overview of SPSS Statistics basics. You can access the tutorial by clicking **Help** and then selecting **Tutorial** from the drop-down menu. You will have an opportunity to explore the tutorial in the exercises at the end of this chapter.

SPSS Statistics Command 4.2: Accessing the Help Menu

Click **Help** → Select the Help option you wish to access (i.e., **Topics**, **Tutorial**, etc.)

These are just some of the Help features available. In addition to the Help menu, assistance is also available in other ways. We will mention some of these as we go on.

Toolbar

In addition to the menu bar, a second common way to communicate with SPSS Statistics is through the use of the *toolbar*. The toolbar is the line of buttons, or tools, running from left to right directly below the menu bar. Although you can use the menu bar and drop-down menus to perform most tasks, sometimes it is easier just to click a button on the toolbar.

You can find out what tasks each button on the toolbar performs by hovering your cursor over the button and waiting a moment until a brief description of the tool pops up onscreen. For instance, if you place your cursor on the button toward the left end of the toolbar that contains a picture of a disk, you will soon see the words "Save File" pop up on the screen. This, of course, lets us know that we can use this tool to save a file.

If you want to take a moment to explore what other tools are available on the toolbar, you can do that now. Otherwise, we will move ahead to the dialog boxes.

Dialog Boxes

Often when you click on one of the toolbars or choose an option from the menu bar, SPSS Statistics responds by opening a dialog box. If, for instance, you click on the **Open Data** tool on the far left-hand side of your toolbar (the tool depicting a partially opened file), SPSS Statistics responds by displaying the Open Data dialog box.

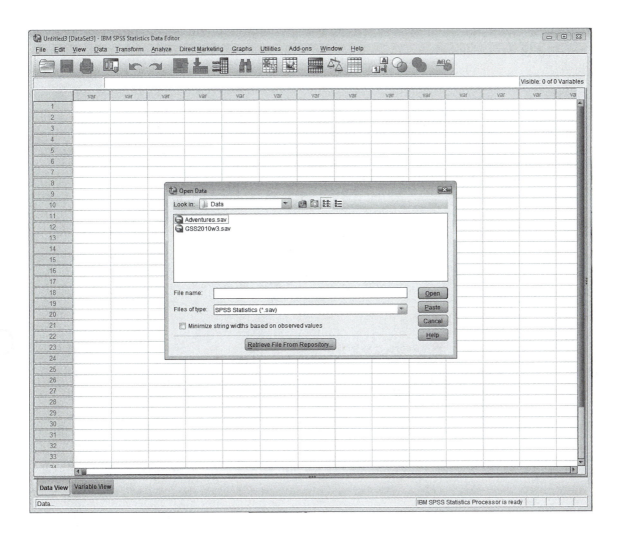

Dialog boxes are important because they tell you what else SPSS Statistics needs to know in order to fulfill your command. In short, they serve as a collection of prompts or hints indicating what other information is necessary in order for SPSS Statistics to comply with your command. Recall when we said earlier that there are other ways to get assistance beyond using the Help option on the menu bar. One of the primary ways to access assistance on SPSS Statistics is through dialog boxes. There are two methods of getting help once you have opened a dialog box.

The first option is to right-click on any control in the dialog box to display a description of the option and directions for its use. The second option is to click on the **Help** button. Most dialog boxes contain a Help button that takes you directly to the Help Topic for that particular dialog box. This Help Topic will provide you with general information and links to related issues.

SPSS Statistics Command 4.3: Getting Help in a Dialog Box

Option 1

Right-click on any control/command

Option 2

Click on **Help** button

Scroll Bars: Moving Through the Data Editor

Before we move on, we want to mention one last set of bars that will make it easier for you to move through the Data Editor: the *scroll bars*.

The horizontal scroll bar, situated across the bottom right side of your screen, allows you to move from left to right, while the vertical scroll bar, located on the far right side of your screen, allows you to move up and down.

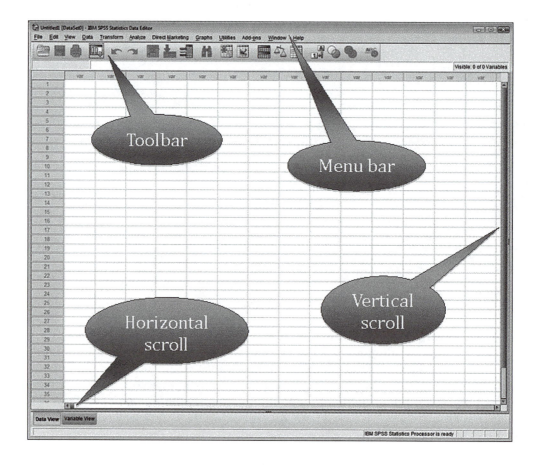

In addition to the scroll bars, you can also use your cursor or the arrow keys on the keyboard to move through the data screen. If you take a moment to experiment with these options, you may notice that you can move the *active cell* (the cell in the screen outlined with thick black lines). Once a cell has been designated as the active cell, you can enter data in it (or modify existing data), something we will focus on in the next demonstration.

SPSS Statistics Command 4.4: Moving Through the Data Screen

Option 1

Use horizontal and/or vertical scroll bars

Option 2

Use cursor → Click

Option 3

Use arrow keys on keyboard

Demonstration 4.3: Entering Data—A Preview

Take a moment now to designate the cell in the upper left-hand corner as the active cell. Once you have done that, you are ready to begin entering data for analysis. In fact, if you decide to conduct your own survey later on, this is how you will enter those data. As a quick preview of this feature, type a **1** and press the **Enter** key on your keyboard.

You have now created the world's smallest data set with one piece of information about one person. The **1** on the left of the screen represents Person 1. If you enter another number as you just did, you will bring Person 2 into existence with one piece of information. Why don't you do that now—enter a **2** for Person 2.

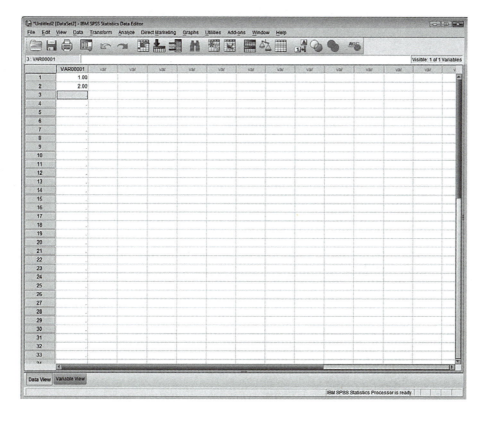

The "VAR00001" at the head of the column in the screen represents the specific information we are storing about each person. It might represent gender, for example. Moreover, a value of 1.00 might mean male and a value of 2.00 might mean female. In that case, we would have indicated that Person 1 is a man and Person 2 a woman.

This is the basic structure of the data sets analyzed by SPSS Statistics. The good news for you is that we've already prepared large data sets for your use, so you won't have to keep entering data like this. However, if you are interested in using SPSS Statistics to create your own data file, you may want to consult Chapter 21, which takes you through this process step-by-step.

Before loading our GSS data, we want to close the small data set we just created. To do that, click **File** → **Close**. At this point, SPSS Statistics will ask you if you want to "save the contents of data editor to Untitled?" which is just a fancy way of asking if you want to save the "world's smallest data set" that we just created. Because we are not interested in saving this data set, click **No**. If we wanted to save the data, we would click **Yes** and follow the instructions in the dialog box.

Demonstration 4.4: Loading a Data Set

At this point, you should once again be looking at an empty screen. But not for long, because we are finally ready to load one of the GSS data sets provided with this book.

Once you have downloaded the Adventures.SAV file from the *Adventures in Social Research* website, you have two options. If you wish to use the menu bar, simply click **File** and then from the drop-down menu choose **Open** and then **Data**. If you want to use the toolbar, all you have to do is click on the **Open Data** tool. Either way, SPSS Statistics will respond by displaying the Open Data dialog box, similar to the one shown below. This box will assist us in selecting the data set we want.

FILE → OPEN → DATA

Now your computer needs to be directed to the Adventures.SAV file. The file is maintained on the *Adventures in Social Research* student study site (http://www.sagepub.com/babbie8estudy/). You may want to burn the SPSS Statistics data files you use to a CD or DVD or perhaps store them on a jump/thumb drive, network drive, or other easily accessible source. By doing that, you can retrieve versions of the data files you have altered and saved and not be limited to just the unaltered data files found on the website. Data manipulation techniques, such as recoding variables, will be addressed later in the book, where the data file will be altered with new variables appended to the file.

Once you download the Adventures.SAV file from the study site, return to SPSS Statistics and click on **File** → **Open Data**. In the dialog box that appears, go to the location on your hard drive or other storage device to which you downloaded the Adventures.SAV file. Then make sure SPSS Statistics-format (or PASW Statistics-format) data files (.sav extension) are displayed in the "Files of type:" drop-down list. If not, simply click on the down arrow and select the suffix for SPSS Statistics data files, **SPSS Statistics (*.sav)**. Now you should see the file you just downloaded. To select **Adventures.SAV,** highlight it and double-click or click the

Open button in the lower right-hand corner of the dialog box. In a few seconds, SPSS Statistics will display the GSS data in the Data View portion of the Data Editor.

SPSS Statistics Command 4.5: Opening a Data File

Option 1: Menu bar

Click on **File → Open Data**

Option 2: Toolbar

Click on **Open Data** tool

Once the Open Data dialog box is displayed:

Click on **Look in field:** → Select the drive that contains your file → Navigate to the folder where you have saved your file → Click the down arrow to choose **SPSS Statistics (*.sav)** from the "Files of type:" drop-down list → Highlight the name of the data file and double-click OR click **Open**

Demonstration 4.5: Raw Data in Data View

Now you should be looking at the data in the Adventures.SAV file. This is the GSS data we will be using for the demonstrations in the body of each chapter. The information in the Data View portion of the Data Editor consists of variables and respondents or cases.

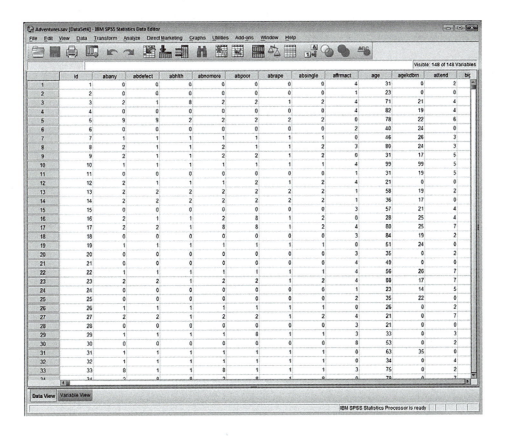

Each row (horizontal) represents a person or *respondent* to the survey (what is sometimes referred to as a *case*). Based on our discussion in Chapter 3, we already know the Adventures.SAV file has 2,044 rows representing each of the 2,044 respondents or cases. You may want to take a few minutes to scroll from the first case or respondent (Row 1) to the last respondent (Row 2044).

Each column (vertical) represents a variable, indicated by the abbreviated variable name in the column heading. This data set contains 150 variables, or columns. If you scroll from left to right across the screen, you can see that this file contains variables ranging from ID and ABANY to ZODIAC and WTSSnr. You can get a brief description of each abbreviated variable name by hovering your cursor over the variable name at the top of each column and waiting for a moment until the description pops up on the screen. For instance, if you place your cursor over the variable name ATTEND, a brief description of the variable that reads "How often R [respondent] attends religious services" will magically appear on your screen.

Finding Variable Information: Values and Labels

Now turn your attention to respondent Record Number 1 and the variable MARITAL. Be careful not to confuse the *record numbers*, which run down the far left side of the screen, with the ID numbers (the actual variable listed in the first column). Initially, they are the same in the Adventures.SAV file, since ID numbers are assigned consecutively, beginning with the number 1. In many types of files, however, they are not the same. Also, when data are sorted, sampled, or split into subsamples, the ID numbers will no longer correspond to the record numbers.

You will notice that respondent Record Number 1 has a 5 in the MARITAL column. This variable reflects respondents' marital statuses (if you forgot that, simply place your cursor over the variable name at the head of the column, and a brief description of the variable will appear on your screen). As you may recall from the list of variables in Appendix A, a 3 (the *numeric value* or code) on MARITAL means the respondent was "never married" (the *value label*). If you didn't recall this, don't worry. There are several ways you can use SPSS Statistics to access the information.

Option 1: Variables Dialog Box

One option is to click on the **Utilities** menu and select **Variables**. Alternatively, you can click on the **Variables icon** on the toolbar (the one with the question mark on it). Either way, SPSS Statistics will respond by opening the Variables dialog box, as shown below.

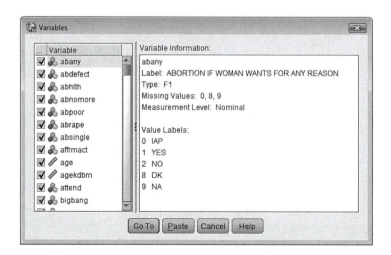

As you can see, the Variables dialog box has two main parts. On the left is a list of the variables contained in our data set. On the right is some information about the highlighted variable ID.

If you don't already see it, click on the variable **ABANY** [Abortion if woman wants for any reason] in the list on the left side of the box. This is the second variable listed. Notice how information about the highlighted variable ABANY is displayed on the right side of the dialog box, including the information we are looking for: numeric values and value labels. This box tells us that for the variable ABANY, a 1 (value) means "yes" (label), a 2 (value) means "no" (label), and so on.

Consequently, if you didn't recall what a value of 3 on MARITAL means, you can easily access the information by scrolling down through the list of variables until you see MARITAL.[2] Now click on **MARITAL**, and this is what you should see.

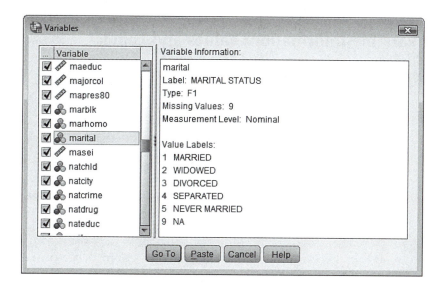

The box on the right side tells us that anyone with a code of 1 is married, anyone with a 2 is widowed, and so forth. You might want to take a minute to explore some of the other variables in the data set, because familiarity with them will be useful later.

When you get bored with the Variables dialog box, there are a couple of ways to leave it. You can either click the **Close** button or the **X.** Either action will produce the same result.

Option 2: Toggling Between Numeric Values and Value Labels

Another way to find out what a numeric value (such as a 5 on MARITAL) means is to click on **View** in the menu bar. Then click on **Value Labels** in the drop-down menu. This command tells SPSS Statistics to change the numeric values in the data screen into descriptive value labels. You will see, for instance, that the numeric value 5 under MARITAL for Respondent 1 has been replaced by a descriptive value label indicating that the respondent was never married.

---

[2]In addition to scrolling, there are several other ways to move through the list of variables on the left side. You can, for instance, click on the up and down arrows on the right-hand side of the variable list. You can also use the arrow keys on your keyboard or type the first letter of the variable's name. The highlight will then move to the first variable in your data set that begins with that letter.

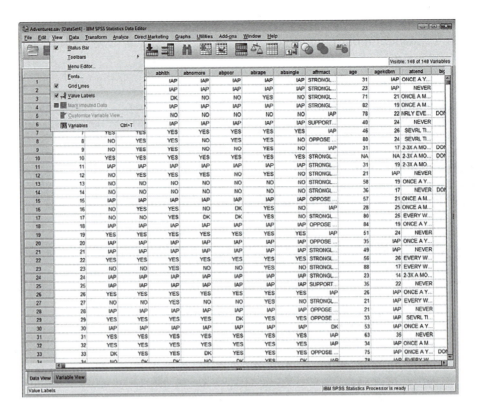

To display numeric values in the data screen once again, all you have to do is click on **View**. You will now see a check mark next to Value Labels in the drop-down menu. Click on **Value Labels** and the check mark will disappear. You should now see the numeric values in the data screen once again. This process of switching back and forth between numeric values and value labels is known as *toggling*.

Option 3: Value Labels Tool

A third way to discover what a numeric value refers to is to click on the button in the toolbar that shows arrows pointing to both an A and a 1, the **Value Labels** tool (second from the right side of the toolbar). You will notice that the numeric values in the data screen have now magically changed into value labels. To change the value labels back into numeric values, all you have to do is click on the **Value Labels** tool once again.

SPSS Statistics Command 4.6: Finding Information on Variables

Option 1: Variables dialog box

Click **Utilities** → **Variables** OR click on the **Variables** icon on the toolbar, and highlight variable name in list on left side

Option 2: Toggling

Click **View** → **Value Labels**

Option 3: Value Labels tool

Click **Value Labels** tool

Demonstration 4.6: Variable View Tab

A fourth way to determine value labels is through the Variable View portion of the Data Editor. So far, we have focused primarily on the Data View screen, which contains the raw data for analysis; however, the Variable View tab also contains important information about the data in your file. In particular, Variable View contains descriptions of the attributes of each variable in your data file.

Click on the **Variable View** tab at the bottom left side of the screen. You will notice that in Variable View, rows represent variables. Take a moment and scroll from the first variable, ID (Row 1), to the last, wtssnr (Row 150).

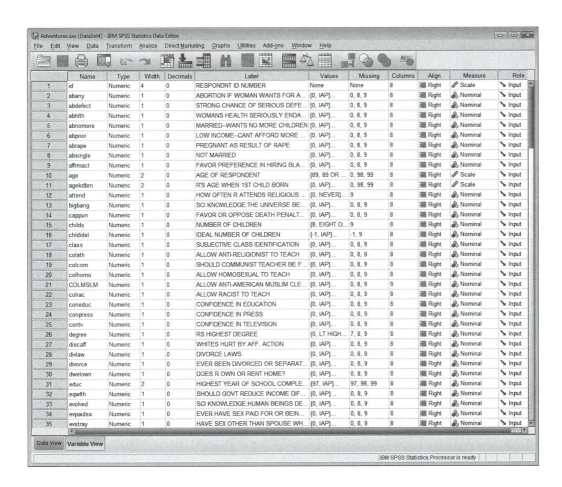

Unlike in Data View, here each column is an attribute associated with the variables. The 11 columns, or attributes, are listed below, followed by a brief description of their contents:

Name   Abbreviated variable name (i.e., ID, ABANY, ABDEFECT)

Type   Data type (e.g., numeric or string)

Width   Number of digits or characters

Decimals   Number of decimal places

Label   Variable labels or description of variable (i.e., respondent ID number; abortion if woman wants for any reason; strong chance of serious defect)

Values   Values and labels (i.e., ABANY: 0 = IAP, 1 = Yes, 2 = No, 8 = DK, 9 = NA)

Missing   Missing values (values designated as "missing")

Columns   Column width (width of column in Data View)

Align   Alignment (alignment of data in Data View, either to right, left, or center)

Measure   Level of measurement (i.e., nominal, ordinal, scale)

Role   The origin of the data: whether it was input (by you or whoever created the file), recoded or calculated from another variable (a target variable), etc.

When in the Variable View, you can add or delete variables and modify the attributes of each variable. We will focus more specifically on how to modify attributes in Variable View later in the text.

For now, it is enough that you are aware that the Values column in the Variable View window is another way to find information about values and labels. Earlier, for instance, we were focusing on the variable MARITAL. In Variable View, identify the cell corresponding with the variable MARITAL (row) and the Values column. Now double-click in the far right side of that cell (intersection between MARITAL and Values) and the Value Labels dialog box will open. Note that rather than double-clicking, you can click once and then click the button with three dots (ellipsis). Once the dialog box opens, you can see that the values and labels for the variable MARITAL are listed. To close the box, click **Close** or the **X** in the upper right-hand corner.

SPSS Statistics Command 4.7:
Finding Information on Variables in Variable View

Option 4: Variable View

In the Variable View portion of the Data Editor . . .

Click on cell corresponding with the variable in question (row) and the Values column

---

Demonstration 4.7: Ending Your SPSS Statistics Session

When you've finished this session, you have a few options. First, you can use the **X** in the upper right-hand corner to close the program.[3] If you click the **X** now, SPSS Statistics will ask if you want to save the contents of the Data Editor. Because the data set is already saved, you can safely click the **No** button now. If you had entered your own data or changed the existing data set, you would definitely want to save the data, in which case the computer would ask you to name the new data set.

You can also end your session by clicking **File** and then **Exit**. Before terminating your session, SPSS Statistics will ask if you want to save the file. Once again, you can click **No** because you have not altered the data file.

---

[3]Don't forget that the Minimize and Maximize buttons to the left of the close button are particularly useful if you are working with SPSS Statistics and other files simultaneously.

SPSS Statistics Command 4.8: Ending Your SPSS Statistics Session

> *Option 1: Close button*
>
> Click on **X** (Close button)
>
> *Option 2: File menu*
>
> Click **File** → **Exit**
>
> Mac OS users, Click **File** → **Quit SPSS (or PASW) Statistics**

Conclusion

In this first encounter, you've learned how to launch SPSS Statistics, load a data set, and explore it. In the next chapter, we'll revisit some of the variables in the data set and see how SPSS Statistics lets us explore more deeply than we've done in this first incursion.

Main Points

- Because SPSS Statistics is generalized, we use SPSS Statistics commands to tell it specifically what to do.
- The SPSS Statistics Data Editor is a screen designed to receive, display, and hold data for analysis.
- The two major components of the Data Editor are Data View and Variable View screens.
- The Data View screen contains the raw data on your file.
- Variable View contains descriptions of the attributes of each of the variables on your data file.
- You can use SPSS Statistics either to enter data or to load an existing data set, such as the data provided on the publisher's website associated with this book (http://www.sagepub .com/babbie8estudy/).
- In Data View, rows represent respondents or cases and columns represent variables.
- In Variable View, each row is a variable and each column is an attribute associated with the variables.

Key Terms

Numeric value	Title bar
Toolbar	Menu bar
Respondent	Variable View
Dialog box	Value label
Menu bar	Active cell
Data Editor	Scroll bars
Record numbers	Toggling
Data View	

SPSS Statistics Commands Introduced in This Chapter

4.1.   Starting an SPSS Statistics Session

4.2.   Accessing the Help Menu

Review Questions

1. What are the two main sections (tabs) in the SPSS Statistics Data Editor?

2. What type of information does the Data View screen contain?

3. What type of information does the Variable View screen contain?

4. In Data View, rows represent _____ and columns represent _____.

5. In Variable View, rows represent _____ and columns represent _____.

6. If a researcher has conducted a survey and wants to enter the raw data into SPSS Statistics, would he or she enter it into the Data View, Variable View, or Utilities View portion of the Data Editor?

7. If you have a question or problem, what is one of the features SPSS Statistics provides to offer you assistance?

8. If you don't recall what a value of a particular variable in your data set means, what three SPSS Statistics commands can you use to find this information?

9. What does "toggling" refer to?

10. If you are using several documents at once, which Windows (or Mac OS) features might you want to take advantage of?

11. If you use SPSS Statistics to enter your own data or change an existing data set, what one thing should you do before ending your SPSS Statistics session?

NAME _____

CLASS _____

INSTRUCTOR _____

DATE _____

To complete the following exercises, load the data file Adventures.SAV (found on the Adventures in Social Research *website: http://www.sagepub.com/babbie8estudy).*

1. What does the variable COLCOM measure? (Hint: In Data View, place cursor over column heading to determine descriptive variable label, OR in Variable View, double-click on cell corresponding with COLCOM and Label column.)

2. Using the variable COLCOM, note the numeric value for each of the following respondents (listed by record number, the column on the far left side of the screen). Then open the Variables dialog box to determine the value label:

	Numeric Value	*Value Label*
Record No. 89	_____	_____
Record No. 837	_____	_____
Record No. 1240	_____	_____

3. What does the variable COLRAC measure?

4. Using the variable COLRAC, note the numeric value for each of the following respondents (listed by record number). Then use the toggle command to determine the appropriate value label:

	Numeric Value	*Value Label*
Record No. 3	_____	_____
Record No. 505	_____	_____
Record No. 1460	_____	_____

5. What does the variable COLHOMO measure?

6. Using the variable COLHOMO, note the numeric value for each of the following respondents (listed by record number). Then use the Value Labels tool to determine the appropriate value label:

	Numeric Value	*Value Label*
Record No. 14	_____	_____
Record No. 233	_____	_____
Record No. 1475	_____	_____

7. What does the variable COLATH measure?

8. Using the variable COLATH, note the numeric value for each of the following respondents (listed by record number). Then access the Variable View tab, identify the cell that corresponds with COLATH and the Values column, and double-click to access the Value Labels dialog box. Use the Value Labels box to determine the appropriate value label.

	Numeric Value	*Value Label*
Record No. 7	_____	_____
Record No. 158	_____	_____
Record No. 1235	_____	_____

9. Select any variable that interests you from the Adventures.SAV file (other than those we have worked with previously in this exercise). Then use SPSS Statistics to find the following variable information:

   a. What is the abbreviated variable name? _____

   b. What does the variable measure (i.e., variable label)? _____

   c. List the numeric values and value labels for the variable.

   _____   _____

   _____   _____

   _____   _____

   _____   _____

	Numeric Value	*Value Label*
	_____	_____
	_____	_____

   d. Choose three respondents and list how they responded to the variable/question item you chose (list case number, numeric value, and value label).

	Case Number	*Numeric Value*	*Value Label*
Case No.	_____	_____	_____
Case No.	_____	_____	_____
Case No.	_____	_____	_____

10. Access the SPSS Statistics Help feature Tutorial.

Hint: Click **Help** → **Tutorial**

Once you have opened the Table of Contents, work your way through the following aspects of the tutorial: **Introduction, Using the Help System,** and **Reading Data.** To exit or close the Tutorial, click on the **X** (Close button).

Chapter 5 **Describing Your Data**

Religiosity

In this chapter, we are going to analyze data. Before we begin, we want to give you a few shortcuts for opening frequently used data files. We also want to show you how to set display options for SPSS Statistics and define values as missing. This will help ensure that what you see on your screen matches the screens in the text.

In order to begin, you'll need to launch SPSS Statistics as described earlier. If you have trouble recalling how to open the program, simply refer back to the discussion in the previous chapter.

Demonstration 5.1: Opening Frequently Used Data Files

If the SPSS or PASW Statistics "What would you like to do?" dialog box is displayed when you open SPSS Statistics, you can access a recently used data file by clicking the button labeled "Open an existing data source" next to the SPSS Statistics icon (the first option displayed).[1] If the Adventures.SAV file is listed, highlight it and click **OK**. This feature is available in prior versions of SPSS software, as shown in the screen from SPSS Statistics version 19 below.

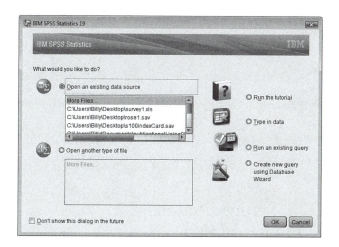

---

[1]If you don't see this box, it is probably because someone using SPSS Statistics before you requested that it not be displayed every time the program is opened. If you prefer this dialog box not be displayed when you open SPSS Statistics, click the "Don't show this dialog in the future" option in the bottom left corner.

Another shortcut for opening a frequently used file is to select **File** on the menu bar and then click **Recently Used Data** in the drop-down menu. Since SPSS Statistics keeps track of files that have been used recently, you may see the Adventures.SAV file listed in the box. To access the file, click **Adventures.SAV.**

If the Adventures.SAV file is not listed, don't worry. You can also open the file by following the instructions given in Chapter 4 for opening a data file.

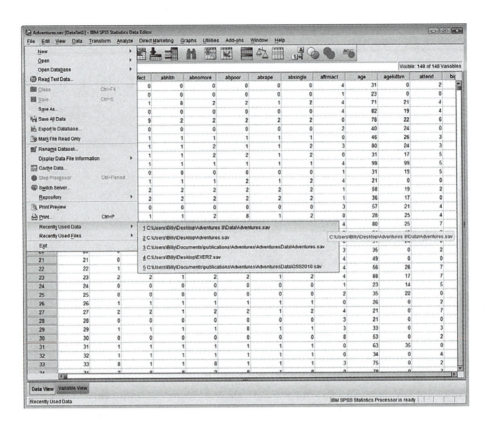

SPSS Statistics Command 5.1:
Shortcut for Opening Frequently Used Data Files

Option 1

In the "What would you like to do now?" dialog box, select **Open an existing data source** → Highlight the name of the file → Click **OK**

Option 2

Click **File** → Select **Recently Used Data** → Click on file name

Demonstration 5.2: Setting Options—
Variable Lists and Output Labels

Before we begin analyzing data, we want to make a few changes in the way SPSS Statistics displays information. This will make it easier for us to tell SPSS Statistics what we want it to do when we start running frequencies.

To set options, simply click on **Edit** (located on the menu bar). Now choose **Options,** and the Options window will open. You will see a series of tabs along the top of the box. Select **General** (if it is not already visible).

Toward the top of the box you will now see the Variable Lists option. In that box, choose **Display names** and **Alphabetical**. Now click **OK** at the bottom of the window. You may get a warning telling you that "Changing any option in the Variable List group will reset all dialog box settings to their defaults, and all open dialogs will be closed." If so, click **OK** once again.

EDIT → OPTIONS

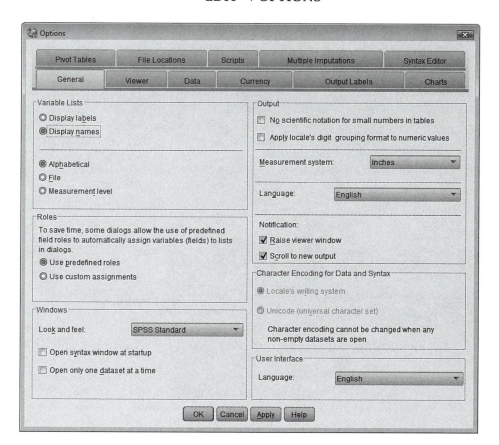

What we have just done is tell SPSS Statistics to display abbreviated variable names alphabetically whenever it is listing variables. This will make it easier for us to find variables when we need them.

SPSS Statistics Command 5.2: Setting Options—
Displaying Abbreviated Variable Names Alphabetically

Click **Edit** → **Options** → **General tab** → **Display names** → **Alphabetical** → **OK** → **OK**

A second change we want to make deals with the way SPSS Statistics displays output (such as frequency distribution tables, charts, graphs, and so on). By following the commands listed below, you can ensure that SPSS Statistics will display both the value and label for each variable you select for analysis, as well as abbreviated variable names and variable labels.

Simply choose **Edit** and **Options** once again. This time, click on the **Output Labels** tab located at the top of the Options window.

At the bottom left-hand corner of this tab, under the heading "Pivot Table Labeling," you will see two rectangles labeled "Variables in labels shown as:" and "Variable values in labels shown as:." Click on the down arrow next to the first rectangle ("Variables in labels shown as:") and choose the third option, **Names and Labels**. Now click on the down arrow next to the second rectangle ("Variable values in labels shown as:") and select **Values and Labels**; then click **OK.**

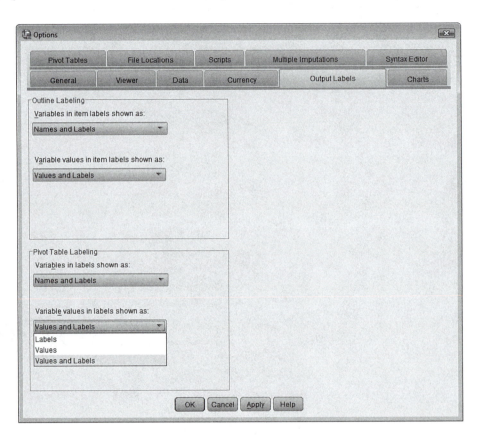

This ensures that when displaying output, both variable names and labels and value names and labels are shown. When we begin producing frequency distributions, you will see why access to all this information is helpful.

SPSS Statistics Command 5.3: Setting Options—Output Labels

Click **Edit** → **Options** → **Output Labels** tab → Click down arrow next to "Variables in labels shown as:" → **Names and Labels** → Click down arrow next to "Variable values in labels shown as:" → **Values and Labels** → **OK**

One last thing we want to do before running frequencies is show you how to set values and labels as "missing." This is a simple process that will once again help ensure that what you see on the screen matches what you see in the text. It is also important because when you begin analyzing data, you will want to be able to define values as missing.

Defining values and labels as missing is a simple process. Since we will be working with our four religious variables frequently throughout this chapter, we will demonstrate this procedure using these variables: RELIG, respondent's religious preference; ATTEND, how often the respondent attends religious services; POSTLIFE, respondent's belief in life after death; PRAY, how often the respondent prays. Click on the **Variable View** tab. Scroll down until you can see the values and labels for our first variable, RELIG, by clicking on the right side of the cell that corresponds with RELIG and the Values column. As you can see, this variable has several categories and labels ranging from 0 = "IAP" (inapplicable), 1 = "Protestant," 2 = "Catholic," all the way to 13 = "Inter-nondenominational," 98 = *DK (don't know)*, and 99 = *NA (no answer)*. When we are running frequencies and conducting other analyses, we may want to exclude those respondents who were not asked the question (IAP), said they didn't know (DK), or had no answer (NA).

Cases may be easily excluded by declaring categories we want to ignore as "missing." In the Variable View tab, click on **Cancel** to close the Value Labels dialog box. Now click on the right side of the cell that corresponds with the variables RELIG and Missing. On opening that box, you can see that you have several options:

- If you do not want any values/labels defined as missing, you can choose the "No missing values" option and click **OK**.
- If you want one to three values/labels set as missing, simply click on the option labeled "Discrete missing values," type the values in the box, and click **OK**.
- If you want a range plus one value defined as missing, you can choose the last option ("Range plus one optional discrete missing value"), type in the range and discrete value, and then click **OK**.[2]

In the case of RELIG, you may already see that the values 0, 98, and 99 have been labeled as missing. If that's the case, then you needn't do the following. If, however, you see only the values 0 and 99, we want to verify that 98 is classified as missing so that we also exclude

---

[2]We include a demonstration involving this option later in the text.

those people who said they don't know their religion. To do this, simply make sure the option labeled "Discrete missing values" is selected, then type **98** in the remaining box, if applicable, and click **OK.**

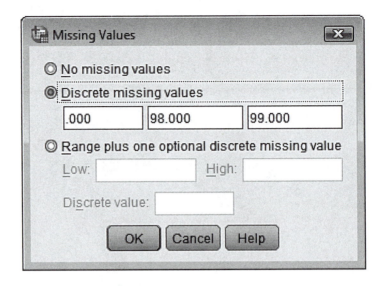

SPSS Statistics Command 5.4:
Setting Values and Labels as "Missing"

Access **Variable View** tab → Click on cell that corresponds with column Missing and row containing appropriate variable

In the Missing Values box, use one of the options available to set variables as missing ("Discrete missing values" or "Range plus one optional discrete missing value") or ensure that there are no missing values ("No missing values")

To make sure you are comfortable setting values as missing, go ahead and set the values for the following variables as missing: ATTEND, 9; POSTLIFE, 0, 8, 9; PRAY, 0, 8, 9. These should already be set in your data file, but it is very important to check. As we move through the text, if you find that your screens don't match those in the book, it may be because you have not set values as a particular value or as "missing." Now that you know how to do that, it should be easy to rectify the problem!

Demonstration 5.3: Frequency Distributions

Now that we have loaded the Adventures.SAV file, set options for SPSS Statistics, and defined selected values as "missing," we are ready to begin looking at some aspects of religious behavior. We will do this by first asking SPSS Statistics to construct a *frequency distribution*. A frequency distribution is a numeric display of the number of times (frequency) and the relative percentage of times each value of a variable occurs in a given sample.

We instruct SPSS Statistics to run a frequency distribution by selecting **Analyze** from the menu bar and then choosing **Descriptive Statistics**, then **Frequencies** in the drop-down menus.

Once you've completed these steps, you should be looking at the following screen:

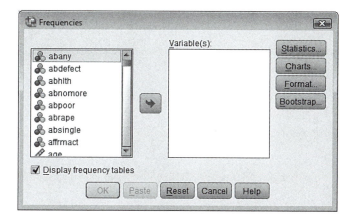

Let's start by looking first at the distribution of religious preferences among the sample. This is easily accomplished as follows: First, use the scroll bar on the right-hand side of the list of variable names to move down until RELIG is visible.

You may notice that it is fairly easy to find RELIG because the abbreviated variable names are displayed in alphabetical order (as opposed to the variable labels, which can be cumbersome to sift through). This is a result of our first task in this chapter, setting the options so SPSS Statistics shows abbreviated variable names alphabetically. If you notice that your variables are not formatted properly, you can revisit that discussion earlier in this chapter—or you can right-click (or if you have a one-button Macintosh mouse, press the **<control>** key while you click), and options will appear for sorting and displaying variable information.

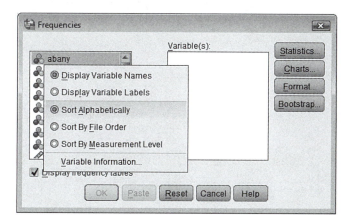

Once you locate RELIG, highlight it and click on the arrow to the right of the list. This will transfer the variable name to the "Variable(s):" field. Alternatively, you can also transfer a variable to the "Variable(s):" field by double-clicking it.

If you transfer the wrong variable, don't worry. You can move the variable from the "Variable(s):" field back to the variable list by highlighting it and clicking on the left-pointing arrow or simply double-clicking the abbreviated variable name.

Once RELIG has been successfully transferred to the "Variable(s):" field, you can display its values and labels by highlighting it and then right-clicking (or with a one-button Macintosh mouse, by holding down the **<control>** key while you click). From the drop-down menu, select **Variable Information**.

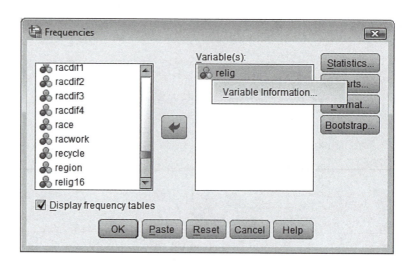

You will now see a box with the name, label, measurement, and value labels for the variable RELIG. Click on the down arrow to the right of the "Value labels:" rectangle to access a complete list of values and labels.

Once you have moved RELIG to the "Variable(s):" field, click **OK**. This will set SPSS Statistics going on its assigned task. Depending on the kind of computer you are using, this operation may take a few seconds to complete. Eventually, a new window called the *SPSS (or PASW) Statistics Viewer* will be brought to the front of the screen, and you should see the following:

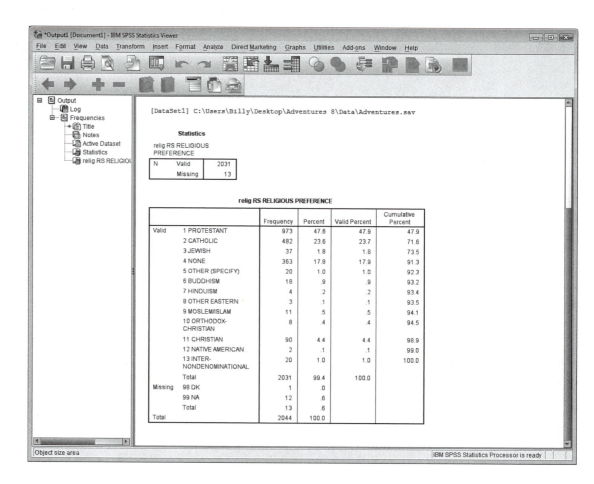

If you don't see the entire table as shown above, don't worry. It's probably because of resolution/size differences between computer screens. To see all the information, simply use the vertical scroll bar.

SPSS Statistics Command 5.5: Running Frequency Distributions

Click **Analyze → Descriptive Statistics → Frequencies**

→ Highlight the abbreviated variable name

→ Click on arrow (pointing toward "Variable(s):" field) OR double-click variable name

→ **OK**

The SPSS Statistics Viewer: Output

You should now be looking at the SPSS Statistics Viewer. After you run a procedure such as a table, graph, or chart, the results are displayed in the Viewer. This window is divided into two main parts or panes: the *Outline pane* (on the left) and the *Contents pane* (on the right). As its name suggests, the Outline pane contains an outline of all the information stored in the Viewer. This outline gives us a complete list of everything we have instructed SPSS Statistics to do in this session. This is useful because if you use SPSS Statistics for several hours to run a large amount of analysis, you can use this feature to navigate through your output. If, for instance, you want to find a specific table, all you have to do is select the name of the table in the Outline and wait a moment. The table will then appear in the Contents pane, that part of the Viewer where charts, tables, and other text output are displayed.

There are several ways to navigate through the SPSS Statistics Viewer. You can use your vertical and horizontal scroll bars, or you can use the arrow keys or "page up" and "page down" keys on your keyboard. For easier navigation, simply click on an item in the Outline pane and your results will be displayed in the Contents pane.

SPSS Statistics Command 5.6:
Navigating Through the SPSS Statistics Viewer

Option 1: Use the vertical or horizontal scroll bars

Option 2: Use the arrow or "page up" and "page down" keys on your keyboard

Option 3: Click on an item in the Outline pane

You can also change the width of the Outline pane by clicking and dragging the border to the right of the pane (the border that separates the Outline and Contents panes). You may want to take a few moments to experiment with each of these options before we move ahead.

SPSS Statistics Command 5.7:
Changing the Width of the Outline Pane

Click and drag border to right of Outline pane

Hiding and Displaying Results in the Viewer

There are several ways to hide a particular table or chart. You can simply double-click on its **book icon** in the Outline pane. Alternatively, you can highlight the item in the Outline pane and then from the menu bar choose **View** and **Hide**. To display it again, simply highlight the item in the Outline pane and select **View** and then **Show** from the drop-down menu. Easier yet, you can highlight the item in the Outline pane and then click on the **Hide** icon (closed book) on the Outlining toolbar. To display it again, simply highlight the item and then click on the **Show** icon (open book) on the Outlining toolbar.

SPSS Statistics Command 5.8: Hiding and Displaying Results in the Viewer

> *Option 1:* Double-click on **book icon** in Outline pane
>
> To display again, double-click on **book icon** again
>
> *Option 2:* Highlight item in Outline pane → Click **View** → **Hide**
>
> To display again, highlight item in Outline pane → Click **View** → **Show**
>
> *Option 3:* Highlight item → Click **Hide** icon (closed book)
>
> To display again, highlight item in Outline pane → Click **Show** icon (open book)

In addition, you can also hide all the results from a procedure by clicking on the box with a minus (−) sign to the left of the procedure name in the Outline pane. Once you do this, you will notice that the minus sign turns into a plus (+) sign. This not only hides the results in your Contents pane but compresses the Outline itself. To display the results again, simply click on the box to the left of the procedure name with the plus (+) sign.

SPSS Statistics Command 5.9: Hiding and Displaying All Results From a Procedure

> Click on box with minus (−) sign to left of procedure name in Outline pane
>
> To display again, click on box with plus (+) sign

Don't forget, you can also hide the SPSS Statistics Viewer itself by using the minimizing and reducing options. There are several more advanced ways of working with SPSS Statistics output and using the SPSS Statistics Viewer.[3] You can, for instance, change the order in which output is displayed, delete results, change the formatting of text within a table, and so on.

We are going to mention some of these more advanced options as we go along. In addition, the lab exercises at the end of this chapter give you an opportunity to work through a part of the SPSS Statistics Tutorial that provides an overview of some of these more advanced options.

Reading Frequency Distributions

Now that we are more familiar with the SPSS Statistics Viewer, let's take a few minutes to analyze the output for the variable RELIG, which is located in the Contents pane.

The small box titled "Statistics" tells us that of the 2,044 respondents in our subsample of the 2010 General Social Survey (GSS), 2,031 gave valid answers to this question, whereas 13 respondents have been labeled as "missing."[3] The larger box below marked "RELIG RS RELIGIOUS PREFERENCE" contains the data we were really looking for.

Let's go through this table piece by piece. As we requested when we changed the options earlier, the first line identifies the variable, presenting both its abbreviated variable name and label. Variable names are the key to identifying variables in SPSS Statistics commands. Earlier versions of SPSS software limited variable names to eight characters, a convention that has been continued by the GSS data analysts. Sometimes it is possible to express the name of a variable clearly in eight or fewer characters (e.g., SEX, RACE), and sometimes the task requires some ingenuity (e.g., CHLDIDEL, for the respondent's ideal number of children).[4]

The leftmost column in the table lists the *numeric values* and *value labels* (recall that we asked SPSS Statistics to display both when we changed the options) of the several categories constituting the variable RELIG. These include (1) Protestant, (2) Catholic, (3) Jewish, (4) None, (5) Other, and so on. As we discussed, the numeric values (sometimes called *numeric codes* or simply *values*) are the actual numbers used to code the data when they were entered. The value labels (or just labels) are short descriptions of the response categories; they remind us of the meaning of the numeric values or codes. As you'll see later on, you can change both kinds of labels if you choose.

The *frequency column* simply tells how many of the 2,044 respondents said they identified with the various religious groups. We see, for example, that the majority (973) said they were Protestant, 482 said they were Catholic, and so forth. Note that in this context, "None" means that some respondents said they had no religious identification; it does not mean that they didn't answer. Near the bottom of the table, we see that only 13 people failed to answer the question.

The next column tells us what percentage of the whole sample each of the religious groups represents. Thus, for example, 47.6% are Protestant, calculated by dividing the 973 Protestants by the total sample of 2,044. Usually, you will want to work with the *valid percentage*, presented in the next column. As you can see, this percentage is based on the elimination of those who gave no answer, so the first number here means that 47.9% of those giving an answer said they were Protestant.

The final column presents the *cumulative percentage*, adding the individual percentages of the previous column as you move down the list. Sometimes this will be useful to you. In this case, for example, you might note that 71.6% of those giving an answer were in traditional Christian denominations, combining the Protestants and Catholics.

Demonstration 5.4: Frequency Distributions—Running Two or More Variables at One Time

Now that we've examined the method and logic of this procedure, let's use it more extensively. As you may have already figured out, SPSS Statistics doesn't limit us to one variable at a time. (If you tried that on your own before we said you could, you get two points for being adventurous. Hey, this is supposed to be fun as well as useful.)

So, return to the Frequencies window with the following actions:

Analyze → Descriptive Statistics → Frequencies

---

[3]Reminder: If your frequency table looks different from the one shown in the book, it may be a result of the "missing" values for the variable RELIG. We ran this frequency distribution with the values 0, 98, and 99 missing. The procedures for defining missing values are outlined earlier in this chapter.

[4]For a brief overview of the rules regarding abbreviated variable names, access the Help feature of SPSS Statistics. Select **Topics** and **Index,** and then search for **Variable Names → Rules**.

If you are doing this all in one session, you may find that RELIG is still in the "Variable(s):" field. If so, you have a few options. You can double-click on **RELIG** to move it back to the variable list. Alternatively, you can highlight it. Once you do that, you will notice that the arrow between the two fields changes direction. By clicking on the arrow (which should now be pointing left), you can return RELIG to its original location. A third option is to simply click the **Reset** button. Choose whichever option appeals to you, and then take a moment to move the RELIG variable back to its original position.

Now let's get some other variables relating to religion. One at a time, highlight and transfer **ATTEND**, **POSTLIFE**, and **PRAY**.[5] When all three are in the "Variables(s):" field, click **OK**.

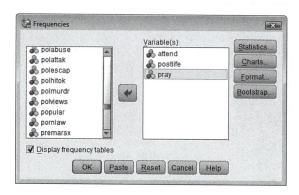

After a few seconds of cogitation, SPSS Statistics will present you with the Viewer window again. You should now be looking at the results of our latest analysis. Bear in mind that if you are doing this all in one session, the results of the latest procedure we ran (a frequency distribution for RELIG) may still be listed in the Outline and displayed in the Contents pane directly above the current procedure. If so, just scroll down until your screen looks like this:

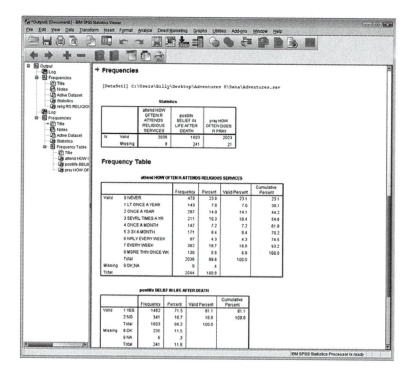

---

[5]Reminder: For the variable ATTEND, the value 9 should be labeled as "missing." For the variables POSTLIFE and PRAY, values 0, 8, and 9 should be labeled as "missing." The procedures for defining values as missing are outlined earlier in this chapter.

SPSS Statistics Command 5.10: Running Frequency Distributions With Two or More Variables

Analyze → **Descriptive Statistics** → **Frequencies** → Double-click on variable name OR highlight variable name → Click right-pointing arrow

Repeat this step until all variables have been transferred to the "Variable(s):" field → **OK**

Take a few minutes to study the new table. The structure of the table is the same as the one we saw earlier for religious preference. This one presents the distribution of answers to the question concerning the frequency of attendance at religious services. Notice that the respondents were given several categories to choose from, ranging from "Never" to "More than once a week." The final category combines those who answered "Don't know" (DK) with those who gave "No answer" (NA).

Notice that religious service attendance is an ordinal variable. The different frequencies of religious service attendance can be arranged in order, but the distances between categories vary. Had the questionnaire asked, "How many times did you attend religious services last year?" the resulting data would have constituted a ratio variable. The problem with that measurement strategy, however, is one of reliability: We couldn't bank on all the respondents recalling exactly how many times they had attended religious services.

The most common response in the distribution is "Never." More than one fifth, or 23.1%, of respondents (valid cases) gave that answer. The most common answer is referred to as the *mode.* This percentage is followed by 18.8% of the respondents, who reported that they attend religious services "Every week."

If we combine the respondents who report attending religious services weekly with the category immediately below them, more than once a week, we might say that roughly 25.6% of our sample report attending religious services at least once a week. If we add those who report attending nearly every week, we see that 29.9% attend religious services about weekly. Combining adjacent values of a variable in this fashion is called *collapsing categories*. It is commonly done when the number of categories is large and/or some of the values have relatively few cases. In this instance, we might collapse categories further, for example:

About weekly	30%
1 to 3 times a month	16%
Seldom	32%
Never	23%

Note: Percentages add to 101% due to rounding of percentages within categories.

Compare this abbreviated table with the original, and be sure you understand how this one was created. Notice that we have rounded off the percentages here, dropping the decimal points. As is the case in this instance, combined percentages can add to slightly more *or* slightly less than 100% due to rounding of the individual categories. Typically, data such as these do not warrant the precision implied in the use of decimal points, because the answers given are themselves approximations for many of the respondents. Later we'll show you how to tell SPSS Statistics to combine categories in this fashion. That will be especially important when we want to use those variables in more complex analyses.

Now let's look at the other two religious variables, beginning with POSTLIFE, as shown on the next page.

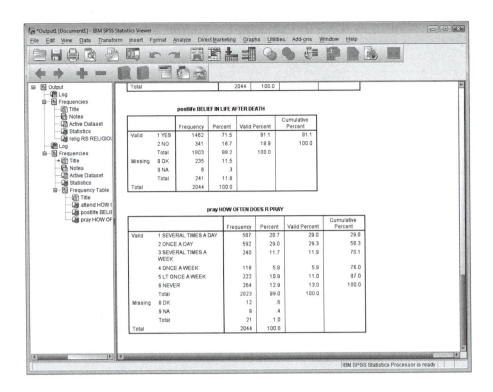

As you can see, significantly fewer categories make up this variable: just "Yes" and "No." Notice that zero respondents are coded ***IAP (inapplicable)***. We know this because the IAP category is not listed. Missing value categories are listed only when there are missing cases to report. This means that zero people were not asked this question. Thus, all 2,044 in the sample were asked this question. There will be a number of questions asked of everyone, such as this; however, due to the split-ballot nature of the GSS data design, many questions were not asked of all respondents. To collect data on a large number of topics, the GSS asks only subsets of the sample for some of the questions. Thus, you might be asked whether you believe in an afterlife but not asked your opinions on abortion. Someone else might be asked about abortion but not about the afterlife. Still other respondents will be asked about both. We will see some of these questions later.

Notice that more than 4 out of 5 American adults believe in an afterlife. Is that higher or lower than you would have predicted? Part of the fun of analyses such as these is the discovery of aspects of our society that you might not have known about. We'll have numerous opportunities for that sort of discovery throughout the remainder of the book.

Now let's look at the final religious variable, PRAY. One quick observation you can make from the distribution is that more than three fourths of Americans pray at least once a week. Is this higher or lower than you were expecting?

Writing Box 5.1

Once you've made an interesting discovery, you will undoubtedly want to share it with others. Often, the people who read our research have difficulty interpreting tables, so it becomes necessary for us to present our findings in prose. A rule of thumb for writing research is to write it so it can be understood by reading either the tables or the prose alone. To aid you in this process, throughout the text we have incorporated "Writing Boxes" using a style appropriate for professional reports or journal articles. In this box, for instance, we present in prose findings from the variables we just examined.

Almost half the respondents in our study (48%) identify themselves as Protestants, while a little less than a quarter (24%) are Roman Catholics. Jews make up only 2% of the sample, and just over 1 in 6 (18%) say they have no religious identification. The rest, as you can see, are spread thinly across a wide variety of other religious persuasions.

Religious identification is not the same as religious practice. For example, while 85% claimed a religious affiliation, only a little more than a quarter (30%) attend religious services about weekly. Another 16% reported attending one, two, or three times a month. Thirty-two percent reported attending services several times a year or less, leaving 23% who never attend religious services.

Religious belief is another dimension of religiosity. In the study, respondents were asked whether they believed in a life after death. The vast majority—4 out of 5 (81%)—said they did.

Finally, respondents were asked how often they prayed, if at all. Responses to this question bear a striking contrast to earlier answers. Recall that 16% said they had no religious affiliation, but 87% of the respondents said they pray *at least* now and then. Clearly, some respondents separate their spirituality from formal religious organizations. More than half of all respondents (58%) said they pray at least once every day.

This just about completes our introduction to frequency distributions. Now that you understand the logic of variables and the values that constitute them and know how to examine them with SPSS Statistics, you may want to spend some time looking at other variables in the data set. You can see them by using the steps we've just gone through.

If you create a frequency distribution for AGE (with values 0, 98, and 99 set as missing), you may notice the following: 0.5% of the sample is 18 years of age, 1.2% is 19, 1.2% is 20, and so forth (by looking in the Valid Percent column.) You may also notice that the table is *not* very useful for analysis. It is not only long (more than 80 values) and cumbersome but difficult to interpret as well. This is due primarily to the fact that AGE is presented here as a continuous, interval/ratio variable.

The Frequencies procedure we just reviewed is generally preferred when you are dealing with discrete variables. If you want to describe the distribution of a continuous variable, such as AGE, you will probably want to do one of two things. One option is to use the Descriptives procedure to display some basic summary statistics, such as common measures of central tendency and dispersion (we also use descriptive statistics when working with discrete variables). Alternatively, you might consider combining adjacent values of the variable (as we did with ATTEND previously) to decrease the number of categories and make the variable more manageable and conducive to a frequency distribution. When we collapse categories on SPSS Statistics, it is called recoding. We will consider the first option (running basic measures of central tendency and dispersion) in this chapter and save the second option (recoding) for a later chapter.

Descriptive Statistics: Basic Measures of Central Tendency and Dispersion

While it is fairly easy to learn how to request basic statistics through SPSS Statistics, it is more difficult to learn which statistics are appropriate for variables of different types and at different levels of measurement. At your command, SPSS Statistics will calculate statistics for variables even when the measure is inappropriate. If, for instance, you ask SPSS Statistics to calculate the mean of the variable SEX, it will dutifully comply and give you an answer of 1.56. Of course, in this instance, the answer is meaningless because it would make no sense to report the average gender for the 2010 GSS as 1.56. What, after all, does that mean?

When it comes to a discrete, nominal variable, such as SEX, the mode, not the mean, is the appropriate measure of central tendency. Since the GSS measures SEX as a dichotomy (having exactly two categories), the mean does tell us that there are more females than males (since males are coded "1" and females are coded "2", the value is closer to 2). It also tells us the proportion of one category to the other. However, for all other nominal variables that are not dichotomous, the value provides no useful immediate information.

Consequently, before running measures of central tendency and dispersion, we want to provide a brief overview of which measures are appropriate for variables of different types (Table 5.1) and at different levels of measurement (Table 5.2).[6] An "X" indicates that the measure is generally considered both appropriate and useful, whereas a long dash [—] indicates that the measure is permissible but not generally considered very useful. For example, Table 5.2 shows that the mode is appropriate for variables at the nominal level, whereas the median, range, and perhaps the mode are appropriate for variables at the ordinal level, and every measure listed is generally appropriate for ratio- and interval-level data. Note the exception that for interval/ratio variables that have skewed distributions, the median is generally a better central tendency measure than the mean.

Table 5.1   Basic Descriptive Statistics Appropriate for Different Types of Variables

Type of Variable	Measures of Central Tendency			Measures of Dispersion		
	Mode	Median	Mean	Range	Interquartile Range	Variance and Standard Deviation
Discrete	X	X				
Continuous		X	X	X	X	X

Table 5.2   Basic Descriptive Statistics Appropriate for Different Levels of Measurement

Level of Measurement	Measures of Central Tendency			Measures of Dispersion		
	Mode	Median	Mean	Range	Interquartile Range	Variance and Standard Deviation
Nominal	X					
Ordinal	—	X		X	X	
Interval/ratio	—	—	X	X	X	X

You are probably already familiar with the three **measures of central tendency** listed above. As we noted in the previous section, the **mode** refers to the most common value (answer or response) in a given distribution. The **median** is the middle category in a distribution, and the **mean** (sometimes imprecisely called the average) is the sum of the values of all the cases divided by the total number of cases.

You may already be familiar with some of the **measures of dispersion** as well. In this section, we will focus on three: range, interquartile range, and standard deviation. While *variance* is also listed in the tables, we'll save it for later in the text. The *range* indicates the distance separating the lowest and highest values in a distribution, whereas the *interquartile range* is the difference between the upper quartile (Q3, 75th percentile) and lower quartile (Q1, 25th percentile), or the range of the middle 50% of the distribution. The *standard deviation* indicates the extent to which the values are clustered around the mean or spread away from it.

---

[6]In addition to the variables' type and level of measurement, other issues may enter into the choice of particular statistics. Additional considerations beyond the scope of this text include (but are not limited to) your research objective and the shape of the distribution (i.e., symmetrical or skewed).

Demonstration 5.5: The Frequencies Procedure

Several of the procedures in SPSS Statistics produce descriptive statistics. To start, we are going to examine descriptive statistics options in the Frequencies procedure.

Open the Frequencies dialog box by choosing **Analyze → Descriptive Statistics → Frequencies.** Then select **Statistics** in the lower portion of the dialog box. The Statistics box that appears allows us to instruct SPSS Statistics to calculate all the measures of central tendency and dispersion listed in Tables 5.1 and 5.2.[7] The Frequencies procedure is generally preferred for discrete variables because, in addition to calculating most basic statistics, it presents you with a frequency distribution table (something not particularly useful for continuous variables), though you do have the option to suppress the frequency table.

To produce descriptive statistics using the Frequencies command, open the Frequencies dialog box. You know how to do this now. You've done it before. If there are any variables listed in the "Variable(s):" window, click **Reset** to clear them. We will use a simple example to show you the commands, and then you can practice on your own with other variables. Enter the variable **SEX**, and then click **Statistics** to the right of the window.

ANALYZE → DESCRIPTIVE STATISTICS → FREQUENCIES

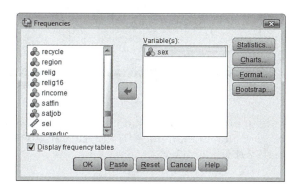

When the Statistics window opens, you will see the measures of central tendency listed in the box in the upper right-hand corner and the measures of dispersion listed in the box in the lower left-hand corner. To select an option, simply click in the box next to the appropriate measure. For our simple example, all we have to do is click the box next to **Mode** (upper right-hand corner). After you have made your selection(s), click **Continue**. You will now be back in the Frequencies box, and you can select **OK**.

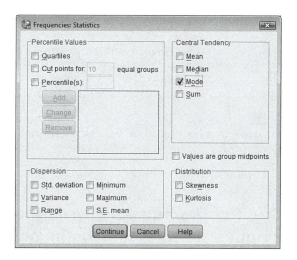

---

[7]To determine the interquartile range, simply select **Quartiles**, then use your output to subtract Q1 from Q3 [Q3 − Q1, or Quartile 3 − Quartile 1].

You should now see your output displayed in the Viewer. Remember, the output should include both the mode and frequencies for the variable SEX. By just looking at the frequencies, you should be able to say quite easily what the mode (the most frequent response or value given) is for this variable. You can check to make sure you are right by comparing your answer with the one calculated by SPSS Statistics.

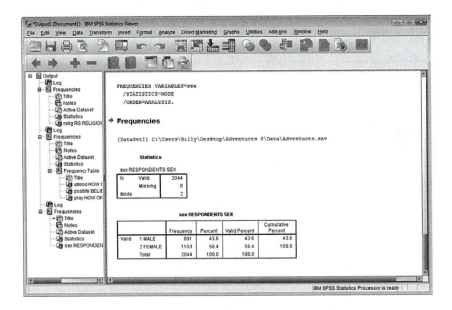

By glancing through our output, we find that the mode in this instance is 2, because there are more females (numeric value 2) than males (numeric value 1) in our data set. There are 1,153 females (56.4%) and 891 males (43.6%).

SPSS Statistics Command 5.11: The Frequencies Procedure—Descriptive Statistics (Discrete Variables)

Click **Analyze** → **Descriptive Statistics** → **Frequencies**

→ Highlight the variable name and click on the right-pointing arrow OR double-click on the variable name

→ Select **Statistics**

→ Choose statistics by clicking in the boxes next to the appropriate measures

Click **Continue** → **OK**

You may want to practice this command by choosing another discrete variable from the Adventures.SAV file, determining the appropriate measure(s) for the variable, and following the procedure listed above.

Demonstration 5.6: The Descriptives Procedure— Calculating Descriptive Statistics for Continuous Variables

We can also instruct SPSS Statistics to calculate basic statistics using the Descriptives procedure. This option is generally preferred when you are working with continuous, interval/

ratio variables or when you do not want to produce a frequency table. Bear in mind, the Descriptives procedure allows you to run all the measures outlined in Tables 5.1 and 5.2 except mode, median, and interquartile range.

If you want to use SPSS Statistics to determine descriptive statistics for AGE, you can do that by clicking **Analyze** in the menu bar. Next, select **Descriptive Statistics** and then **Descriptives** from the drop-down menus. You should now be looking at the Descriptives dialog box, as shown below.

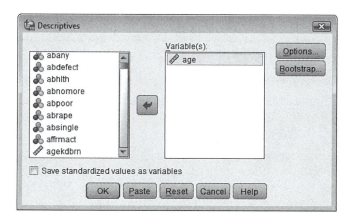

If there are any variables displayed in the "Variable(s):" box, go ahead and click **Reset**. Then move **AGE** from the variable list on the left side to the "Variable(s):" box, either by double-clicking it or by highlighting it and clicking on the arrow between the two fields. Once you have done that, AGE will appear in the "Variable(s):" field. Now select **Options** at the bottom of the window. This opens the Options window, which lets you specify what statistics you want SPSS Statistics to calculate.

The **Mean, Standard deviation, Minimum,** and **Maximum** options may already be selected. If not, select those options by clicking in the boxes next to them. You can also request **Range,** although in this case, it is not necessary because calculating the measure by hand is just as easy. Now that you have specified which statistics you want SPSS Statistics to run, click **Continue** and then **OK** in the Descriptives box.

Descriptive Statistics

	N	Minimum	Maximum	Mean	Std. Deviation
age AGE OF RESPONDENT	2041	18	89	47.97	17.678
Valid N (listwise)	2041				

We can see that the mean age of respondents in this study is 47.97. As we already know, this was calculated by adding the individual ages and dividing that total by 2,041, the number of people who gave their ages. Also, note that the Valid N is 2,041, three short of our total sample of 2,044. So where did those other three other people go? Did we lose them? No, they opted out of the question by giving no answer (NA). Theoretically, one or more of them could have responded that they didn't know their own age (DK), but that was not the case here. (You can always confirm this by looking at the full frequency distribution.)

We can also see that the minimum age reported was 18 and the maximum 89. The distance between these two values is the range, in this case, 71 years (89 − 18).[8] The standard deviation tells us the extent to which the individual ages are clustered around the mean or spread away from it. It also tells us how far we would need to go above and below the mean to include about two thirds of all the cases. In this instance, about two thirds of the 2,041 respondents have ages between 30.292 and 65.648 (47.97 ± 17.678). Later, we'll discover other uses for the standard deviation.

SPSS Statistics Command 5.12: The Descriptives Procedure— Descriptive Statistics (Continuous Variables)

Click **Analyze** → **Descriptive Statistics** → **Descriptives**

Highlight variable name and double-click OR click on the right-pointing arrow

→ Click **Options**

→ Choose statistic(s) by clicking in the box next to the appropriate measure(s)

→ Click **Continue**

→ **OK**

---

[8]Reminder: For the variable AGE, the values 0 and 99 should be labeled as "missing." The procedure for defining values as missing is outlined earlier in this chapter.

You may want to practice using the Descriptives command by choosing other continuous variables and running appropriate statistics. If you run into any problems, simply review the steps outlined above.

Demonstration 5.7: Printing Your Output (Viewer)

If you would like to print your output, you have several options. First, make sure that the SPSS Statistics Viewer is the *active window* (i.e., it is the window displayed on your screen). From the menu, choose **File** and then **Print**. You can control the items that are printed by selecting either **All visible output** (allows you to print only the items currently displayed in the Contents pane, as opposed to hidden items, for instance) or **Selection** (prints only items currently selected in the Outline or Contents panes). Then click **OK**.

If, for instance, you want to print only a selected item (such as the frequency distribution for the variable ATTEND), you click directly on the table or the name of the table in the Outline. Once you have done this, you can access the Print Preview option to preview what will be printed on each page. Conversely, you can select output you don't want to print and remove it by pressing the **Delete** key.

To access Print Preview, simply select **File** and then **Print Preview** from the drop-down menu. When you are ready to return to the SPSS Statistics Viewer, select **Close**.

You can then click **File**, **Print**, **Selection,** and **OK** to print your output.

SPSS Statistics Command 5.13: Printing Your Output (Viewer)

Make sure SPSS Viewer is active window

To select only certain items to print: Highlight item in Contents or Outline pane → Click **File** → **Print** → **Selection** → **OK**

To print all visible output: **File** → **Print** → **All visible output** → **OK**

SPSS Statistics Command 5.14: Print Preview

File → **Print** → **Preview**

Demonstration 5.8:
Adding Headers/Footers and Titles/Text

Headers and footers are the information printed at the top and bottom of the page. When you are in the SPSS Statistics Viewer and ready to print output, you may want to enter text as headers and footers. This is particularly useful for those of you working in a classroom or laboratory setting and using shared printers.

You can add headers and footers by selecting **File** and then **Page Setup** from the drop-down menu. Now click on **Options** and add any information you want in the Header and/

or Footer fields (such as your name, the date, title for the table, etc.). Then click **OK** and **OK** once again to return to the SPSS Statistics Viewer.

You may want to use the Print Preview option discussed earlier to see how your headers and/or footers will look on the printed page.

SPSS Statistics Command 5.15: Adding Headers and Footers

Click **File** → **Page Setup** → **Options**

→ Add information in Header and/or Footer fields

→ **OK** → **OK**

In addition to adding headers and footers, you can also add text and titles to your output by clicking on the place in either the Outline or Contents pane where you wish to add the title/text. Now simply click on the **Insert Title** or **Insert Text** tool on the toolbar or click **Insert** → **New Title** or **New Text** in the menu bar. SPSS Statistics will then display an area for you to add your title or text.

After inserting your text or title, you may want to use the Print Preview option to see how your text/title will appear on the printed page.

SPSS Statistics Command 5.16: Adding Titles/Text

Click on area in Contents or Outline pane where you wish to add title or text →

Click on **Insert Title** OR **Insert Text** tool on the toolbar

→ Add text or titles OR click **Insert** on the menu bar

→ **New Title** OR **New Text**

→ Add text or title

Demonstration 5.9: Saving Your Output (Viewer)

That's enough work for now. But before we stop, let's save the work we've done in this session so we won't have to repeat it all when we start up again.

Any changes you have made to the file (such as defining values as missing) and any output you have created is being held only in the computer's volatile memory. That means if you leave SPSS Statistics right now, the changes and output will disappear. To avoid this, simply follow the instructions here. This section walks you briefly through steps involved in saving output. The next section gives you instructions for saving changes to your data file.

If you are writing a term paper that will use the frequencies and other tables created during this session, you can probably copy portions of the output and paste it into your word-processor document. To save the Output window, first make sure it is the window frontmost on your screen, and then select **Save As** under the **File** menu, or choose the **Save File** tool on the toolbar. Either way, SPSS Statistics will present you with the following window:

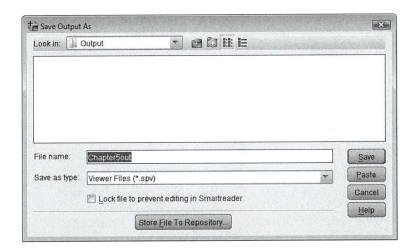

In this example, we've decided to save our output to the Documents folder on our computer's hard drive under the name "Chapter5out" (which SPSS Statistics thoughtfully provided as a default name). If you are using a public or shared computer, you may want to save the file to a portable drive or accessible Internet location (e.g., thumb drive, USB drive, network drive).

As an alternative, you might like to use a name such as "Out1022" to indicate that it was the output saved on October 22, or some similar naming convention. Either way, insert your drive if necessary, change the file name if you wish, choose the appropriate location in the "Look in:" field, and then click **Save**.

SPSS Statistics Command 5.17: Saving Your Output (Viewer)

> Make sure the SPSS Viewer/Output is the active window. Click **File Save As** OR click on **Save File** tool
>
> → Choose appropriate drive
>
> → Name your file
>
> → Click **Save**

Demonstration 5.10: Saving Changes to Your Data Set

If you have made no changes to labels, missing values, etc., then there is no need to resave the data file; the one you already accessed today is up to date. If you did make changes, we will want to maintain the labels we set for our variables as missing, so let's go ahead and save the data file. If you are sharing a computer with other users, you should save Adventures .SAV on a removable or network drive for safekeeping. If you are using your own personal computer, you can save your work on your computer's hard drive.

Keep in mind that whenever you save a file in SPSS Statistics (whether you are saving output or your data set), you are saving only what is visible in the active window. To save your data set, you need to make sure the Data Editor is the active window. You can do that by either clicking on **Adventures.SAV** in the taskbar or choosing **Window** in the menu bar and then selecting **Adventures.SAV** from the drop-down menu. Then, go to the **File** menu and select **Save As**. From the "Save in:" drop-down menu at the top of the Save As window, select a location for your data to be saved—either the USB drive, network drive, or the computer's hard drive.

Type the name for your file in the "File Name:" window. Although you can choose any name you wish, in this case, we will use "AdventuresPLUS" to refer to the file that contains our basic Adventures file with any changes you might have made today. Then click **Save**. Now you can leave SPSS Statistics with the **File** → **Exit** command. If you open this file the next time you start an SPSS Statistics session, it will have all the new, added labels.

SPSS Statistics Command 5.18: Saving Changes Made to an Existing Data Set

Make sure the Data Editor is the active window → Click **File** → **Save As**

→ Select appropriate drive

→ Name your file

→ Click **Save**

Conclusion

You've now completed your first interaction with data. Even though this is barely the tip of the iceberg, you should have begun to get some sense of the possibilities that exist in a data set such as this. Many of the concepts with which social scientists deal are the subjects of opinion in everyday conversations.

A data set such as the one you are using in this book is powerfully different from opinion. From time to time, you probably hear people make statements such as these:

Almost no one goes to church anymore.

Americans are pretty conservative by and large.

Most people are opposed to abortion.

Sometimes opinions such as these are an accurate reflection of the state of affairs; sometimes they are far from the truth. In ordinary conversation, the apparent validity of such assertions typically hinges on the force with which they are expressed and/or the purported wisdom of the speaker. Already in this book you have discovered a better way of evaluating such assertions. In this chapter, you've learned some of the facts about religion in the United States today. The next few chapters take you into the realms of politics and attitudes toward abortion.

Main Points

- In this chapter, we began our discussion of univariate analysis by focusing on two basic ways of examining the religiosity of respondents to the 2010 General Social Survey (GSS): running frequency distributions and producing basic descriptive statistics.
- Before embarking on our analysis, we reviewed a shortcut for opening a frequently used data file and learned to change the way SPSS Statistics displays information and output, as well as how to define certain values as "missing."
- The Frequencies command can be used to produce frequencies for one or more variables at a time.
- When we request tables, charts, or graphs on SPSS Statistics, the results appear in the SPSS Statistics Viewer.

- There are two main parts, or panes, in the Viewer: Outline (on the left) and Contents (on the right).
- The Outline pane lists the results of our output for the entire session.
- The Contents pane presents the results of our analyses.
- Frequency distributions are more appropriate for discrete, rather than continuous, variables.
- When working with continuous variables, a better option for describing your distribution is either to produce descriptive statistics or to reduce the number of categories by recoding.
- Certain measures of central tendency and dispersion are appropriate for variables of different types and at different levels of measurement.
- There are several ways to run measures of central tendency and dispersion on SPSS Statistics. We reviewed two: the Frequencies procedure and the Descriptives procedure.
- When you are ready to end your SPSS Statistics session, you may want to save and/or print your output so you can use it later.
- Before exiting SPSS Statistics, don't forget to save any changes you made to the existing data set.

Key Terms

Frequency distribution	Frequency column
Median	Measures of central tendency
SPSS (or PASW) Statistics Viewer	Valid percentage
Mean	Measures of dispersion
Outline pane	Cumulative percentage
Range	Collapsing categories
Contents pane	Interquartile range (IQR)
DK (don't know)	NA (no answer)
Numeric value	Active window
Standard deviation	IAP (inapplicable)
Value label	Mode
Variance	

SPSS Statistics Commands Introduced in This Chapter

5.1.   Shortcut for Opening Frequently Used Data Files

5.2.   Setting Options—Displaying Abbreviated Variable Names Alphabetically

5.3.   Setting Options—Output Labels

5.4.   Setting Values and Labels as "Missing"

5.5.   Running Frequency Distributions

5.6.   Navigating Through the SPSS Statistics Viewer

5.7.   Changing the Width of the Outline Pane

5.8.   Hiding and Displaying Results in the Viewer

5.9.   Hiding and Displaying All Results From a Procedure

5.10. Running Frequency Distributions With Two or More Variables

5.11. The Frequencies Procedure—Descriptive Statistics (Discrete Variables)

5.12. The Descriptives Procedure—Descriptive Statistics (Continuous Variables)

Review Questions

1. Analysis of one variable at a time is often referred to as what type of analysis?

2. What are the two parts of the SPSS Statistics Viewer called?

3. What is a frequency distribution?

4. In a frequency distribution, what information does the column labeled "frequency" contain?

5. In a frequency distribution, what information does the column labeled "valid percent" contain?

6. Is a frequency distribution generally preferred for continuous or discrete variables? Why?

7. List the measures of central tendency and dispersion appropriate for variables at each of the following levels of measurement:
   a. Ordinal
   b. Nominal
   c. Interval/ratio

8. List the measures of central tendency and dispersion appropriate for each of the following types of variables:
   a. Discrete
   b. Continuous

9. Calculate the mode, median, mean, and range for the following distribution: 7, 5, 2, 4, 4, 0, 1, 9, 6.

10. If Q1 (Quartile 1) = 25 and Q3 (Quartile 3) = 40, what is the interquartile range for the distribution?

11. Is the standard deviation a measure of central tendency or dispersion?

12. In general terms, what information does the standard deviation tell us?

13. Explain why, for certain GSS variables, a large number of the responses may be labeled "missing," coded as IAP.

14. In order to save or print output, which SPSS Statistics window should be the "active window"?

15. In order to save your data file, which SPSS Statistics window should be the "active window"?

NAME _____

CLASS _____

INSTRUCTOR _____

DATE _____

To complete the following exercises, you need to open the Adventures.SAV or the AdventuresPLUS.SAV file.

Produce and analyze frequency distributions and appropriate measures of central tendency for the variables RACE, HEALTH, and TEENSEX. Then use your output to answer Questions 1 through 6.

Before you begin, make sure to define values for the following variables as "missing":

RACE: 0

HEALTH: 0, 8, 9

TEENSEX: 0, 8, 9

1. (RACE) The largest racial grouping of respondents to the 2010 GSS was _____, with _____%. The second largest grouping was _____, with _____%.

2. Which measure of central tendency is most appropriate to summarize the distribution of RACE and why? List the value of that measure in the space provided.

3. (HEALTH) _____% of respondents to the 2010 GSS reported that they are in good health. This was followed by _____% of respondents who reported being in excellent health, whereas about _____% of respondents reported being in either fair or poor health.

4. Which measure of central tendency is most appropriate to summarize the distribution of HEALTH and why? List the value of that measure in the space provided.

5. (TEENSEX) Nearly _____% of respondents to the 2010 GSS believe that sex before marriage, particularly when it comes to teens 14 to 16 years of age, is either always wrong or almost always wrong. Only _____% think it is not wrong at all, whereas _____% think it is sometimes wrong.

6. Which measure of central tendency is most appropriate to summarize the distribution of TEENSEX and why? List the value of that measure in the space provided.

Use the Descriptives procedure to examine the variables EDUC and TVHOURS. Then use your output to answer Questions 7 and 8.

Before you begin, make sure to define values for the following variables as "missing":

EDUC: 97, 98, 99

TVHOURS: 1, 98, 99

7. (EDUC) The mean number of years of education of respondents to the 2010 GSS is _____, and two thirds of respondents report having between _____ and _____ years of education.

8. (TVHOURS) Respondents to the 2010 GSS report watching an "average" of _____ hours of television a day, with two thirds of them watching between _____ and _____ hours of television per day.

9. Are respondents to the 2010 GSS generally satisfied that the government is doing enough to halt the rising crime rate (NATCRIME) and deal with drug addiction (NATDRUG)? (Hint: Consider the type and level of measurement for each variable before using either the Frequencies or Descriptives command to produce frequency tables and/or descriptive statistics. For the variables NATCRIME and NATDRUG, values 0, 8, and 9 should be set as "missing.")

Choose three variables you are interested in from the Adventures.SAV (or AdventuresPLUS.SAV) file. Then based on the variable's type and level of measurement, use either the Descriptives or Frequencies command to produce a frequency table and/or appropriate descriptive statistics.

List the variable name and the values you set as "missing" below. Then describe your findings in the space provided (Questions 10–12).

10. Abbreviated variable name _____

Values labeled "missing" _____

Written analysis:

11. Abbreviated variable name _____

 Values labeled "missing" _____

Written analysis:

12. Abbreviated variable name _____

 Values labeled "missing" _____

Written analysis:

13. Access the SPSS Statistics Help feature Tutorial.

Hint: Click **Help** → **Tutorial** → Then work your way through the following sections: "Using the Data Editor" and "Working With Output."

Chapter 6   **Presenting Your Data in Graphic Form**

Political Orientations

Now let's turn our attention from religion to politics. Some people feel so strongly about politics that they joke about it being a religion. The General Social Survey (GSS) data set has several items that reflect political issues. Two are key political items: POLVIEWS and PARTYID. These items will be the primary focus in this chapter. In the process of examining these variables, we are going to learn not only more about the political orientations of respondents to the 2010 GSS but also how to use SPSS Statistics to produce and interpret data in graphic form.

You will recall that in the previous chapter, we focused on a variety of ways of displaying univariate distributions (frequency tables) and summarizing them (measures of central tendency and dispersion). In this chapter, we are going to build on that discussion by focusing on several ways of presenting your data graphically.

We will begin by focusing on two charts that are useful for variables at the nominal and ordinal levels: bar charts and pie charts. We will then consider two graphs appropriate for interval/ratio variables: histograms and line charts.

Graphing Data With Direct "Legacy" Dialogs

SPSS Statistics gives us a variety of ways to present data graphically. Clicking the Graphs option on the menu bar will give you three options for building charts: Chart Builder, Interactive, and Legacy Dialogs. These take you to three chart-building methods developed by SPSS over the past 20 years. Chart Builder is the most recent. The Interactive method preceded the Legacy Dialogs.

We have chosen to introduce you to graphing data with the Legacy Dialogs because the dialogs guide new users through the creation of a graph a step at a time. The newer methods require a user to be fairly familiar with graphics issues and nomenclature. While the Chart Builder and Interactive methods make greater use of the Microsoft Windows and Apple Macintosh graphic interfaces, we think you will appreciate starting with this established method. After you develop a sense of graph making's issues and options, we encourage you to try out the Chart Builder and Interactive methods.

Demonstration 6.1: Frequency Table—POLVIEWS

We'll start our examination of political orientations with POLVIEWS. Before we get started, make sure to define the values 0, 8, and 9 as "missing." Then use the Frequencies command to find out what this variable measures.

ANALYZE → DESCRIPTIVE STATISTICS → FREQUENCIES

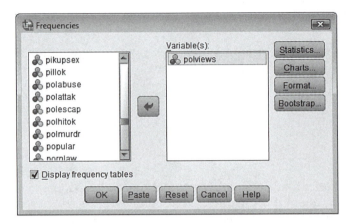

Take a few minutes to examine this table. As you can see, POLVIEWS taps into basic political philosophy, ranging from "extremely liberal" to "extremely conservative." As you might expect, most people are clustered near the center, with fewer numbers on either extreme.

polviews THINK OF SELF AS LIBERAL OR CONSERVATIVE

		Frequency	Percent	Valid Percent	Cumulative Percent
Valid	1 EXTREMELY LIBERAL	76	3.7	3.9	3.9
	2 LIBERAL	259	12.7	13.1	17.0
	3 SLIGHTLY LIBERAL	232	11.4	11.8	28.7
	4 MODERATE	746	36.5	37.8	66.5
	5 SLGHTLY CONSERVATIVE	265	13.0	13.4	80.0
	6 CONSERVATIVE	315	15.4	16.0	95.9
	7 EXTRMLY CONSERVATIVE	80	3.9	4.1	100.0
	Total	1973	96.5	100.0	
Missing	8 DK	61	3.0		
	9 NA	10	.5		
	Total	71	3.5		
Total		2044	100.0		

Bar Chart: POLVIEWS

Sometimes, the information in a univariate analysis can be grasped more quickly if it is presented in graphic form rather than in a table of numbers. Without looking back at the table you just created, take a moment to think about the distribution of political orientations for respondents to the 2010 GSS. You may recall that most respondents were clustered near the center, but do you remember the relative sizes of the different groups? Was the "moderate" group a little bigger than the others or a lot bigger? Sometimes, a graphic presentation of such data sticks in your mind more than a table of numbers.

In this section, we are going to focus on one basic procedure to construct a simple *bar chart* for POLVIEWS. Bar charts display the same type of information as frequency tables do (the number or percentage of cases in a category). The difference is that bar charts display this information graphically rather than in a table.

SPSS Statistics offers an easy method for producing a simple bar chart. Under the **Graphs** menu, select **Legacy Dialogs**, then **Bar**, and the Bar Charts dialog box will open. This box will give you an opportunity to select the kind of graph you would like: simple, clustered, or stacked. Because we have only one variable, we want to choose the **Simple** type.

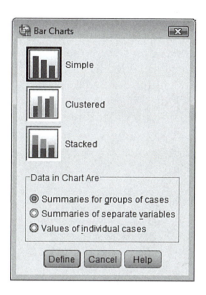

Probably that's the one already selected, but you can click it again to be sure. Then, click the **Define** button.

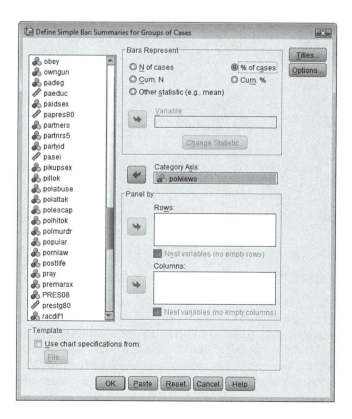

The next window allows you to specify further the kind of bar chart you would like, including a specification of the variable(s) to be graphed. As a start, let's find POLVIEWS in the variable list and highlight it. Now click the right-pointing arrow next to the "Category Axis:" box. This lets SPSS Statistics know that you want to construct the bar chart with the categories of POLVIEWS (extremely liberal, liberal, slightly liberal, and so on) on the *horizontal* **(x)** *axis*.

The "Bars Represent" box at the top of the window allows you to specify a format for the *vertical* **(y)** *axis* of the bar chart. In other words, you can choose to display "N of cases" (frequencies), "% of cases" (percentages), "Cum. N" (cumulative frequencies), or "Cum. %" (cumulative percentages). For now, let's click on % **of cases.**

Next, click on the **Options** button in the upper right-hand corner. After you do this, you will be presented with a screen that allows you to select how missing values will be treated. Because we are not interested in cases coded as "missing," make sure that the check mark next to the line that says "Display groups defined by missing values" is NOT showing. If the box has a check in it, simply click on the box to make the mark disappear, and then select **Continue**. If the box is already blank, click **Continue**.

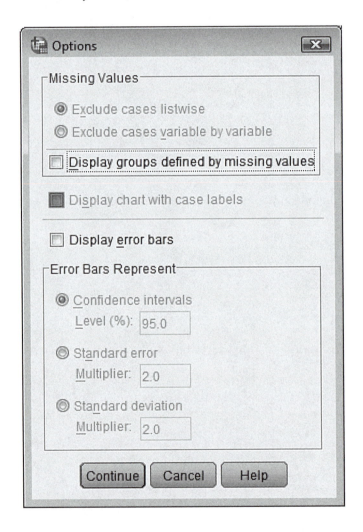

You should now be back in the Define Simple Bar dialog box. Before we tell SPSS Statistics to produce the chart, we want to draw your attention to one option you may find useful. The **Titles** button, located in the upper right-hand corner, opens a dialog box that allows you to specify titles, subtitles, and/or footnotes to define your chart further. While we are not going to do that now, you may want to keep this option in mind, particularly if you are planning to prepare charts for a presentation or inclusion in a paper, report, or publication.

Now that you are ready to instruct SPSS Statistics to produce the chart, click **OK**. It may take SPSS Statistics a few seconds to construct the bar graph to your specifications, depending on the speed of your computer, but in a short time, your chart will appear in the SPSS Statistics Viewer.

If you take a moment now to compare your bar chart to the frequency table you ran for POLVIEWS, it should become clear why graphic presentations are sometimes preferred to tables and why they can be more powerful. Viewing the data in graphic form makes it easy to see that most people are "middle-of-the road" and very few people are extreme in their political views. Graphic presentations are useful because they often do a better job of communicating the relative sizes of the different groups than does a table of numbers. Chances are, for instance, that after analyzing this chart, you will have a more vivid memory of the distribution of political views in the United States.

Moreover, it should also be clear that bar charts are essentially just the graphic or visual equivalent of a frequency distribution—equivalent in the sense that they convey essentially the same information in a different format. Each category of the variable is represented by a bar whose height is proportional to the percentage of the category.

SPSS Statistics Command 6.1: Simple Bar Chart

Click **Graphs** → **Legacy Dialogs** → **Bar**

→ **Simple** → **Define**

    → Highlight the variable name

    → Click arrow pointing to the "Category Axis:" box

    → Select option for vertical (*y*) axis in "Bars Represent" box

→ Click **Options**

    → Make sure there is NOT a check mark next to the "Display groups defined by missing values" option → Click **Continue** → **OK**

Demonstration 6.2: SPSS Statistics Chart Editor

After studying your chart, you may decide you would like to change the appearance (i.e., color, style, etc.) of this graphic. The good news is that SPSS Statistics allows you to do this quite easily in the Chart Editor. To start the Chart Editor, place your cursor on your bar chart and double-click. You should now see the Chart Editor, which provides you with a variety of options.

The Chart Editor can be very useful, particularly if you are planning to present, publish, or otherwise share your analysis. While we can't cover nearly all the possibilities available, we will introduce you to a few fun and useful options. Remember, once you are familiar with how to access the Chart Editor, you can experiment on your own.

Once the Chart Editor is open, you can change the color of your chart. Single-click on any bar in the chart to select all the bars. Then select the **Show Properties** window on the Chart Editor toolbar (the third from the left). Alternatively, you can also click on **Edit** → **Properties**.

Once the Properties dialog box opens, click on the **Fill & Border** tab. To change either the color or borders of the bars, select **Fill** or **Border** → the color of your choice → **Apply**. You can see that this dialog box also allows you to edit your chart in other ways. If, for instance, you want to insert a pattern, you can simply click on the down arrow next to **Pattern** → select the pattern of your choice → **Apply**. You can see that the Properties box also allows you to make other changes to your chart. If you want to change the style of your bar, for instance, you can click on the tab labeled **Depth and Angle**. In the upper left side, you will see that you have the option of choosing either "Flat," "Shadow," or "3-D Effect." Go ahead and select **3-D Effect**, then click **Apply,** and you will see how it changes the appearance of your chart.

The Properties box enables you to edit your chart in other ways as well. Take a few minutes to examine each of the tabs, and then once you are done, click **Close** or **X**. The Chart Editor has many other useful aspects yet to explore. The last option we will mention for now also happens to be one of the most useful. You can add percentages to each bar on your chart by clicking on the **Show Data Labels** icon (the one that looks like a bar chart, second from the left, in the row immediately above the chart). Alternatively, you can select **Elements** → **Show Data Labels**. As you can see, both methods display the percentages for each bar.

The Data Value Labels tab of the Properties dialog box allows you to customize this display. Take a few moments to experiment with some of those options, such as customizing the label position. When you are done, click on **Close** or the **X**. Remember, if at a later date you want to remove the percentages, choose **Elements** → **Hide Data Labels** or click on the **Hide Data Labels** icon (second from the left, immediately above the chart).

These are only a few of the Chart Editor's capabilities. You should experiment with the Chart Editor features on your own. Remember, once you are done, you can exit the Chart Editor by selecting **X** or **File** → **Close** from the drop-down menu.

Familiarity with the Chart Editor makes it easy to modify SPSS's charts for use in black-and-white publications, color overhead transparencies, and computer presentations. We will try to do our part by mentioning some other options as we go along. In addition, in Lab Exercise 7.1, you will be given an opportunity to work with an aspect of the SPSS Tutorial that focuses on creating and editing charts.

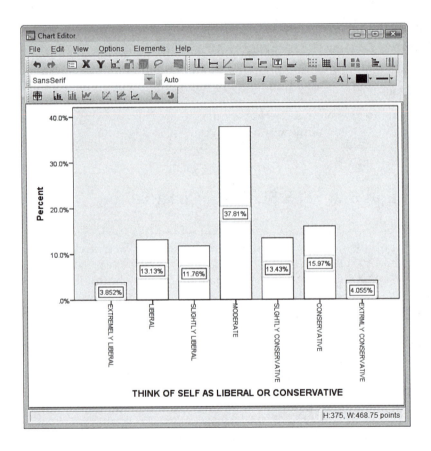

SPSS Statistics Command 6.2: SPSS Statistics Chart Editor

Double-click on chart

→ Click on appropriate icon in SPSS Chart Editor to make necessary changes

Before we turn to another basic chart useful for nominal and ordinal variables, let's take a few minutes to practice creating a simple bar chart. We will use the POLVIEWS variable again, but this time, construct your chart so the bars represent the number of cases as opposed to the percentage of cases. (Hint: Select **N of cases** in the "Bars Represent" box.) By following the instructions listed in SPSS Statistics Command 6.1, you should get the following chart.

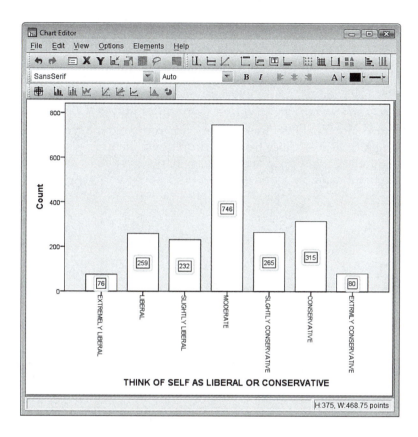

Demonstration 6.3: Frequency Table—PARTYID

Another basic indicator of a person's political orientation is found in the party with which she or he tends to identify. Let's turn now to the variable PARTYID.

Before we get the Frequencies for this variable, make sure to define the values 8 and 9 as "missing." Once you have done that, go ahead and run Frequencies for PARTYID. You should get the following results:

partyid POLITICAL PARTY AFFILIATION

		Frequency	Percent	Valid Percent	Cumulative Percent
Valid	0 STRONG DEMOCRAT	348	17.0	17.2	17.2
	1 NOT STR DEMOCRAT	348	17.0	17.2	34.3
	2 IND,NEAR DEM	265	13.0	13.1	47.4
	3 INDEPENDENT	360	17.6	17.8	65.1
	4 IND,NEAR REP	197	9.6	9.7	74.9
	5 NOT STR REPUBLICAN	277	13.6	13.7	88.5
	6 STRONG REPUBLICAN	184	9.0	9.1	97.6
	7 OTHER PARTY	49	2.4	2.4	100.0
	Total	2028	99.2	100.0	
Missing	9 NA	16	.8		
Total		2044	100.0		

Pie Chart: PARTYID

Pie charts are another common way of presenting nominal and ordinal data graphically. Like bar charts, pie charts depict the same type of information found in a frequency table graphically—the differences in frequencies or percentages among categories of a given variable. However, in this case, the information is not displayed on the *x*- and *y*-axes but as segments of a circle or, as the name suggests, slices of a pie.

A *pie chart* is simply a graphic display of data that depicts the differences in frequencies or percentages among categories of a nominal or ordinal variable. The categories are represented as pieces of pie whose segments add up to 100%.

To create a pie chart, simply select **Graphs, Legacy Dialogs**, and then **Pie** from the drop-down menu. Once the Pie Charts dialog box opens, you will see that by default, SPSS Statistics has selected the option that is appropriate for our purposes: "Summaries for groups of cases." To accept this and open the Define Pie dialog box, click **Define**.

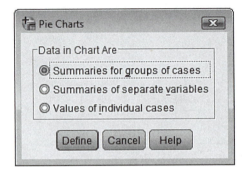

The Define Pie dialog box may look somewhat familiar to you. That is because it is similar to the Define Simple Bar dialog box we used previously. In this case, the main difference is that after highlighting the variable **PARTYID** on the left, you need to click on the arrow pointing to the "Define Slices by:" box.

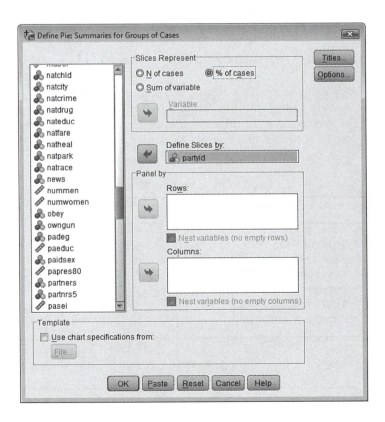

The rest of the steps are probably so familiar that you will not need much instruction, but just in case, follow the steps below.

In the "Slices Represent" box at the top, select **% of cases**. Choose **Options** and turn off the "Display groups defined by missing values" option by clicking in the box to the left and removing the check mark. Click **Continue** to return to the Define Pie box. Notice that similar to the Define Bar box, you can use the **Titles** option to insert titles, subtitles, and/or text. When you are ready to run your pie chart, click **OK.**

You will see that by default, SPSS Statistics does not display the percentages for each slice. Nevertheless, it appears that the largest slices are "Strong Democrat," "Not Strong Democrat," "Independent," and "Not Strong Republican."

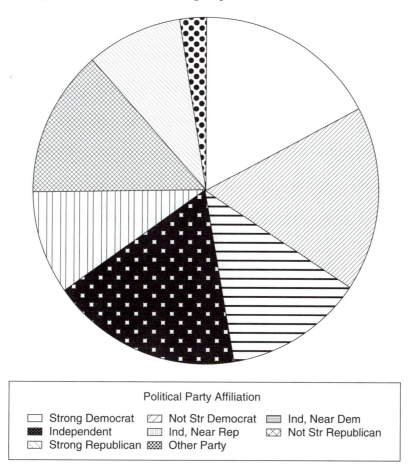

Political Party Affiliation

	Strong Democrat		Not Str Democrat		Ind, Near Dem
	Independent		Ind, Near Rep		Not Str Republican
	Strong Republican		Other Party		

SPSS Statistics Command 6.3: Pie Chart

Click **Graphs** → **Legacy Dialogs** → **Pie**
→ **Define**

   → Highlight variable name

   → Click arrow pointing to "Define Slices by:"

   → Select desired option in "Slices Represent" box

   → Click **Options**

      → Click on check mark next to "Display groups defined by missing values" to remove it → Click **Continue**

   → Click **OK**

We can request percentages for each slice by accessing the SPSS Statistics Chart Editor. To do that, once again click on the **Show Data Labels** icon or select **Elements** → **Show Data Labels**. As you can see, the percentages are now displayed in each slice of the pie chart.

To close the Chart Editor, select **File** and then **Close** or click **X**.

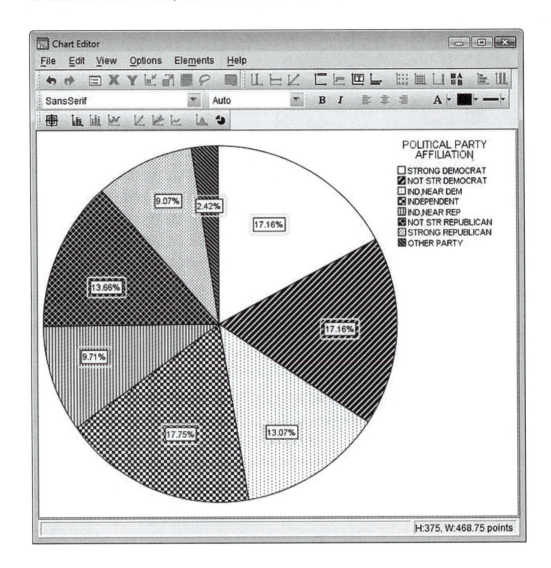

SPSS Statistics Command 6.4: Accessing Pie Options (SPSS Statistics Chart Editor)

Double-click on pie chart in SPSS Statistics Viewer

→ Select **Data Labels** icon OR **Elements** → **Show Data Labels**

Remember that in addition to adding (or deleting) percentages and values for your pie chart, you can also change the color and text, "explode" a slice for emphasis, and so on. The ability to vary fill patterns, for example, is especially useful when preparing charts for reproduction in black and white. To experiment with this, access the SPSS Statistics Chart Editor once again.

Once again, open the Properties dialog box by selecting the **Show Properties Window** icon or **Edit → Properties**. Now select the **Fill & Border** tab. This tab should look familiar because we used it to change the color of the bar chart. You can use this option once again to edit the colors of the pie chart or use the "Pattern" option to experiment with fill patterns. Once you are done experimenting, close the Chart Editor.

Before we experiment with other variables, let's take a few minutes to practice creating a pie chart. We will use the PARTYID variable again, but this time, construct your chart so the slices represent the number of cases as opposed to the percentage of cases. (Hint: Select **N of cases** in the "Slices Represent" box.) By following the instructions listed in SPSS Statistics Command 6.3, you should get the following chart.

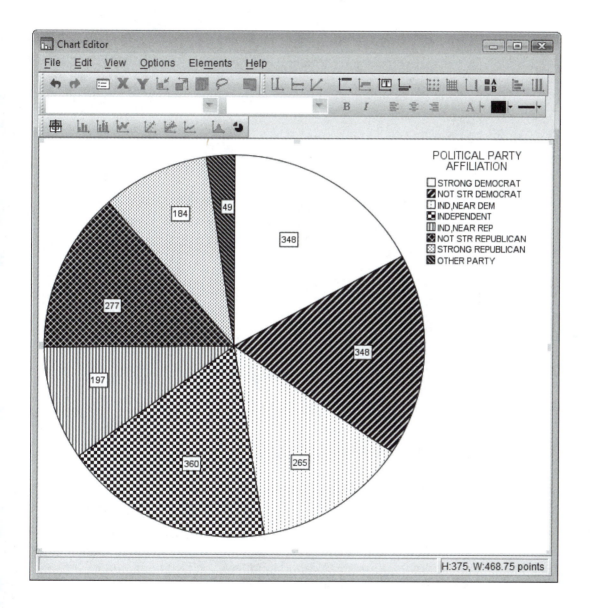

Note that you can also change the type of labels directly in the SPSS Statistics Chart Editor. In the Properties window, as seen below, click the **Data Value Labels** tab, and options are presented for what kinds of labels are "Displayed" and "Not Displayed." You can move a label type (e.g., percentage) from the "Displayed" box to the "Not Displayed" box to remove it from the graph or vice versa.

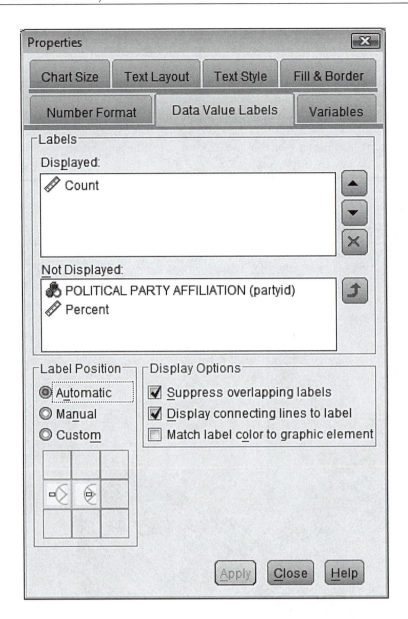

Demonstration 6.4: Political Attitudes

The GSS data set contains other variables that also tap into people's political orientations. For instance, GUNLAW measures how people feel about the registration of firearms. This has been a controversial issue in the United States for a number of years, involving, on one hand, Second Amendment guarantees of the right to bear arms and, on the other hand, high rates of violent crime, often involving firearms. CAPPUN measures whether respondents favor or oppose capital punishment, another topic associated with political attitudes.

As you can see, there is no lack of ways to explore people's political outlooks in the data set. We're going to focus on some of these items in later sections of the book. You should take some time now to explore some of them on your own. Take capital punishment, for example. How do you think the American people feel about this issue? Do you think most are in favor of it or most are opposed? This is your chance to find out for yourself. If you have any interest in political matters, you should enjoy this exercise.

You may have your own personal opinion about extramarital sex or homosexuality, but do you have any idea how the general population feels about such things? How about the death penalty and permits to purchase firearms? Take a moment to think about how the general public feels about these issues.

Be sure to set appropriate values for each variable as "missing" before you proceed. We defined 0, 8, and 9 as "missing" for all five variables (HOMOSEX, PREMARSX, XMOVIE, CAPPUN, and GUNLAW).

You can check to see if you were correct by running charts and graphs for each variable. Once you have done that, compare your findings to those in Writing Box 6.1.

Writing Box 6.1

Here's where the sample stood on a number of controversial issues. There is overwhelming support (75%) for requiring people to obtain police permits in order to buy a gun. About two thirds (68%) support the death penalty for persons convicted of murder.

Three questions had to do with sexual matters, and levels of support differed widely among the three items. Asked whether premarital sex was always wrong, almost always wrong, sometimes wrong, or not wrong at all, 53% chose the last of these, saying it is always all right. Interestingly, 22%—the next most popular response—said it was always wrong, pointing to a polarization of opinions on this topic.

The same question was asked with regard to homosexuality: "sexual relations between two adults of the same sex." Again, opinions were polarized, but the skew was toward disapproval. More than 45% said it was always wrong, while just over 42% said it was not wrong at all.

Finally, respondents were asked whether they had attended an X-rated movie during the past year. About one quarter (25%) said they had.

Demonstration 6.5: Histogram—AGE

In the next few sections, we will examine two other types of graphs appropriate for interval/ratio variables: histograms and line charts.

Like bar charts, *histograms* have two axes, the vertical (*y*) axis and the horizontal (*x*) axis. The categories of the variable are displayed along the horizontal axis, while frequencies or percentages are displayed along the vertical axis. Unlike bar charts, the categories of a histogram are displayed as contiguous bars (bars that touch each other). Both the height and width of each bar are proportional to the frequency or percentage of cases in each category. On a histogram, the sum of the areas covered by all the contiguous bars is 1, or 100%, if proportions or percentages are graphed. If frequencies are graphed, the sum of the areas is the number of cases.

Here's how SPSS Statistics can be instructed to produce a histogram for an interval/ratio variable. Because the political items we have been focusing on in this chapter are nominal and/or ordinal, we need to turn our attention to another issue. For now, let's focus on AGE (respondents' ages at the time of the interview), a measure that you may suspect is at least peripherally related to political orientations and party identification.

Before we instruct SPSS Statistics to create a histogram, make sure the values 98 and 99 are defined as "missing" for the variable AGE. Now select **Graphs**, **Legacy Dialogs**, and then **Histogram** from the drop-down menu. Highlight the variable **AGE** and then either double-click or click the arrow pointing toward the "Variable:" field. For this chart, we don't have to worry about missing values because SPSS Statistics automatically removes them from display.

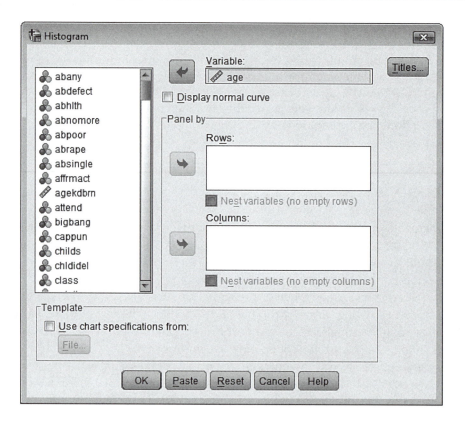

You will notice that once again, you have the option of adding titles, subtitles, and/or footnotes (by clicking the **Titles** button) or displaying a line graph of the normal distribution across your histogram. But we won't do those things now; instead, we will go ahead and run the histogram by clicking **OK**.

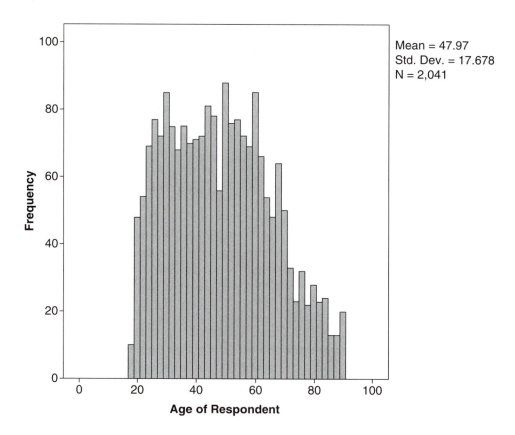

As with bar charts, the horizontal (*x*) axis shows categories of AGE that SPSS Statistics has automatically created. Depending on the range of the variable, SPSS Statistics computes an interval width for each category. For AGE, SPSS Statistics created 33 intervals each with a width of about 2 years. The numbers under the bars of the histogram indicate ages. As with the bars on a bar chart, the bars on the histogram are proportional to the number of cases in the intervals.

SPSS Statistics Command 6.5: Histogram

Click **Graphs** → **Legacy Dialogs** → **Histogram**

→ Double-click on variable name OR highlight variable and click on arrow pointing toward the variable field

→ Click **OK**

The Chart Editor can be used to customize your histogram. Take a few moments to experiment with this option and see if you can improve on the way SPSS Statistics made your chart.

Demonstration 6.6: Line Chart—INCOME

Total family income (INCOME) is another interval/ratio variable somewhat related to political orientations and party identification. While it is possible to represent this distribution using a histogram, we are going to consider another type of graph you are probably already familiar with, a *line chart* (sometimes also called a *frequency polygon*).

A line chart is similar to a histogram and bar chart in that the categories for the variable are displayed on the horizontal (*x*) axis and the frequency or percentage is situated on the vertical (*y*) axis. Unlike the bar chart or histogram, however, the line chart has a single line running from the far left to the far right of the graph, which connects points representing the frequency or percentage of cases for each category of the variable.

Line charts are particularly useful in showing the overall shape or distribution of variables with a large number of values or categories. Consequently, while bar and pie charts are generally used to display discrete variables, histograms and line charts are most often used to display the distribution of continuous, interval/ratio variables.

Before we produce our line chart, make sure the values 13, 98, and 99 for INCOME are defined as "missing." To begin, click **Graphs** → **Legacy Dialogs** once again. This time, however, select **Line,** and the Line Charts dialog box will open, as shown below.

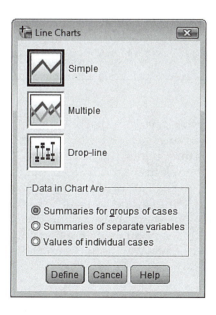

On the left side of the box, you see three types of graphs listed: simple, multiple, and drop-line. Because we are charting only one variable, we want to choose **Simple**, which is probably already highlighted, but you can click on it once again just to be sure. Now click the **Define** button, and the Define Simple Line dialog box will open.

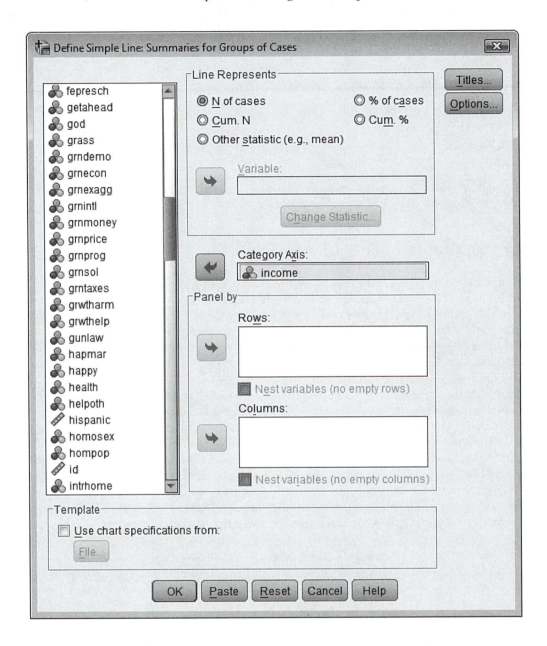

On the left side of the box, you can see the list of variables. Scroll down the list until you find **INCOME,** and highlight it. Now click on the arrow next to the "Category Axis:" box to transfer the variable name. This time, under "Line Represents," choose **N of cases** as opposed to **% of cases** (it may already be selected), and then click the **Options** button to make sure there is not a check mark next to the "Display groups defined by missing values" option. After you have done that, click **Continue,** and you will be back in the Define Simple Line box. As in the previous chart dialog boxes, the **Titles** option can be used to add titles, subtitles, and/or footnotes to the chart. You can explore that option now or just click on **OK** if you are ready to produce your line chart. Once you do that, you should see the following line chart displayed in your Viewer.

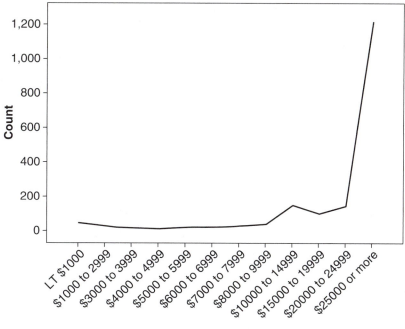

Total Family Income

Take some time to review your line chart. You can see how respondents' reported annual income ranged from those who received less than $1,000 to those who received more than $25,000 in 2010. You may want to take note of the shape of the distribution and whether the sample is concentrated in a specific area or fairly evenly spread out. (Hint: Part of the explanation lies in the distribution of income in the United States. The other part lies in the coding of this variable: The last category includes all respondents making $25,000 or more per year.)

SPSS Statistics Command 6.6: Simple Line Chart

Click **Graphs** → **Legacy Dialogs** → **Line**

→ **Simple** → **Define**

→ Highlight the variable name → Click arrow pointing toward the "Category Axis:" box → Click **Options** →

Click on the box next to the "Display groups defined by missing values" option to make sure it is NOT checked

→ Click **Continue**

→ **OK**

As before, you can use the Chart Editor to change the appearance of your line graph. You can, for instance, change the style or color of your new chart, as well as experiment with several other options available through the Chart Editor.

Some Guidelines for Choosing a Chart or Graph

In this chapter, we have focused on four common and useful charts. The following tables present a summary of the discussion regarding which charts are appropriate for variables of different types (Table 6.1) and at different levels of measurement (Table 6.2). Bear in mind that you can also use SPSS Statistics to produce a number of additional types of charts and graphs. To get a sense of the options available, click **Graphs** on the toolbar. You can see that, in addition to the four options discussed, you have the opportunity to work with many other types of charts.

Table 6.1   Charts Appropriate for Variables of Different Types

	Type of Variable	
	Discrete	Continuous
Bar chart	X	
Pie chart	X	
Histogram		X
Line chart		X

Table 6.2   Charts Appropriate for Variables at Different Levels of Measurement

	Level of Measurement		
	Nominal	Ordinal	Interval/Ratio
Bar chart	X	X	
Pie chart	X	X	
Histogram			X
Line chart			X

While it is fairly easy to learn how to produce charts on SPSS Statistics or any other statistical package, it is more difficult to learn which types of graphs are appropriate for variables of different types and at different levels of measurement. Remember, at your command, SPSS Statistics will produce any chart or graph you request, regardless of whether it is meaningful or appropriate.

Saving and Printing Your Charts

That is enough work for now. If you are interested in saving or printing any of your charts or if you want to save the changes we made to the Adventures.SAV (or AdventuresPLUS. SAV) file (e.g., defining certain values as missing), you can do so by following the commands for saving and printing output discussed at the end of the previous chapter.

Conclusion

Politics is a favorite topic for many Americans, often marked by the expression of unsubstantiated opinions. Now you are able to begin examining the facts of political views. In later chapters, we'll move beyond describing political orientations and start examining why people have the political views they have.

Main Points

- In this chapter, we focused on two key political items: POLVIEWS and PARTYID.
- POLVIEWS taps into basic political philosophy, whereas PARTYID measures political party affiliation.
- A graph or chart can sometimes communicate the relative sizes of different groups or the shape of the distribution of a variable more powerfully than can a table of numbers.
- You can use SPSS Statistics to produce two charts appropriate for discrete, nominal, or ordinal variables: simple bar charts and pie charts.
- Bar and pie charts are the graphic equivalent of frequency tables, best used for variables with a limited number of categories.
- You can also use SPSS Statistics to produce two charts appropriate for continuous, interval/ratio variables: histograms and line charts.
- Line charts and histograms are most useful for variables with numerous response categories.

Key Terms

Bar chart

Histogram

Horizontal (*x*) axis

Line chart

Vertical (*y*) axis

Pie chart

SPSS Statistics Commands Introduced in This Chapter

6.1. Simple Bar Chart

6.2. SPSS Statistics Chart Editor

6.3. Pie Chart

6.4. Accessing Pie Options (SPSS Statistics Chart Editor)

6.5. Histogram

6.6. Simple Line Chart

Review Questions

1. What does the variable POLVIEWS measure?

2. What does the variable PARTYID measure?

3. If you wanted to produce a chart for a discrete variable, which type of chart would be your best option: bar chart, line chart, or histogram?

4. If you wanted to produce a chart for a variable with many categories or values, which type of chart would be your best option: bar chart, pie chart, or line graph?

5. Why are graphic presentations of data useful?

6. Name one reason why you might make a pie chart as opposed to a line chart for the variable PARTYID.

7. Name two measures of political attitudes (other than POLVIEWS and PARTYID) contained on Adventures.SAV.

8. List the types of chart(s)/graph(s) generally recommended for variables at the following levels of measurement:
   a. Nominal
   b. Ordinal
   c. Interval/ratio

9. List the types of chart(s)/graph(s) generally recommended for variables of the following types:
   a. Discrete
   b. Continuous

10. A researcher creates a problematic, overcategorized bar chart for the variable AGE (measured in years). Which of the following is NOT a solution to this problem?
    a. Combine categories of the variable AGE (recode), and produce an appropriate chart for the new variable.
    b. Produce a pie chart for the variable AGE.
    c. Produce a line chart for the variable AGE.

SPSS STATISTICS LAB EXERCISE 6.1

NAME _____

CLASS _____

INSTRUCTOR _____

DATE _____

To complete the following exercises, access the Adventures.SAV file.

1. (NEWS) 0, 8, and 9 should be defined as "missing."

   a. What does the variable NEWS measure (hint: variable label)?

   b. What is the variable's level of measurement? _____

   c. List one type of chart appropriate for this variable: _____

   d. Produce the chart listed in response to Question 1c above, and then, based on your output, fill in the blanks below.

About _____% of respondents read the newspaper once a week or more, whereas about ____% read the newspaper less than once a week. Only ____% of the sample said they never read the newspaper.

2. (TVHOURS) 1, 98, and 99 should be defined as "missing."

   a. What does the variable TVHOURS measure?

   b. Would you describe the variable as discrete or continuous? _____

   c. List two charts appropriate for this variable: _____ and _____

   d. Produce one of the charts listed in response to Question 2c above that will allow you to answer the question below. (Hint: When you run a histogram, SPSS Statistics automatically calculates two statistics that are useful for answering this question—mean and standard deviation.)

Using the standard deviation, we would expect two thirds of respondents to watch between ____ hour(s) and _____ hours of television per day.

Questions 3 through 5: After you have produced and analyzed the charts for CONPRESS and CONTV, briefly describe the distribution in the spaces provided below. You may want to note the largest and smallest categories for the items, as well as any other interesting or pertinent information.

3. (CONPRESS) 0, 8, and 9 should be defined as "missing."

   a. List one type of chart appropriate for this variable: _____

   b. Description of chart:

4. (CONTV) 0, 8, and 9 should be defined as "missing."

    a. List one type of chart appropriate for this variable: _____

    b. Description of chart:

5. (EDUC) 97, 98, and 99 should be defined as "missing."

    a. List one type of chart appropriate for this variable: _____

    b. Produce the chart listed in response to Question 5a above, and then describe the distribution below:

6. How do American adults feel about sex education in public schools? Produce and analyze a chart for the variable SEXEDUC (0, 8, and 9 should be defined as "missing"), then describe your findings below:

7. Do Americans tend to think sex before marriage for teens 14 to 16 is appropriate or inappropriate? Produce and analyze a chart for the variable TEENSEX (0, 8, and 9 should be defined as "missing"), then describe your findings below:

8. Choose three variables in your Adventures.SAV file related to personal sexual behavior, and then list the abbreviated variable names and variable labels below:

        Abbreviated Variable Name   *Variable Label*

Variable 1 _____

Variable 2 _____

Variable 3 _____

9. Produce and analyze charts for the three terrorism preparedness variables listed in response to Question 8 above. Based on your findings, how would you describe the sexual behavior of Americans?

Chapter 7   **Recoding Your Data**

Religiosity and Political Orientations

At the end of our discussion of frequency distributions, we mentioned that frequency tables are not particularly useful when you are working with continuous variables or any variables with a larger number of response categories. As an alternative, we suggested you may want to use basic measures of central tendency and dispersion to describe a distribution. That's what we did with AGE in Chapter 5.

But often, instead of using statistics to describe a variable, we want to use a set of meaningful groupings that suit our research. Even when we know our respondents' exact ages, it may be more useful to group them in categories such as young, middle, and older. The groups are formed by *collapsing categories* of the variable and recoding them. Those of you who have worked with frequency distributions may recognize that what we are doing is similar to the creation of a grouped frequency distribution. Only, in this case, it is much easier because we have SPSS Statistics to assist in the process.

SPSS Statistics can be instructed to combine adjacent categories using the Recode command. *Recoding* is a technique that allows us to combine or group two or more categories of a variable together in order to simplify the process of analysis. When we recode, we can take a variable such as ATTEND with nine valid categories and combine them to create four categories (or fewer). Recoding is a useful technique to master for several reasons. In addition to helping us create easier-to-read tables and identify patterns in responses, it also enables us to group continuous data into categories so we don't present our readers with an excessively long table. If you don't believe this could be a problem, try imagining the size of a table relating our Adventures.SAV file variables EDUC and AGE!

Recoding takes several steps. While it may take a little time to get used to the process of recoding, if you follow along with the demonstrations below and practice them using the lab exercises, you should get the hang of it quickly. If you have trouble recalling the steps at first, don't worry, you will have plenty of opportunities to practice recoding in later chapters and exercises.

Demonstration 7.1: Modifying Variables With Recode—ATTEND → CHATT

Throughout this chapter, we will use a number of variables from the Adventures.SAV file that you have worked with in previous chapters, beginning with ATTEND. You may recall that ATTEND measures how often the respondent attends religious services. To help refresh your memory, go ahead and run a frequency distribution for this variable. You should get a table that matches the one below.

attend HOW OFTEN R ATTENDS RELIGIOUS SERVICES

		Frequency	Percent	Valid Percent	Cumulative Percent
Valid	0 NEVER	470	23.0	23.1	23.1
	1 LT ONCE A YEAR	143	7.0	7.0	30.1
	2 ONCE A YEAR	287	14.0	14.1	44.2
	3 SEVRL TIMES A YR	211	10.3	10.4	54.6
	4 ONCE A MONTH	147	7.2	7.2	61.8
	5 2-3X A MONTH	171	8.4	8.4	70.2
	6 NRLY EVERY WEEK	87	4.3	4.3	74.5
	7 EVERY WEEK	382	18.7	18.8	93.2
	8 MORE THN ONCE WK	138	6.8	6.8	100.0
	Total	2036	99.6	100.0	
Missing	9 DK,NA	8	.4		
Total		2044	100.0		

The recoding process will allow us to make this variable more manageable by combining adjacent categories. The result will be a new variable with four, as opposed to nine, response categories. When we are done, you will be able to compare the frequency for the new variable with the frequency table you just ran and see the difference firsthand.

To begin this process, select **Transform**, and then select **Recode** from the drop-down menu. SPSS Statistics now asks if you want to replace the existing values of the variables with the new, recoded ones. Select **Into Different Variables**, because we are going to assign a new name to the new, recoded values.[1] SPSS Statistics now presents you with the following screen, in which you can describe the recoding you want.

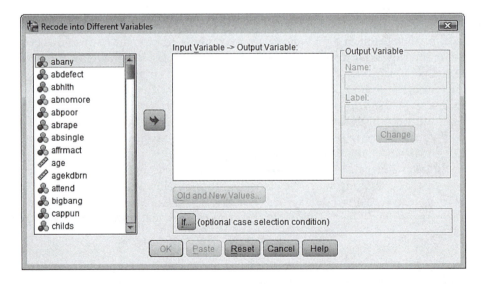

---

[1]Because the student version of SPSS Statistics limits the number of variables that can be used to 50, if you are using an older student version, you might want to save the recode under the same name. It's okay to do this if you save the modified data set later under a different name (e.g., Adventure01.SAV). That way, you can reopen one of the modified data sets or you can retrieve the original data in its unrecoded form. Note that the Sample.SAV file does not use all 50 allowed variable slots, so there is room to add a few recoded variables and save to a usable data file.

In the variable list at left, find and select **ATTEND**. To transfer ATTEND to the "Input Variable → Output Variable:" field, either double-click on ATTEND or click on the arrow to the right of the variable list.

Notice that you need to tell SPSS Statistics what you would like to name the new, recoded variable. You can accomplish this easily in the section of the window called "Output Variable." Let's name the recoded variable "CHATT" (for CHurch ATTendance). Type **CHATT** into the space provided for the output variable name, and then label the variable (perhaps "Recoded Church Attendance"). When you are done, click the **Change** button. As you can see in the middle field, SPSS Statistics will now modify the entry to read "ATTEND → CHATT."

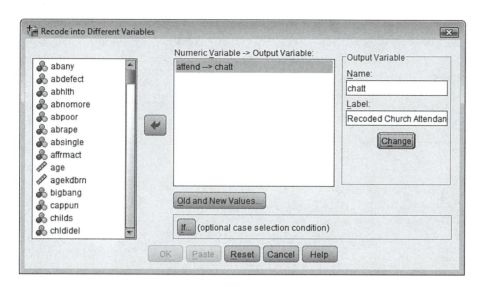

Thus far, we have created a new variable, but we haven't entered any data into it. We initiate this final step by clicking the **Old and New Values** button. Now, SPSS Statistics presents you with the following window.

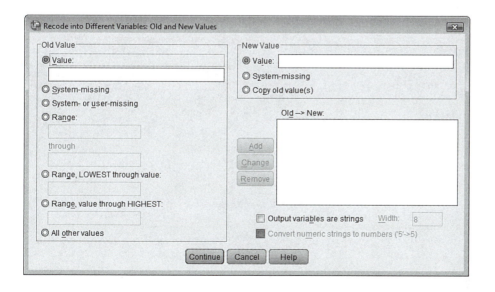

The left side of this window provides us with several options for specifying the old values we want to recode. The first, which SPSS Statistics has selected as a default, lets us specify a single value on the original variable, such as 8 ("More than once a week"). A more efficient option for our present purposes is found farther down the list, letting us specify a range of values. (Remember, to find the numerical codes assigned to ATTEND, you can use one of the several options discussed earlier.)

In our manual collapsing of categories on this variable earlier, you'll recall that we combined the values 6 ("Nearly every week"), 7 ("Every week"), and 8 ("More than once a week"). We can accomplish the same thing now by clicking the first **Range** button and entering 6 and 8 in the two boxes.

At the top of the right side of the window, notice that there is a space for you to enter the new value for this combination of responses. Let's recode it 1. Enter that number in the field provided, as shown below.

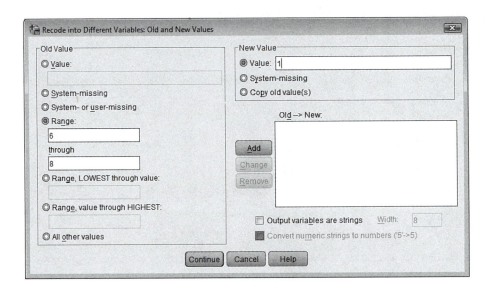

Once you've added the recode value, notice that the **Add** button just below it is activated. Where it was previously grayed out, the button is now illuminated and available for use. Click it.

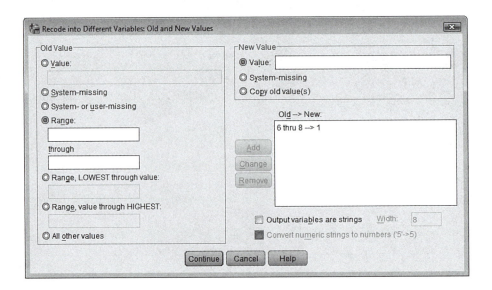

This action causes the expression "6 thru 8 → 1" to appear in the field. We've given SPSS Statistics part of its instructions. Now let's continue.

Click **Range** again, and now let's combine values 4 ("Once a month") and 5 ("Two to three times a month"). Give this new combined category the value of 2. Click **Add** to add it to the list of recodes. Now combine categories 1 ("Less than once a year"), 2 ("Once a year"), and 3 ("Several times a year"). Recode the new category as 3, and **Add** it to the list. Finally, let's recode 0 ("Never") as 4. On the left side of the window, use the **Value** button to accomplish this. Enter 0 there, and enter 4 as the new value. Click on **Add**.

To tidy up our recoding, we could have SPSS Statistics maintain the missing data values of the original variable. We would accomplish this by clicking **System- or user-missing** as an old value (on the left side) and **System-missing** as the new value (on the right side), and then clicking **Add**.[2] Although it is good practice to consciously recode every category, in this case it is not necessary. Any cases not covered by the range of the old values will be undefined and treated as missing values.

Your Recode window should now look like this:

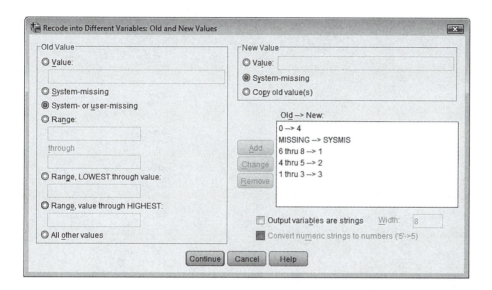

As we wrap up, we should repeat that there are no hard-and-fast rules for choosing which categories to combine in a recoding process such as this. There are, however, two rules of thumb to guide you: one logical, the other empirical.

First, there is sometimes a logical basis for choosing cutting points at which to divide the resulting categories. In recoding AGE, for example, it is often smart to make one break at 21 years (the traditional definition of adulthood) and another at 65 (the traditional age of retirement). In the case of church attendance (attendance at religious services), our first combined category observes the Christian norm of weekly church attendance.

The second guideline is based on the advantage of having sufficient numbers of cases in each of the combined categories, because a very small category will hamper subsequent analyses. Ideally, each of the combined categories should have roughly the same number of cases.

How do you suppose we'd continue the recoding process? Click **Continue**, you say? Hey, you may be a natural at this. Do that. This takes you back to the Recode into Different Variables window. Now that you've completed your specification of the recoding of this variable, all that remains is to click **OK** at the bottom of the window. SPSS Statistics may take a few seconds now to accomplish the recoding you've specified.

Go to the Data View portion of the Data Editor window now. To see your new variable, scroll across the columns of the window until you discover CHATT in the last column used

---

[2]For a brief overview of the difference between "system- or user-missing" and "system-missing," right-click on either term and after a moment, a brief explanation will display on your screen.

thus far.[3] Notice the values listed in the column. Case No. 1 has a value of 3.00 on the new variable, Case No. 2 has a value of 4.00, Case No. 3 has a value of 2.00, Case No. 4 has a value of 2.00, and so on.

Now find ATTEND. Notice that Case No. 1 has a 2 in ATTEND. That's correct because everyone with a 6, 7, or 8 in the original variable was recoded as a 1 in the new variable, CHATT. Similarly, notice that Case No. 2 has a 0 in ATTEND. That, too, is correct because everyone with a 0 in the original variable was recoded as a 4. You can check a few more people if you want to verify that the coding was accomplished as we instructed. This is a good idea, by the way, to ensure that you haven't made a mistake. (Presumably, SPSS Statistics doesn't make computational mistakes.) The next step in recoding is to define your new, recoded variable CHATT. We can do this by accessing the Variable View tab. Once you are in Variable View, scroll to the bottom of your screen until you see CHATT listed in the last row, as shown below.

	Name	Type	Width	Decimals	Label	Values	Missing	Columns	Align	Measure	Role
121	relig	Numeric	2	0	RS RELIGIOUS...	{0, IAP}...	0, 98, 99	8	Right	Nominal	Input
122	relig16	Numeric	2	0	RELIGION IN ...	{0, IAP}...	0, 98, 99	8	Right	Nominal	Input
123	rincome	Numeric	2	0	RESPONDENT...	{0, IAP}...	13 - 99, 0	8	Right	Nominal	Input
124	satfin	Numeric	1	0	SATISFACTIO...	{0, IAP}...	0, 8, 9	8	Right	Nominal	Input
125	satjob	Numeric	1	0	JOB OR HOUS...	{0, IAP}...	0, 8, 9	8	Right	Nominal	Input
126	sei	Numeric	4	1	RESPONDENT...	{-1.0, IAP}...	-1.0, 99.8, 9...	8	Right	Scale	Input
127	sex	Numeric	1	0	RESPONDENT...	{1, MALE}...	0	8	Right	Nominal	Input
128	sexeduc	Numeric	1	0	SEX EDUCATI...	{0, IAP}...	0, 8, 9	8	Right	Nominal	Input
129	sexfreq	Numeric	1	0	FREQUENCY ...	{-1, IAP}...	-1, 0, 9	8	Right	Nominal	Input
130	SEXORNT	Numeric	1	0	SEXUAL ORIE...	{0, IAP}...	0, 8, 9	9	Right	Nominal	Input
131	sexsex	Numeric	1	0	SEX OF SEX P...	{0, IAP}...	0, 8, 9	8	Right	Nominal	Input
132	sexsex5	Numeric	1	0	SEX OF SEX P...	{0, IAP}...	0, 8, 9	8	Right	Nominal	Input
133	sibs	Numeric	2	0	NUMBER OF B...	{-1, IAP}...	-1, 98, 99	8	Right	Scale	Input
134	socbar	Numeric	1	0	SPEND EVENI...	{-1, }...	0, 8, 9	8	Right	Nominal	Input
135	socfrend	Numeric	1	0	SPEND EVENI...	{-1, }...	0, 8, 9	8	Right	Nominal	Input
136	spanking	Numeric	1	0	FAVOR SPAN...	{0, IAP}...	0, 8, 9	8	Right	Nominal	Input
137	teensex	Numeric	1	0	SEX BEFORE ...	{0, IAP}...	0, 8, 9	8	Right	Nominal	Input
138	thnkself	Numeric	1	0	TO THINK FOR...	{0, IAP}...	0, 8, 9	8	Right	Nominal	Input
139	trdunion	Numeric	1	0	WORKERS NE...	{0, IAP}...	0, 8, 9	8	Right	Nominal	Input
140	tvhours	Numeric	2	0	HOURS PER D...	{-1, IAP}...	-1, 98, 99	8	Right	Scale	Input
141	union	Numeric	1	0	DOES R OR S...	{0, IAP}...	0, 8, 9	8	Right	Nominal	Input
142	usewww	Numeric	1	0	R USE WWW ...	{0, IAP}...	0, 8, 9	8	Right	Nominal	Input
143	VOTE08	Numeric	1	0	DID R VOTE IN ...	{0, IAP}...	0, 8, 9	8	Right	Nominal	Input
144	workhard	Numeric	1	0	TO WORK HARD	{0, IAP}...	0, 8, 9	8	Right	Nominal	Input
145	wrkstat	Numeric	1	0	LABOR FORC...	{0, IAP}...	0, 9	8	Right	Nominal	Input
146	wrkwayup	Numeric	1	0	BLACKS OVE...	{0, IAP}...	0, 8, 9	8	Right	Nominal	Input
147	wwwhr	Numeric	3	0	WWW HOURS...	{-1, IAP}...	-1, 998, 999	8	Right	Scale	Input
148	xmarsex	Numeric	1	0	SEX WITH PE...	{0, IAP}...	0, 8, 9	8	Right	Nominal	Input
149	xmovie	Numeric	1	0	SEEN X-RATE...	{0, IAP}...	0, 8, 9	8	Right	Nominal	Input
150	zodiac	Numeric	2	0	RESPONDENT...	{0, IAP}...	0, 98, 99	8	Right	Nominal	Input
151	chatt	Numeric	8	0	Recoded Churc...	None	None	10	Right	Nominal	Input
152											
153											
154											
155											

Remember that the Variable View tab stores important information about each of our variables. Since we just created the variable CHATT, we need to specify its characteristics.

Let's begin with the column labeled "Decimals." Simply point and click on the cell that corresponds with the Decimals column and the variable CHATT. Notice that up and down arrows appear on the right side of the cell. Because decimal points are of no use to us in this situation, let's get rid of them by using the down arrow to change the number of decimal places to 0. Now move to the column labeled "Width" directly to the left. To do that, click on the cell that corresponds with the Width column and the CHATT row. Because our recoded variable contains codes that range from 1 to 4, the width need only be set at 1—though leaving it set at 8 will not make any discernible differences for the purposes of analysis covered in this book. Again, if you wish, use the down arrow to change the width to 1.

---

[3]If you used the same variable name, look at ATTEND.

Recall that when we began the process of recoding, we not only named our new variable CHATT, but we labeled it "Recoded Church Attendance" (in the Recode into Different Variables dialog box). Consequently, you do not have to change any of the information in the Label column for the variable CHATT. However, if you want to view the entire label or if you want to change the label, simply double-click on the cell that corresponds with the Label column and the CHATT row. You will notice that the width of the cell increases so you can better view the entire label and, if you desire, modify the variable label.

The next column, labeled "Values," is important because it allows us to store labels for the values of our new variable. To do this, simply click on the right side of the cell that corresponds with the Variables column and the CHATT variable. You will notice that the Value Labels box appears on the screen, as shown below.

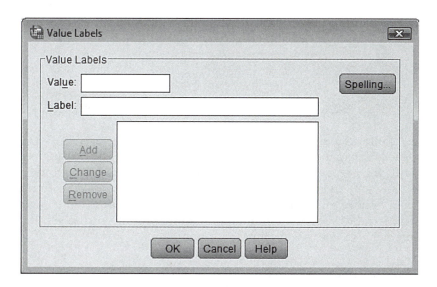

To add the values and labels for our new variable, simply type "1" in the "Value:" field. Recall that this value represents people who originally scored 6 ("Nearly every week"), 7 ("Every week"), and 8 ("More than once a week"). Let's call this new, combined category "About weekly." Type that description in the "Label:" field, and click **Add**. You can see that the information now appears in the field below, telling us that 1 = "About weekly." Now enter the remaining value labels as indicated below. Once you have entered all the value and label information (and your screen looks like ours below), click **OK**.

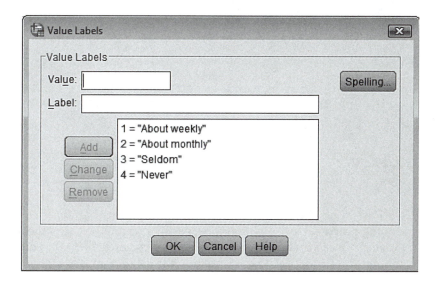

If you accidentally add the wrong information, don't worry. You can delete the information by highlighting it and clicking **Remove**. Alternatively, as shown below, you can correct the value or label and then click the **Change** button, which will become usable once you begin typing something different into the "Value:" or "Label:" field.

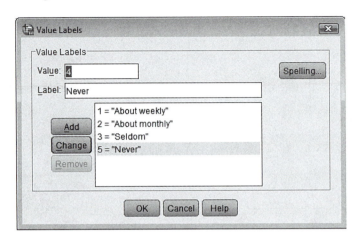

The last bit of information of concern to us at this point is the variable's level of measurement. We can easily select the appropriate level of measurement by clicking on the cell that corresponds with the Measure column (the second-to-last column on the far right side of your Variable View tab) and the CHATT row. Notice that once you point and click on that cell, three options appear, representing each of the levels of measurement: nominal, ordinal, and scale (interval/ratio). Use the down arrow and your mouse to select the appropriate level of measurement (in this case, ordinal).

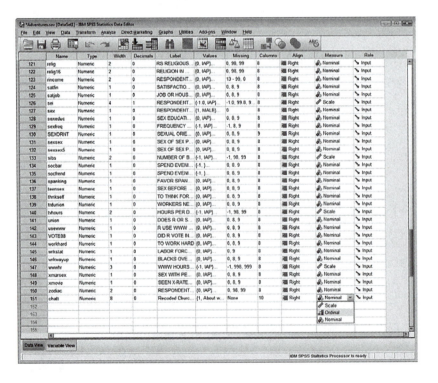

Now that we have used the Variable View tab to define our variable, let's review the results of our recoding process. We can do this most easily through the use of the now familiar Frequencies command. Once you have accessed the Frequencies dialog box, scroll down the variable list. You'll see that CHATT is now included in the list (which is in alphabetical order).

Once you have found CHATT, go ahead and run a frequency distribution for the variable CHATT. Your table should be similar to the one shown in the screen below.

chatt Recoded Church Attendance

		Frequency	Percent	Valid Percent	Cumulative Percent
Valid	1 About weekly	607	29.7	29.8	29.8
	2 About monthly	318	15.6	15.6	45.4
	3 Seldom	641	31.4	31.5	76.9
	4 Never	470	23.0	23.1	100.0
	Total	2036	99.6	100.0	
Missing	System	8	.4		
Total		2044	100.0		

Notice how much more manageable the recoded variable is. Now we can use the recoded variable in our later analyses.

Summary of Steps Involved in Recoding

Step 1: Preparing to Recode a Variable

- Identify which variable you want to recode (e.g., ATTEND).
- Run a frequency distribution for the variable you want to recode to see its current coding (as we did with ATTEND).
- Figure out which values you want to combine to create the new variable (as we did with ATTEND). Keep in mind the loose guidelines for recoding discussed earlier. You should also consider the level of measurement for your new recoded variable, because it will impact the type of statistical analyses you can perform.

Many of us find it helpful to make a list of the old values and new values with their labels before we start the recoding process. Here's what we did for ATTEND.

Old Variable ATTEND		New Variable CHATT	
Value	Label	Value	Label
0	Never	4	Never
1	Less than once a year	3	Seldom
2	Once a year	3	
3	Several times a year	3	
4	Once a month	2	About monthly
5	Two to three times a month	2	
6	Nearly every week	1	About weekly
7	Every week	1	
8	More than once a week	1	

- Decide on a name and label for your new (recoded) variable (i.e., CHATT for CHurch ATTendance, "Recoded Church Attendance").

Step 2: Recoding

Use the SPSS Statistics Recode command to create your new variable (i.e., **Transform Recode → Into Different Variables**).

Step 3: Checking Your Recode in the Data Editor

Check your new variable in the Data View portion of the Data Editor to make sure your recoding was correct.

Step 4: Defining Your Variable

Define your new variable using the Variable View tab (i.e., decimals, width, label, values, measure).

Step 5: Running a Frequency Distribution

Run a frequency distribution for your new variable to ensure the recoding was done correctly.

Demonstration 7.2: Recoding AGE → AGECAT

The value of recoding is especially evident in the case of continuous interval/ratio variables such as age and education. Because these types of variables have so many categories, they are totally unmanageable in some forms of analysis. Fortunately, we can recode scale variables as easily as we just recoded ATTEND.

Why don't we go ahead and recode the variable AGE, because in addition to giving us more practice with recoding, it will allow us to take advantage of an additional feature in the recoding process.

If you follow the steps listed above, you will note that in accordance with Step 1, we have already identified the variable we want to recode (AGE). Now take a moment to run a frequency distribution for AGE so you can see the categories we want to combine. Feel free to look back at the instructions given earlier in the chapter if you have any problem recalling how to run a frequency distribution.

In this case, we want to collapse this multitude of categories into four more manageable categories as follows:

Old Variable AGE	New Variable AGECAT	
Values	Values	Labels
18 to 20	1	Under 21
21 to 39	2	21 to 39
40 to 64	3	40 to 64
65 to 89	4	65 and older

Note in this instance that the oldest respondent to the 2010 General Social Survey (GSS) was 89 years old. A value of 99 in this case refers to "NA," or people who did not answer the question, rather than the actual age of a respondent. In this case, the value of 99 for the variable AGE should be labeled as "missing." Go ahead and do that, or verify that it has already been done, before continuing with the recode. If you have any problems setting values as "missing," see the discussion in the previous chapter.

We will call our new, recoded variable AGECAT (to represent AGE CATegories) and label the variable "Recoded AGE." Now we are ready to use the Recode command to create our new variable. To do that, simply select **Transform** → **Recode** → **Into Different Variables** again. Notice that the Recode into Different Variables window still shows our recoding of ATTEND. Clear the boards by clicking **Reset** at the bottom of the window.

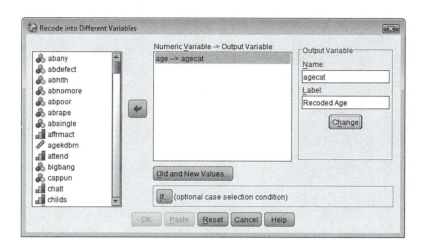

Then, select **AGE** and move it to the "Numeric Variable → Output Variable:" window. Name the new variable "AGECAT," add the variable label, and click **Change**. Now you can select **Old and New Values** to tell SPSS Statistics how to recode.

In recoding AGE, we want to make use of the range option again, but for our first recode, check the **Range, LOWEST through value** option. This will ensure that our youngest category will include the youngest respondents. Enter 20 in the box, specify the new value as 1, and click **Add**.

You may also want to take advantage of the other range option, **Range, value through HIGHEST**. When you do this, however, beware! Although the lowest and highest range specifications are handy, they must be used with care. If we hadn't run a frequency distribution for AGE and checked on our missing values, we could easily specify "65 through highest" as the range and not notice that in this case 99 refers to "NA." As a result, we may run the risk of including 10 people who didn't specify their ages as 65 years and older.

Now do what you have to do to create the remaining recode instructions indicated below.

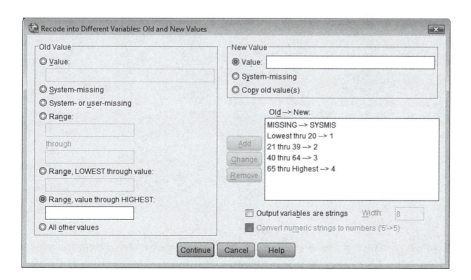

Click **Continue** to return to the main Recode window and then **OK** to make the recoding changes.

Now go to the Data View portion of the Data Editor and you should see your new, recoded variable, AGECAT, in the rightmost column. Check it out.

To complete the recode process, follow Steps 3 and 4. Check your new variable AGECAT against the old variable AGE in Data View, and then define AGECAT in Variable View. When defining your variable, remember to set your decimals to 0 and your width to 1. Then list the values and labels for your recoded variable AGECAT and select the appropriate level of measurement (in this case, ordinal), as shown below.

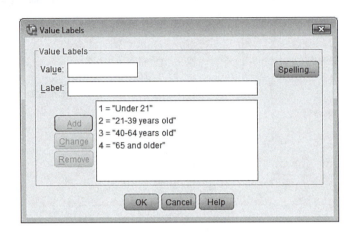

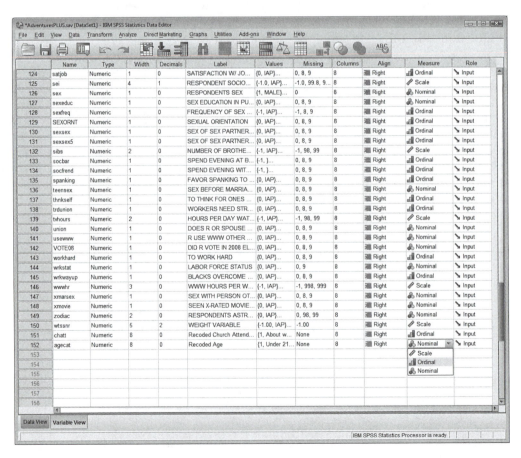

Once you've completed the recoding and labeling, check the results of your labors by following Step 5 and running a frequency distribution for your new variable AGECAT.

Notice that AGE may still be in the list of variables to be analyzed. You can place AGE back in the list of abbreviated variable names by using one of the methods we discussed previously (clicking **Reset**, or double-clicking or highlighting **AGE** and selecting the left-pointing arrow). If you fail to do this, SPSS Statistics will simply calculate and report the frequencies on AGE again, along with any other variables selected.

agecat Recoded Age

		Frequency	Percent	Valid Percent	Cumulative Percent
Valid	1 Under 21	58	2.8	2.8	2.8
	2 21-39 years old	680	33.3	33.3	36.2
	3 40-64 years old	910	44.5	44.6	80.7
	4 65 and older	393	19.2	19.3	100.0
	Total	2041	99.9	100.0	
Missing	System	3	.1		
Total		2044	100.0		

Writing Box 7.1

The largest percentage of respondents to the 2010 GSS is between the ages of 40 and 64 (45%). This is followed by 33% of respondents who are 21 to 39. Only 19% of respondents are retirement age (65 and older), while less than 3% are under 21 years of age.

SPSS Statistics Command 7.1: Recoding a Variable

Click **Transform** → **Compute** → **Into Different Variables**

→ Highlight name of variable to be recoded

→ Click on right-pointing arrow OR double-click on variable name →

In "Output Variable" field, type name and label of new variable → Click **Change**

→ Click **Old and New Values**

→ Choose option for specifying old values you want to recode (i.e., **Range**, etc.) → Enter old value codes → Enter new value recodes → Click **Add** → (Repeat this step until all old values are recoded) → Click **Continue** → **OK**

To define new, recoded variable: Select **Variable View** tab → Identify new, recoded variable in last row → Click on cells corresponding with new, recoded variable and the following columns to change **Width, Decimals, Values,** and **Measure**

Demonstration 7.3: Recoding POLVIEWS → POLREC

Now that we have recoded ATTEND and AGE, let's turn to two of the political variables we examined in the previous chapter. You may recall that in Chapter 6 we examined the respondent's basic political philosophy using the variable POLVIEWS. We also looked at the respondent's party identification using the variable PARTYID. You may also recall that both of these variables had seven or more valid categories. In the following two demonstrations, we will use the Recode command to reduce the number of categories for these variables to a more manageable size.

We will begin with the variable POLVIEWS. Go ahead and run a frequency table for this variable. Your results should be the same as those below.

polviews THINK OF SELF AS LIBERAL OR CONSERVATIVE

		Frequency	Percent	Valid Percent	Cumulative Percent
Valid	1 EXTREMELY LIBERAL	76	3.7	3.9	3.9
	2 LIBERAL	259	12.7	13.1	17.0
	3 SLIGHTLY LIBERAL	232	11.4	11.8	28.7
	4 MODERATE	746	36.5	37.8	66.5
	5 SLGHTLY CONSERVATIVE	265	13.0	13.4	80.0
	6 CONSERVATIVE	315	15.4	16.0	95.9
	7 EXTRMLY CONSERVATIVE	80	3.9	4.1	100.0
	Total	1973	96.5	100.0	
Missing	8 DK	61	3.0		
	9 NA	10	.5		
	Total	71	3.5		
Total		2044	100.0		

As you can see, POLVIEWS has seven valid categories ranging from "Extremely liberal" to "Extremely conservative." If you simply wanted a measure of liberals versus conservatives, you might decide to recode the variable into just three categories: Liberal, Moderate, and Conservative.

Let's go ahead and do that. We will create a new variable called "POLREC" by recoding POLVIEWS as follows:

1 through 3 → 1   *Liberal*

4 → 2   *Moderate*

5 through 7 → 3   *Conservative*

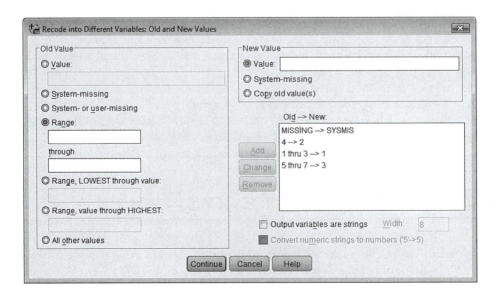

Once you have successfully recoded POLVIEWS, remember to access the Variable View tab and set the decimals, width, values, and appropriate level of measurement for your new, recoded variable POLREC. In this case, assign new labels to the values of POLREC as follows:

1 = "Liberal"

2 = "Moderate"

3 = "Conservative"

To see the results of our recoding, we repeat the Frequencies command with the new variable POLREC.

polrec Recoded Political Views

		Frequency	Percent	Valid Percent	Cumulative Percent
Valid	1 Liberal	567	27.7	28.7	28.7
	2 Moderate	746	36.5	37.8	66.5
	3 Conservative	660	32.3	33.5	100.0
	Total	1973	96.5	100.0	
Missing	System	71	3.5		
Total		2044	100.0		

If your screen looks like the one shown above, congratulations! You have mastered the art of recoding and are ready to move ahead. If not, don't worry, you may just need to go back and review the steps outlined earlier.

Demonstration 7.4: Recoding PARTYID → PARTY

As a final demonstration, let's go ahead and recode the variable PARTYID. You may recall that **PARTYID** contains eight categories ranging from "Strong Democrat" to "Other party." If you want to refresh your memory, go ahead and run a frequency table for this variable, as shown below.

partyid POLITICAL PARTY AFFILIATION

		Frequency	Percent	Valid Percent	Cumulative Percent
Valid	0 STRONG DEMOCRAT	348	17.0	17.2	17.2
	1 NOT STR DEMOCRAT	348	17.0	17.2	34.3
	2 IND,NEAR DEM	265	13.0	13.1	47.4
	3 INDEPENDENT	360	17.6	17.8	65.1
	4 IND,NEAR REP	197	9.6	9.7	74.9
	5 NOT STR REPUBLICAN	277	13.6	13.7	88.5
	6 STRONG REPUBLICAN	184	9.0	9.1	97.6
	7 OTHER PARTY	49	2.4	2.4	100.0
	Total	2028	99.2	100.0	
Missing	9 NA	16	.8		
Total		2044	100.0		

As the frequency table shows, this variable probably has more answer categories than we will be able to manage easily. Consequently, we should recode PARTYID to create a new, less cumbersome measure of party identification called "PARTY."

It makes sense to combine the first two categories: the "strong" and "not strong" Democrats (0 and 1). Similarly, we will want to combine the corresponding Republican categories (5 and 6). Two of the categories, however, need a little more discussion: the two Independent groups who said, when pressed by interviewers, that they were "near" one of the two parties.

Should we combine those near the Democrats with that party, for example, or should we combine them with the other Independents? There are a number of methods for resolving this question. For now, however, we are going to choose the simplest method. As we continue our analyses, it will be useful if we have ample numbers of respondents in each category, so we will recode with an eye to creating roughly equal-sized groups. In this instance, that means combining the three Independent categories into one group (2, 3, and 4). So let's recode as follows:

0 through 1 → 1   *Democrat*

2 through 4 → 2   *Independent*

5 through 6 → 3   *Republican*

7 → 4   *Other*

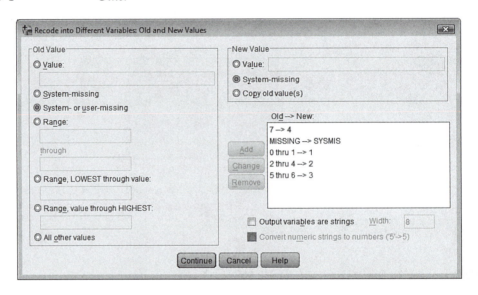

Then, label PARTY as follows: 1 = "Democrat," 2 = "Independent," 3 = "Republican," 4 = "Other."

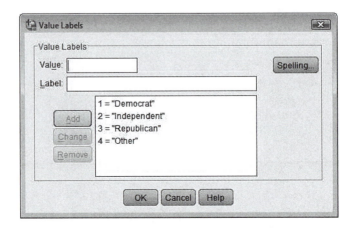

Enter and execute these commands now. Once you've done so, we'll be ready to create a frequency table and either a pie or bar chart for the new variable, as shown below.

party Recoded PartyID

		Frequency	Percent	Valid Percent	Cumulative Percent
Valid	1 Democrat	696	34.1	34.3	34.3
	2 Independent	822	40.2	40.5	74.9
	3 Republican	461	22.6	22.7	97.6
	4 Other	49	2.4	2.4	100.0
	Total	2028	99.2	100.0	
Missing	System	16	.8		
Total		2044	100.0		

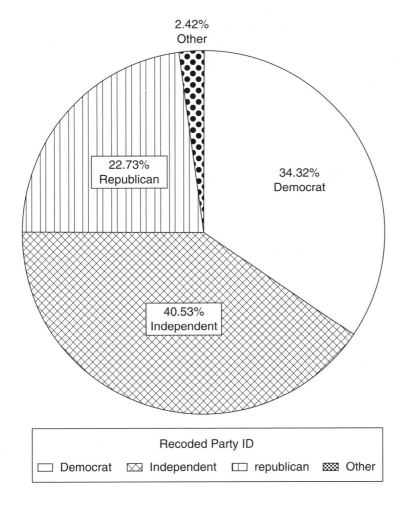

Writing Box 7.2

Whereas more than one third of respondents (34%) to the 2010 GSS are Democrats, only a bit more than a quarter (29%) describe themselves as "liberal." On the other hand, 23% of respondents identify with the Republican Party, but almost 34% describe themselves as "conservative." More than one third of the respondents say they are moderate (38%), and a similar number describe themselves as Independents (41%).

Demonstration 7.5: Saving Changes to Your Data Set

That's enough work for now. But before we stop, let's save the work we've done in this session so we won't have to repeat it all when we start up again. The recoding you've done so far is being held only in the computer's volatile memory. That means if you leave SPSS Statistics right now, all the recoding changes will disappear. Because we will want to use the recoded variables CHATT, AGECAT, POLREC, and PARTY later, a simple procedure will save us time at our next session.

If you are sharing a computer with other users, you should save Adventures.SAV as "AdventuresPLUS.SAV" on a removable or network drive for safe keeping. If you are using your own personal computer, you can save your work on your computer's hard drive. Keep in mind that whenever you save a file in SPSS Statistics (whether you are saving output or your data set), you are saving only what is visible in the active window.

To save your data set, you need to make sure the Data Editor is the active window. You can do that by either clicking on **Adventures.SAV** in the taskbar or choosing **Window** in the menu bar and then selecting **Adventures.SAV** from the drop-down menu. Then, go to the **File** menu and select **Save As**. From the "Save in:" drop-down menu at the top of the Save As window, select a location for your data to be saved—either the USB drive, network drive, or the computer's hard drive.

Type the name for your file in the "File Name" field. Although you can choose any name you wish, in this case, we will use "AdventuresPLUS.SAV" to refer to the file that contains our basic Adventures.SAV file with the changes we made today using the instructions in Chapter 7 (i.e., four newly recoded variables). Then click **Save**. Now you can leave SPSS Statistics with the **File** → **Exit** command. If you open this file the next time you start an SPSS Statistics session, it will have all the new, recoded variables.

SPSS Statistics Command 7.2: Saving Changes Made to an Existing Data Set

> Make sure the Data Editor is the active window
>
> → Click **File** → **Save As** → Select appropriate drive → Name your file → **Save**

Conclusion

Social scientists have a variety of options when it comes to working with data. While in the previous two chapters we examined variables relating to religious and political attitudes, we were somewhat restricted. In this chapter, we learned that while the researchers who collected the GSS data opted to distinguish between people who go to church two to three times a month and people who go to church nearly every week, using SPSS Statistics we can combine adjacent categories and create a new, more manageable variable. Similarly, while some researchers may want to know the exact age of respondents, others may want to combine categories of this variable to create age range categories.

Thankfully, the software we are using allows us to do this fairly easily. By recoding "into different variables," we have the luxury of examining the variables in their original format and combining categories to create a wholly new, more accessible version as well.

Main Points

- When working with continuous variables, one option often used to describe your distribution is to reduce the number of categories by recoding.
- Modifying variables in SPSS Statistics by combining adjacent categories is called collapsing categories.
- Recoding is particularly useful when you are dealing with continuous variables or variables with many response categories.
- There are no rules for deciding which categories to combine when you are recoding. There are, however, two guidelines—one logical and the other empirical.
- Before exiting SPSS Statistics, don't forget to save any changes you made to the existing data set.

Key Terms

Collapsing categories   Recoding

SPSS Statistics Commands Introduced in This Chapter

7.1.   Recoding a Variable

7.2.   Saving Changes Made to an Existing Data Set

Review Questions

1. What are the two general guidelines you should keep in mind when deciding how to combine adjacent categories for recoding?

2. It makes most sense to consider recoding if a variable has which of the following characteristics?
   a. It is a nominal variable
   b. It is a discrete variable
   c. It is a continuous variable
   d. None of the above

3. It makes most sense to consider recoding if a variable has which of the following characteristics?
   a. It has a small number of response categories
   b. It has a larger number of response categories
   c. It has a large number of missing data
   d. None of the above

4. If you were going to recode the variable CHILDS, which measures the respondent's ideal number of children, how might you collapse (or combine) the following categories?
   a. None
   b. One
   c. Two
   d. Three
   e. Four
   f. Five

g. Six

h. Seven or more

i. As many as you want

5. If you were going to recode the variable PRAY, which measures how often the respondent prays, how might you collapse (or combine) the following categories?

a. Several times a day

b. Once a day

c. Several times a week

d. Once a week

e. Less than once a week

f. Never

6. After recoding a variable and going back to the Data Editor window, we recommend that you access the Variable View tab because it allows you to accomplish what specific task(s)?

7. After recoding, which SPSS Statistics procedure gives you the best way to check your new, recoded variable?

8. In order to save your data file, which SPSS Statistics window should be the active window?

NAME _____

CLASS _____

INSTRUCTOR _____

DATE _____

To complete the following exercises, you should open your AdventuresPLUS.SAV file.

1. What does the variable GRNTAXES measure? (Hint: Check the variable label.)

2. Produce a frequency distribution for the variable GRNTAXES and use your findings to fill in the blanks below. Before you begin, make sure you define the values 0, 8, and 9 as "missing."

Valid Percentage

1 Very willing _____

2 Fairly willing _____

3 Neither willing nor unwilling _____

4 Not very willing _____

5 Not at all willing _____

3. Now follow these steps to recode the variable GRNTAXES.

■ Recode the variable GRNTAXES to create a new variable, RECGRTAX (recoded GRNTAXES), as follows:

Old Variable GRNTAXES		*New Variable RECGRTAX*	
Values	*Labels*	*Values*	*Labels*
1	*Very willing*	*1*	*Willing*
2	*Fairly willing*	*1*	
3	*Neither willing nor unwilling*	*2*	*Neither*
4	*Not very willing*	*3*	*Not willing*
5	*Not at all willing*	*3*	

Use the Recode command to create your new, recoded variable RECGRTAX.

■ Define your new variable.

Run a frequency distribution for the variable RECGRTAX, then list the valid percentages for each category below.

Valid Percentage

1 Willing _____

2 Neither willing nor unwilling _____

3 Not willing _____

4. Print out your frequency table, attach it to this sheet, and then write a brief analysis of your table below.

5. What does the variable NEWS measure? (Hint: Check the variable label.)

6. Produce a frequency distribution for the variable NEWS and use your findings to fill in the blanks below. Before you begin, make sure you define the values 0, 8, and 9 as "missing."

Valid Percentage

1 Every day _____

2 Few times a week _____

3 Once a week _____

4 Less than once a week _____

5 Never _____

7. Now follow these steps to recode the variable NEWS.

■ Recode the variable NEWS to create a new variable, RECNEWS (recoded NEWS), as follows:

Old Variable NEWS		New Variable RECNEWS	
Values	Labels	Values	Labels
1	Every day	1	Few times a week or more
2	Few times a week	1	
3	Once a week	2	Once a week or less
4	Less than once a week	2	
5	Never	3	Never

Use the Recode command to create your new, recoded variable RECNEWS.

■ Define your new variable.

Run a frequency distribution for the variable RECNEWS, and then list the valid percentages for each category below.

Valid Percentage

1 Few times a week or more _____

2 Once a week or less _____

3 Never _____

8. Print out your frequency table, attach it to this sheet, and then write a brief analysis of your table below.

9. What does the variable EDUC measure? (Hint: Check the variable label.)

10. Produce a frequency distribution for the variable EDUC. Print out your frequency distribution and attach it to this sheet. Before you begin, make sure you define the values 97, 98, and 99 as "missing."

11. Now follow these steps to recode the variable EDUC.

   ■ Recode the variable EDUC to create a new variable, EDCAT (recoded EDUC), as follows:

Values	Old Variable EDUC Values	New Variable EDCAT Labels
0 to 11	1	Less than high school education
12	2	High school graduate
13 to 15	3	Some college
16	4	College graduate
17 to 20	5	Graduate studies and beyond

   Use the Recode command to create your new, recoded variable EDCAT.

   ■ Define your new variable.

Run a frequency distribution for the variable EDCAT, and then list the labels and valid percentages for each category below.

Labels   *Valid Percentage*

1 _____   _____

2 _____   _____

3 _____   _____

4 _____   _____

5 _____   _____

12. Print out your frequency table, attach it to this sheet, and then write a brief analysis of your table below.

13. Choose one variable from the AdventuresPLUS.SAV file that you want to recode. Then fill in the information requested below.

    a. Name of the variable being recoded: _____

    b. Run a frequency table for this variable, attach a copy to this sheet, and describe the frequency table below.

    c. List the categories of the variable and indicate how you will collapse them below.

    d. Name of the new variable you are creating: _____

    e. Complete the steps for recoding the variable.

    f. Run a frequency distribution for your new variable, attach it to this sheet, and describe the table below.

14. Save your recoded variables so we can refer back to them later. Save the changes using the file name "AdventuresPLUS.SAV."

15. Access the SPSS Statistics Help feature Tutorial, and work your way through the section titled "Modifying Data Values."

Chapter 8   Creating Composite Measures

Exploring Attitudes Toward Abortion in More Depth

Now that you've had a chance to become familiar with univariate analysis, we are going to add a little more sophistication to the process.

You may recall that in Chapter 1, we discussed abortion. The 2010 General Social Survey (GSS) contains several measures of American attitudes toward this issue. We are going to explore these attitudes in more depth in this chapter. We will begin by identifying these variables and running frequencies. Then we are going to learn how to combine the multiple measures of abortion into a single variable, called a *composite measure*.

Demonstration 8.1: Identifying the Seven Abortion Variables—File Info

To start the process, we will instruct SPSS Statistics to produce frequency tables for the abortion attitude variables contained in the file AdventuresPLUS.SAV (and Adventures.SAV). If you do not recall the names of the seven variables used to measure abortion attitudes, you can review the list of variables in the codebook (Appendix A). Alternatively, you may view the variables with SPSS Statistics by selecting **Utilities** and clicking on **Variables**. In a moment, information about the variables on your Adventures.SAV file should appear.

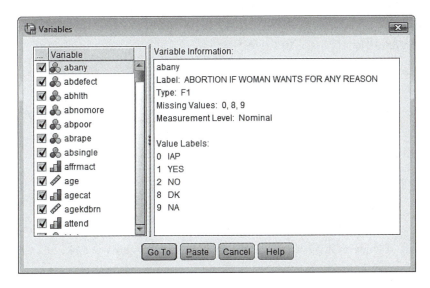

With the file arranged alphabetically, you can see the seven abortion variables contained in the Adventures.SAV file at the beginning of the list: ABANY, ABDEFECT, ABHLTH, ABNOMORE, ABPOOR, ABRAPE, and ABSINGLE.

SPSS Statistics Command 8.1: Identifying Variables—File Info

Click **Utilities** → **Variables**

Demonstration 8.2: Running Frequencies for Several Variables at Once[1]

Once you have identified the names of the abortion attitude variables, you can easily instruct SPSS Statistics to run frequencies for all the items at once. If the variables are clustered together (as the abortion opinion items in our Adventures.SAV file are), you can select the entire cluster by clicking on the first and last variables while holding down the **Shift** key and then clicking the right-pointing arrow to transfer them to the "Variable(s):" field.

If the variables are not clustered together, you can transfer them all at once by holding down the <Ctrl> (**Control**) key (or <**Apple**> key for Macintosh computers) on your keyboard as you click the name of each item. After all the items have been highlighted, click on the right-pointing arrow to transfer them to the "Variable(s):" field.

You may want to experiment with both of these options. Ultimately, however, you want to transfer all seven abortion items to the "Variable(s):" field, as shown below.

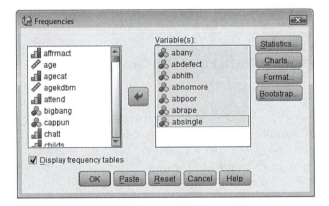

SPSS Statistics Command 8.2: Running Frequencies for Several Variables (Not Clustered)[2]

Analyze → **Descriptive Statistics** → **Frequencies**

→ Press and hold <**Ctrl**> (**Control**) key, or <**Apple**> key for Macintosh computers, on the keyboard as you click on the names of the variables

→ Click right-pointing arrow to transfer selected variables to "Variable(s):" field

→ **OK**

---

[1]For each of the seven abortion variables, the values 0, 8, and 9 should be defined as "missing."

[2]To run frequencies for several variables at once when the variables are clustered, see the discussion in Chapter 5.

Once you have given SPSS Statistics the command that will result in the frequency distributions for the various abortion items, you should get the results in your Output window. The tables discussed below indicate the different levels of support for abortion under various circumstances.

Items With the Highest Levels of Support

It would seem at first glance that very high percentages of the general public support a woman's right to an abortion if the woman's health is at risk, if the pregnancy is a result of rape, and if there is a strong chance of serious birth defects. As the tables below show, the large majority of respondents (75%–87%) to the 2010 GSS said they would support abortion in each of these cases.[3]

abhlth WOMANS HEALTH SERIOUSLY ENDANGERED

		Frequency	Percent	Valid Percent	Cumulative Percent
Valid	1 YES	1066	52.2	87.1	87.1
	2 NO	158	7.7	12.9	100.0
	Total	1224	59.9	100.0	
Missing	0 IAP	763	37.3		
	8 DK	35	1.7		
	9 NA	22	1.1		
	Total	820	40.1		
Total		2044	100.0		

abrape PREGNANT AS RESULT OF RAPE

		Frequency	Percent	Valid Percent	Cumulative Percent
Valid	1 YES	987	48.3	79.9	79.9
	2 NO	248	12.1	20.1	100.0
	Total	1235	60.4	100.0	
Missing	0 IAP	763	37.3		
	8 DK	28	1.4		
	9 NA	18	.9		
	Total	809	39.6		
Total		2044	100.0		

abdefect STRONG CHANCE OF SERIOUS DEFECT

		Frequency	Percent	Valid Percent	Cumulative Percent
Valid	1 YES	921	45.1	75.2	75.2
	2 NO	304	14.9	24.8	100.0
	Total	1225	59.9	100.0	
Missing	0 IAP	763	37.3		
	8 DK	32	1.6		
	9 NA	24	1.2		
	Total	819	40.1		
Total		2044	100.0		

[3]In each instance, the percentages have been rounded.

Items With Less Support

When you examine the other three items, however, you will find a significant decrease in levels of support for women getting an abortion. As the tables below show, support for a woman getting an abortion if she is not married, has low income and cannot afford any more children, or doesn't want any more children range between 43% and 48%.[4]

absingle NOT MARRIED

		Frequency	Percent	Valid Percent	Cumulative Percent
Valid	1 YES	524	25.6	42.7	42.7
	2 NO	704	34.4	57.3	100.0
	Total	1228	60.1	100.0	
Missing	0 IAP	763	37.3		
	8 DK	36	1.8		
	9 NA	17	.8		
	Total	816	39.9		
Total		2044	100.0		

abpoor LOW INCOME--CANT AFFORD MORE CHILDREN

		Frequency	Percent	Valid Percent	Cumulative Percent
Valid	1 YES	560	27.4	45.5	45.5
	2 NO	671	32.8	54.5	100.0
	Total	1231	60.2	100.0	
Missing	0 IAP	763	37.3		
	8 DK	34	1.7		
	9 NA	16	.8		
	Total	813	39.8		
Total		2044	100.0		

abnomore MARRIED--WANTS NO MORE CHILDREN

		Frequency	Percent	Valid Percent	Cumulative Percent
Valid	1 YES	587	28.7	48.0	48.0
	2 NO	636	31.1	52.0	100.0
	Total	1223	59.8	100.0	
Missing	0 IAP	763	37.3		
	8 DK	36	1.8		
	9 NA	22	1.1		
	Total	821	40.2		
Total		2044	100.0		

[4]In each instance, the percentages have been rounded.

Unconditional Support for Abortion

Finally, let's see what proportion of the population would support a woman having unrestricted freedom to choose an abortion for any reason.

abany ABORTION IF WOMAN WANTS FOR ANY REASON

		Frequency	Percent	Valid Percent	Cumulative Percent
Valid	1 YES	537	26.3	43.7	43.7
	2 NO	693	33.9	56.3	100.0
	Total	1230	60.2	100.0	
Missing	0 IAP	763	37.3		
	8 DK	34	1.7		
	9 NA	17	.8		
	Total	814	39.8		
Total		2044	100.0		

Notice that about the same proportion (44%) supports a woman's unrestricted freedom to choose abortion as supported by the specific situations described in ABNOMORE (48%), ABPOOR (46%), and ABSINGLE (43%). Yes, that's right: A slightly smaller percentage of respondents support the choice of abortion for women who are not married than do those respondents who would support unrestricted freedom to choose abortion. Can you think of any reasons why the results are such?

Support for Abortion: An Overview

Let's construct a table that summarizes those tables we've just examined. It is often useful to bring related tables such as these together in an abbreviated table.

The tables suggest that attitudes toward abortion fall into three basic groups. There is a small minority of no more than 13% who are opposed to abortion under absolutely any circumstances whatsoever. We conclude this because 87% would support abortion if the woman's life were seriously endangered—the highest level of support from the scenarios given.

Another group, about 44% of the population, would support a woman's free choice of abortion for any reason. The remainder of the population would support abortion in only a few circumstances involving medical danger and/or rape. (See Table 8.1 for a summary of abortion support across all variables.)

Table 8.1 Variations in Levels of Support for Abortion

Percentage Who Support a Woman's Right to Choose Abortion Because . . .	
the woman's health would be seriously endangered	87%
the pregnancy resulted from rape	80%
there is a strong sense of a serious defect	75%
a married woman wants no more children	48%
the woman is unmarried	43%
the woman is too poor to afford any more children	46%
the woman wants it for any reason	44%

While the separate abortion items were helpful, they made it difficult for us to get a clear picture of how Americans feel about abortion overall. Consequently, to explore attitudes toward abortion further, it may be useful for us to have a single variable, a composite measure, that captures the complexities of attitudes toward abortion overall—in other words, a single variable that takes into account the complexity of attitudes by capturing the three major positions on this issue.

Now that we have raised the issue, you will probably not be surprised to learn that SPSS Statistics allows us to create such a variable or composite measure made up of multiple indicators of a single concept. By employing the *Count command*, we can use the scores on two or more variables to compute a new variable that summarizes attitudes toward a complicated issue or ambiguous concept such as abortion, sexual permissiveness, religiosity, prejudice, and so on. We will begin this chapter with a fairly simple two-item index indicating anti-abortion attitudes and then move to a somewhat more sophisticated measure indicating pro-abortion attitudes.

Index: A Form of Composite Measure

An *index* is a form of composite measure, composed of more than one indicator of the variable under study. The score on a multiple-choice quiz is an example of a very simple index. Each question is an "item" that indicates some of the student's knowledge of the subject matter. Together, all the items form an index of the student's knowledge. Just as we would not think that a single-question quiz would give us a very accurate assessment of a student's knowledge, so it is when we measure attitudes. We do a better job measuring respondents' attitudes with multiple items.

In addition, there are two other advantages to using an index. First, they include multiple dimensions of the subject under study. In this case, an index composed of ABDEFECT and ABSINGLE will combine two aspects of the debate over abortion rather than being limited to only one (e.g., the impact of birth defects). Second, composite measures tap into a greater range of variation between the extremes of a variable. If we were simply to use one of the abortion items, we would distinguish two groups of respondents: the pros and the antis. Combining two items will allow us to distinguish three groups. A later example will extend this range of variation even further.

ABORT Index

As we noted, the Count command allows us to create a new summary variable based on information from existing variables. The new variable can then be saved in your data set and used like any other variable.[5]

The Count command scans a list of variables for each respondent and counts the number of variables containing a particular code. For instance, the abortion variables have been consistently coded 1 to indicate approval of abortion and 2 to indicate disapproval of abortion. When the time comes, we will instruct Count to count either the number of 1s or the 2s, depending on whether we want our index to reflect overall approval or disapproval of abortion.

Our first index will be based on the two variables mentioned above: ABDEFECT and ABSINGLE. As you probably recall, ABDEFECT measures respondents' support for abortion if there is a chance of serious defect in the baby, whereas ABSINGLE measures respondents' support for abortion if the woman is not married and does not want to marry the "man." You may want to refer to the codebook (Appendix A) to find the exact wording of each question. If you do that, you will notice that both items are coded as follows:

---

[5]If you follow the demonstrations and exercises in this chapter, you will be asked to compute a handful of new variables. Anyone using the student version of SPSS Statistics should be sensitive to its limitation to a maximum of 50 variables in a data file.

Value/Code—Value Label

0—IAP (missing value)

1—Yes

2—No

8—DK (missing value)

9—NA (missing value)

ABORT Index Scores

We are going to treat ABDEFECT and ABSINGLE as if they were items on a quiz designed to find out how much our respondents disapprove of abortion. On this quiz, we will count only the "right" answers, those who answered "No." Consequently, a respondent who answered "No" to both questions will get 2 points. Someone answering "No" to only one question and "Yes" to the other will get 1 point, and a respondent answering "Yes" to both will get a score of 0. On this quiz, then, the higher the score, the more respondents disapprove of abortion.

We will use the Count command to instruct SPSS Statistics to create a new variable (ABORT) based on respondents' scores on ABDEFECT and ABSINGLE. Our index (ABORT) will be made up of three possible scores, as discussed above and shown below.

0—Respondent approves of abortion in both cases (if there is a chance of a serious birth defect and if the woman is single).

1—Respondent approves of abortion in one circumstance (primarily birth defects) but not the other.

2—Respondent disapproves of abortion in both cases (if there is a chance of a serious birth defect and if the woman is single).

When we are done, if a respondent had a 1 ("Yes") on both ABDEFECT and ABSINGLE, his or her score will be a 0 on our new index. If a respondent had a 1 ("Yes") on ABDEFECT and a 2 ("No") on ABSINGLE (or vice versa), his or her score will be a 1 on our new index. And if a respondent had a 2 ("No") for both component variables, his or her score will be 2 on our index.

Demonstration 8.3: ABORT Index

Before we get started, you should check to make sure the values 0, 8, and 9 are defined as "missing" for both ABDEFECT and ABSINGLE.

Once you have done that, go ahead and click **Transform** in the menu bar and then select **Count Values within Cases** in the drop-down menu. This will open the Count Occurrences of Values within Cases dialog box, as shown below.

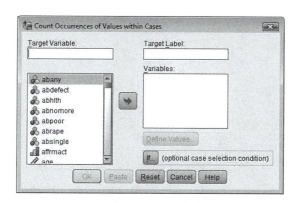

In the upper left-hand corner of the box, you will see a field labeled "Target Variable:" where you can type the name of your new variable (ABORT). To the right, you will see another field labeled "Target Label:" where you can type a descriptive label for your new variable, such as "Simple Abortion Index."

Once you have typed the variable name and label, transfer both ABDEFECT and ABSINGLE from the variable list to the "Variables:" field. Remember, you can do this by highlighting the variable name and then clicking on the right-pointing arrow. Alternatively, you can transfer both variables at the same time by using the <**Ctrl**> key (<**APPLE**> key for Macintosh users) on your keyboard. Whichever method you choose, your screen should look like this once you are done.

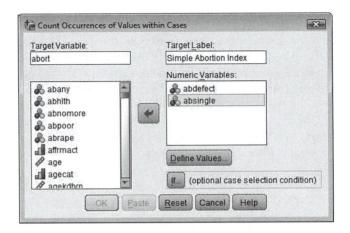

So far, all we have done is tell SPSS Statistics the variables for which specified values will be counted. We have not, however, told SPSS Statistics which value(s) to count. To do that, we need to open the Count Values within Cases: Values to Count dialog box by clicking on the **Define Values** button. Use this dialog box to tell SPSS Statistics which value(s) of ABDEFECT and ABSINGLE to count. For this index, we want SPSS Statistics to count the value of 2, or the number of respondents who said they do not support a woman's right to have an abortion in either case (if there is a chance of a birth defect and if the woman is single). To do this, type **2** in the "Value:" field on the left, and then click the **Add** button. You should now see that the value of 2 is displayed in the large "Values to Count:" field.

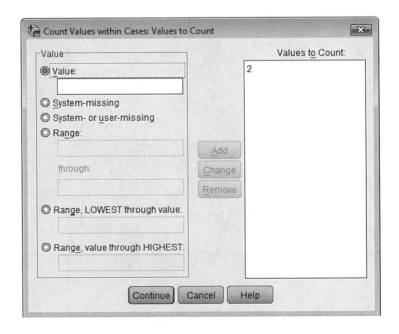

Now click **Continue** to return to the Count Occurrences dialog box.

We are almost ready to set SPSS Statistics on its task. Before we do that, however, we have to deal with missing values. Remember, before we started we made sure the values of 0 (IAP), 8 (DK), and 9 (NA) for ABDEFECT and ABSINGLE were defined as "missing." Since we do not want SPSS Statistics to count respondents with missing codes, we have to tell it to count only respondents who answered 1 ("Yes") or 2 ("No") on both questions.

In order to remove the people with missing value codes from consideration, click the **If** button next to the parenthetical text "optional case selection condition." This opens the Count Occurrences: If Cases dialog box.

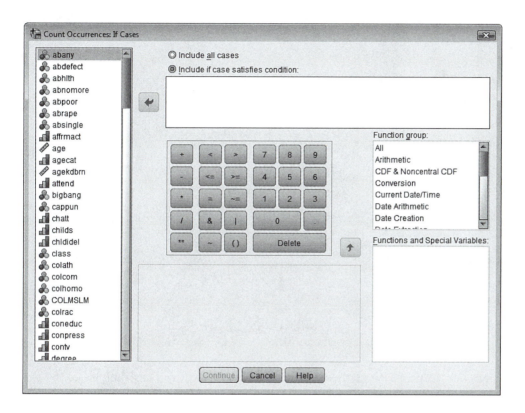

Begin by clicking the round button next to the "Include if case satisfies condition:" field. Now you can either type or paste the following expression in the "Include if case satisfies condition:" field: NMISS (abdefect, absingle) = 0. You can do this by clicking in the field and then using your keypad to type the expression. Alternatively, you can paste the expression by following the directions below.

- Scroll through the "Functions and Special Variables:" list on the bottom right-hand side of the dialog box until you locate the "NMISS (variable[,. . .])" function. This function[6] is used to count the number of variables with missing value codes. Once you have located "NMISS (variable[,. . .])," highlight it and click on the arrow to the left of the list to move it to the empty field.
- The blinking question mark (?) shows that SPSS Statistics is waiting for you to specify the names of the variables. To do this, simply highlight **ABDEFECT** in the variable list, and then transfer it with the right-pointing arrow. You will notice that ABDEFECT has replaced the blinking question mark.

---

[6]SPSS Statistics contains an extensive library of functions for data manipulation and statistical computations that cannot be accomplished with procedures found on the Data and Transform drop-down menus. Detailed information about functions may be found by clicking **Help** → **Index** and looking under "functions."

- Type a **comma** (,), and then follow the same process to transfer the variable ABSINGLE.
- Now you can use the calculator below to finish the expression by clicking = 0.

The expression, as shown below, tells SPSS Statistics to count only when the values 1 or 2 are present for both variables. If respondents answered 0, 8, or 9 for either variable, the index ABORT will be set to "missing."

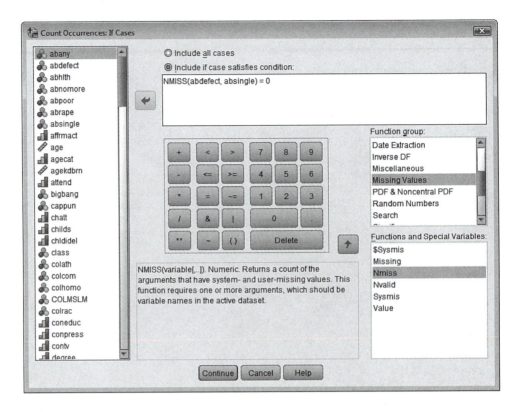

Now click **Continue** and then **OK** in the Count Occurrences dialog box to set SPSS Statistics going on its assigned task.

You should now be back in the Data View portion of the Data Editor. Scroll to the far right side and you will see your new ABORT variable in the last column. You may notice that many respondents (Case Nos. 1, 2, and 4, for instance) have a dot (.) as opposed to a value for ABORT. That is okay; it is just the way SPSS Statistics tells you that this person had a missing value (0, 8, or 9) on either ABDEFECT or ABSINGLE and therefore did not receive a score on our new index.

Demonstration 8.4: Defining ABORT

You may also notice that unlike most other variables on our data file, the scores for ABORT include decimal places. If this is true, it is because the default number of decimals for your copy of SPSS Statistics has been set to two. The unused decimal places may be discarded when you define ABORT.

To do that, simply click on the **Variable View** tab. Once you are in the Variable View window, scroll down until you see ABORT in the last row. Now define your new index, similar to the way we defined new, recoded variables in the previous chapter. In this case, do the following:

- Set the decimals at 0.
- Set the width at 1.
- You should already have a variable label, but you can change that if you prefer.

- Set the level of measurement to ordinal.
- Add values and labels as shown below:

0 = "Yes/Approve" (in both cases)
1 = "Conditional Support" ("Yes" in one case, "No" in one case)
2 = "No/Disapprove" (in both cases)

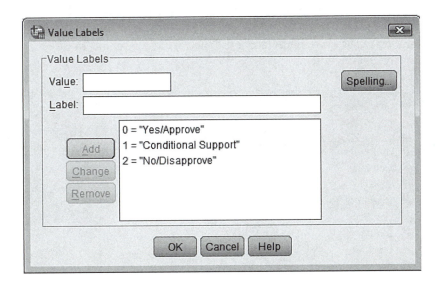

Demonstration 8.5: Checking New Index—Comparing Scores on Old and New Variables

After you have defined your new variable, go back to the Data View portion of the Data Editor and locate ABORT in the far right-hand column. You will see that the decimal places are gone and each respondent has one of three scores: 0, 1, or 2.

Now we want to check our new index or verify that we created it correctly. One simple way to determine whether we constructed our index properly is to compare a few respondents' scores on the original variables (ABDEFECT and ABSINGLE) to their scores on our index (ABORT). For instance, determine how the four respondents (listed below by case/row numbers) answered ABDEFECT and ABSINGLE. We have done it ourselves and listed their responses below.

	ABDEFECT	*ABSINGLE*	*Expected Index Score*	*Actual ABORT Score*
Case No. 1	0	0	Missing	Missing
Case No. 3	1	2	1	1
Case No. 7	1	1	0	0
Case No. 13	2	2	2	2

Now, without looking at your data file, determine what score each respondent should have gotten on your new ABORT index. If we constructed the index correctly, their expected score should match their actual score on ABORT. As you can see above, Case No. 13 responded "No" (2) on ABDEFECT and "No" (2) on ABSINGLE, so he or she should get a score of 2 on our new index, ABORT. Similarly, since Case No. 7 has a 1 for ABDEFECT and ABSINGLE, she or he should be scored as a 0 on our new index, ABORT.

Now go ahead and determine the expected ABORT scores for Cases/Rows No. 1 and No. 3, as well. Once you have done that, compare the expected index scores to the actual ABORT scores. If you constructed the index properly, the expected and actual scores should match.

Demonstration 8.6: Running Frequencies for ABORT

An even better way to see if we really accomplished what we set out to do is to run frequencies for the variable ABORT.

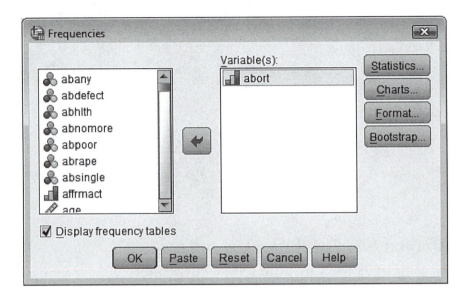

Once you do that, you should see the following table:

abort Simple Abortion Index

		Frequency	Percent	Valid Percent	Cumulative Percent
Valid	0 Yes/Approve	510	25.0	42.9	42.9
	1 Conditional Support	384	18.8	32.3	75.2
	2 No/Disapprove	295	14.4	24.8	100.0
	Total	1189	58.2	100.0	
Missing	System	855	41.8		
Total		2044	100.0		

As the table shows, 510 people, or 43% of respondents (valid percent), scored 0 on the index, meaning that they approved of abortion in both instances. Conversely, 295 people, or 25% of respondents, scored 2, which means that they disapproved of abortion in both cases. In the middle category, we found 32% (384 respondents) who approved in one case but not the other. It should also be noted that a considerable number of people (855) were excluded on the basis of missing data; that, however, does not impact the valid percentages.

Congratulations! You've just created a composite index. We realize you may still be wondering why that's such good news. After all, it wasn't your idea to create the thing in the first place!

SPSS Statistics Command 8.3: Creating a Simple Index Using Count

---

Creating Index

Click Transform → **Count**

→ Type new variable name in "Target Variable:" field

→ Type new variable label in "Target Label:" field

→ Transfer variables for which specified values will be counted to "Variables:" field

→ Click **Define Values**

    → Insert value(s) to be counted → Click **Add**

→ **Continue**

→ Click **If**

    → Click **Include if case satisfies condition:** → Type or paste expression (e.g., "NMISS (abdefect, absingle) = 0")

→ Click **Continue** → **OK**

Defining New Variable

Access **Variable View** tab → Scroll to new variable in last row

→ Set **type**, **width**, **values**, **measure**

Checking New Variable

→ Compare expected and actual index scores

→ Run Frequencies for new variable

ABINDEX: Index Based on Six Abortion Variables

While our first index was created from only two of the abortion items, we can easily create a more thorough index using more items. In addition to providing us with a wider range of scores, having more items will increase our confidence in the index. As with quizzes, we'd have more confidence in a six-item quiz than in a two-item quiz. For our new index, we will use all the abortion items *except* ABANY, supporting a woman's unrestricted choice.

For this index, we will ask SPSS Statistics to count the number of times respondents approve (1 = "Yes") of abortion. Unlike the previous index, the more supportive respondents are of abortion, the higher the scores will be. Respondents who disapprove of abortion will have lower scores. The ABINDEX scores will range from 0 (disapprove of abortion in all instances) to 6 (approve of abortion in all cases), with some respondents falling in between depending on their support/disapproval in various circumstances.

Usually, before beginning we would ask you to make sure the appropriate values (0, 8, and 9) for each variable are defined as "missing." Since we used these variables in the previous demonstrations, they should be ready to use in the next one—although no one would be hurt if you double-checked.

Demonstration 8.7: ABINDEX

Since we are following the same basic procedures we used previously, the steps should be somewhat familiar to you. As a guide, we've outlined the process in brief below.

- Click **Transform** → **Count Values within Cases.**
- Click **Reset** to remove all previous settings.

- Type new variable name (ABINDEX) in "Target Variable:" field.
- Type a label for new variable (descriptive variable label of your choice) in "Target Label:" field (e.g., "Six-Item Abortion Index").
- Transfer six abortion variables (ABDEFECT, ABHLTH, ABNOMORE, ABPOOR, ABRAPE, ABSINGLE) to "Variables:" field[7] → Click **Define Values** → Click button next to **Value**: option.

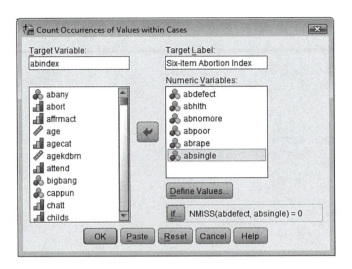

- Type **1** → Click **Add** → Click **Continue** *(because we are counting the "Yes" responses).*

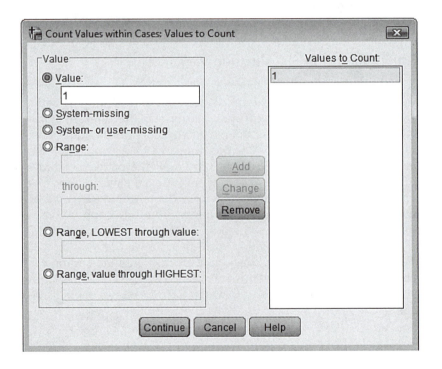

- Click **If**
- Click button next to "Include if case satisfies condition:" option.

---

[7]If the previous index ABORT is in the variable list, be careful you do NOT transfer it to the "Numeric Variables:" field as well.

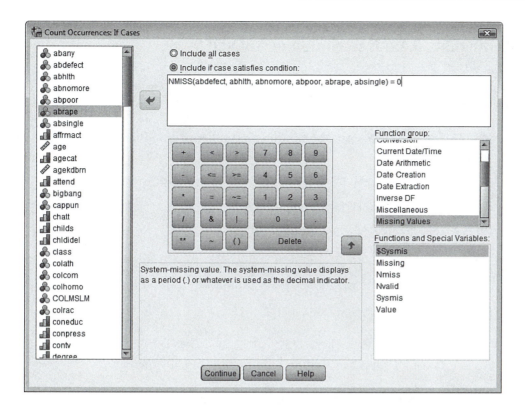

- Type or paste expression "NMISS (abdefect, abhlth, abnomore, abpoor, abrape, absingle) = 0."
- Click **Continue** → **OK**.

Now all we have to do is tidy up our new variable, ABINDEX, by accessing the Variable View tab and setting the decimals, width, values (0 = no support; 6 = high support), and level of measurement (measure).

Once you have done that, you may want to return to the Data View portion of the Data Editor and check your new index by comparing the expected and actual ABINDEX scores for a few respondents/cases chosen at random.

Demonstration 8.8: Running Frequencies

Better yet, let's go ahead and run a frequency distribution for ABINDEX. Here's what you should find.

abindex Six-item Abortion Index

		Frequency	Percent	Valid Percent	Cumulative Percent
Valid	0 No Support	104	5.1	9.3	9.3
	1	80	3.9	7.2	16.5
	2	128	6.3	11.5	27.9
	3	194	9.5	17.4	45.3
	4	89	4.4	8.0	53.3
	5	81	4.0	7.3	60.5
	6 High Support	441	21.6	39.5	100.0
	Total	1117	54.6	100.0	
Missing	System	927	45.4		
Total		2044	100.0		

> ## Writing Box 8.1 Description of ABINDEX
>
> Respondents' attitudes toward abortion have been summarized in the form of an index, ABINDEX. Scores on the index, which range from 0 to 6, are based on responses to six items asked in the questionnaire, each of which indicates some degree of support for a woman's right to choose an abortion (or opposition to it). The six items were presented as different circumstances in which you might approve or disapprove of abortion as a choice:
>
> Strong chance of serious defect (ABDEFECT)
>
> Woman's health seriously endangered (ABHLTH)
>
> Married—wants no more children (ABNOMORE)
>
> Low income—can't afford more children (ABPOOR)
>
> Pregnant as a result of rape (ABRAPE)
>
> Not married (ABSINGLE)
>
> *Note: GSS variable names for items are presented in parentheses.*
>
> Respondents were given one point for each circumstance they felt justified letting a woman choose abortion. Thus, respondents who scored 6 on the index (40% of the sample) felt that each of the circumstances was sufficient justification. Those who scored 0 on the index (a little more than 9% of the sample) were unconditionally opposed to abortion in any and all circumstances. A person who scored 3 on the index felt that three of the circumstances warranted abortion but the other three did not.

That is enough for now. However, before ending this session, make sure you save both of your new variables (ABORT and ABINDEX) with an appropriate filename, such as "AdventuresPLUS.SAV," so we can find and use them later.

It is important to reiterate that anyone using the student version of SPSS Statistics (currently only available as an older version, 18 and earlier) has to be sensitive to the fact that it is limited to 50 variables. Consequently, you may need to follow the instructions in Chapter 5 regarding saving your data. For those of you using the full version of SPSS Statistics, this is not an issue.

Conclusion

In this chapter, we've seen that it is often possible to measure social scientific concepts in a number of different ways. Sometimes, the data set contains a single item that does the job nicely. Measuring gender by asking people for their gender is a good example. In other cases, the mental images that constitute our concepts (e.g., religiosity, political orientations, prejudice) are varied and ambiguous. Typically, no single item in a data set provides a complete representation of what we have in mind.

Often, we can resolve this problem by combining two or more indicators of the concept into a composite index. As we've seen, SPSS Statistics offers the tools necessary for such data transformations. If you continue your studies in social research, you will discover that there are many more sophisticated techniques for creating composite measures. However, the simple indexing techniques you have learned in this chapter will serve you well in the analyses that lie ahead.

Main Points

- Concepts can be measured in a number of different ways.
- In some cases, a single item is enough to measure a concept.
- In the case of a controversial and emotionally charged issue such as abortion, where opinion is varied and intricate, no one item is capable of capturing the complexity of opinion.
- In such cases, composite measures made up of multiple indicators of a single concept can be very useful.
- There are a variety of advantages to using composite measures.
- An index is a form of composite measure.
- The Count Values within Cases command allows us to create a new summary variable or index based on the scores of existing variables.
- We created one fairly simple index (ABORT) and one more-complicated index (ABINDEX) with Count. Both indices were designed to measure overall feelings about abortion.
- After creating variables, it is important to check your work by running frequencies.
- When saving your new, recoded variables, those working with the student version of SPSS Statistics must be aware of the limitation to 50 variables; those working with the full versions of SPSS Statistics are NOT limited in terms of the number of variables they can save.

Key Terms

Composite measure   Index
Count command

SPSS Statistics Commands Introduced in This Chapter

8.1.   Identifying Variables—File Info

8.2.   Running Frequencies for Several Variables (Not Clustered)

8.3.   Creating a Simple Index Using Count

Review Questions

1. What is a composite measure?

2. What is an index?

3. Why do researchers create indices? What are the advantages of using an index?

4. In this chapter, we created indices using which command: Analyze, Compute, Count, or Recode?

5. What dialog box do we use to specify which value(s) of certain variables are to be counted?

6. After creating an index, you access the Variable View tab for what purpose(s)?

7. Can indices be based on the scores of more than two variables?

8. Can the procedure used to create indices in this chapter be used if the variables are not coded the same way?

9. What is one disadvantage of using Count Values within Cases to create an index?

10. In this chapter, we created two indices that helped summarize American attitudes toward abortion. Setting this issue aside for a moment, what other complicated issues or ambiguous concepts are you personally interested in investigating further? Identify three issues or concepts, at least one of which can be investigated using your Adventures .SAV data file.

NAME _____

CLASS _____

INSTRUCTOR _____

DATE _____

In this exercise, you will be given an opportunity to create three indices. In order to complete the exercises, you should access the AdventuresPLUS.SAV file we asked you to create at the end of SPSS Statistics Lab Exercise 7.1. If you did not save the recoded variables, don't worry; when we use them in later exercises, you can simply go back and recode them following the commands in SPSS Statistics Lab Exercise 7.1. At the end of this exercise, we will ask you to save the indices you create on your AdventuresPLUS.SAV file; so if you are working in a lab setting, make sure you have a jump drive handy or access to a network drive or cloud.

INDEX 1: POLIN

The first index is based on two variables contained in your AdventuresPLUS.SAV (or Adventures .SAV) data file, POLABUSE and POLATTAK. Simply follow the steps below and supply the information requested.

1. What does the variable POLABUSE measure?

2. What does the variable POLATTAK measure?

3. List the values and labels for POLABUSE and POLATTAK.

Value   *Label*

_____   _____

_____   _____

_____   _____

_____   _____

Now define 0, 8, and 9 as "missing" for the variables POLABUSE and POLATTAK. Our new variable/index will be named POLIN (for POLice INdex). POLIN scores are as follows:

0—"Approve/Yes" in both cases

1—"Conditional Approval/Yes" in one case but not the other

2—"Disapprove/No" in both cases

4. The _____ [higher/lower] the index score, the *more the respondent agrees* that there are situations in which he or she would approve of a police officer striking an adult male citizen.

5. The _____ [higher/lower] the index score, the *less the respondent agrees* that there are situations in which he or she would approve of a police officer striking an adult male citizen.

Now compute new variable/index POLIN. (Hint: Click **Transform → Count.**) Then define your new variable. (Hint: Access the **Variable View** tab, then set decimals, width, values/labels, and measure.)

6. Access the **Data View** tab. Use the information in the matrix to complete the first two columns indicating the values for the variables POLABUSE and POLATTAK (Case Nos. 1–3, 6–8).

After completing the first two columns, use the information to fill in the expected index score for each case.

Now compare the expected index score to the actual POLIN score for each case. Based on this "test," use the space below to indicate whether your index appears, by this method at least, to have been completed properly, or whether you need to go back and reconstruct the index.

	Value on	*Value on*	*Expected*	*Actual*
	POLABUSE	*POLATTAK*	*Index Score*	*POLIN Score*

Case No. 1 _____

Case No. 2 _____

Case No. 3 _____

Case No. 6 _____

Case No. 7 _____

Case No. 8 _____

Run frequencies for POLIN, and attach your output to this sheet.

7. Of those who answered,

_____% received a score of 0 on the index.

_____% received a score of 1 on the index.

_____% received a score of 2 on the index.

INDEX 2: POLFORCE

Now it is time to compute an index on your own. Create a more complex index (POLFORCE) based on the following four variables: POLABUSE, POLATTAK, POLESCAP, and POLMURDR. Be sure to define 0, 8, and 9 as "missing" for each variable before you begin.

8. After you construct and define your index, run frequencies and attach your output to this sheet. Then describe your findings below. What does this index tell us about Americans' attitudes toward the use of force by police officers?

INDEX 3: CONSTRUCTING AN INDEX OF YOUR CHOICE

Choose variables from the Adventures.SAV (or, of course, AdventuresPLUS.SAV) file and create an index of your choice. When you are finished, answer the following questions:

9. What variables did you choose and why?

10. After you construct and define your index, run frequencies and attach your output to this sheet. Then describe your findings below. After you are done, be sure to save the indexes you created in this lab exercise on your AdventuresPLUS.SAV file so we can use them in later analyses.

Chapter 9   **Suggestions for Further Analysis**

In the preceding chapters, we've given you a number of research possibilities to begin exploring based on the variables contained in your AdventuresPLUS.SAV (and Adventures.SAV) file. The topics we have focused on so far include religion, politics, and abortion. In the event that you've exhausted those possibilities and want to look beyond them, here are some additional topics for you to consider.

Desired Family Size

One of the major social problems facing the world today is that of overpopulation. A brief summary of population growth on the planet should illustrate what we mean.

Year	Population	Doubling Time
0	0.25 billion	—
1650	0.50 billion	1,650 years
1850	1.00 billion	200 years
1930	2.00 billion	80 years
1975	4.00 billion	45 years
1994	5.64 billion	39 years
1998	5.93 billion	49 years
2002	6.27 billion	54 years
2006	6.53 billion	60 years

These data show several things. For example, the world's population has increased more than 20-fold since the beginning of the Christian era. More important, until recently the rate of growth has been increasing steadily. This is most easily seen in the rightmost column above, showing what demographers call the ***doubling time***. It took 1,650 years for the world's population to increase from a quarter of a billion people to half a billion. The time required to double dramatically shortened until 1994. As of 2006, the doubling time increased to about

60 years. Even with the slight increase in doubling time, however, today's children can expect to live out their older years in a world with more than 12 billion people.

This astounding increase in the pace of population growth has been caused by the fact that during most of human existence, extremely high death rates have been matched by equally high birth rates. During the past few generations, however, death rates have plummeted around the world because of improved public health measures, medical discoveries, and improved food production.

The current pace of population growth simply cannot go on forever. Although scientists may disagree on the number of people the planet can support, there is simply no question that there is a limit. At some point, population growth must be slowed even more and stopped—perhaps even reversed.

There are two ways to end population growth: Either death rates can be returned to their former high levels, or birth rates can be reduced. Because most of us would choose the latter solution, demographers have been very interested in variables that measure desired family size.

Demonstration 9.1: Respondents' Ideal Family Size (CHLDIDEL)

Your data set contains a variable, CHLDIDEL, that presents responses to the question, "What do you think is the ideal number of children for a family to have?" If every family had only two children, then births and deaths would eventually roughly balance each other out, producing a condition of population stabilization, or zero population growth (excluding the effect of migration). What percentage of the population do you suppose chose that as the ideal? Some favored larger families, and others said they thought only one child was the ideal.

Why don't you find out what the most common response was by running a frequency distribution, as shown below, as well as the appropriate measures of central tendency.[1]

chldidel IDEAL NUMBER OF CHILDREN

		Frequency	Percent	Valid Percent	Cumulative Percent
Valid	0	6	.3	.4	.4
	1	42	2.1	3.0	3.5
	2	672	32.9	48.7	52.1
	3	346	16.9	25.1	77.2
	4	123	6.0	8.9	86.1
	5	24	1.2	1.7	87.8
	6	7	.3	.5	88.3
	7 SEVEN+	6	.3	.4	88.8
	8 AS MANY AS WANT	155	7.6	11.2	100.0
	Total	1381	67.6	100.0	
Missing	-1 IAP	614	30.0		
	9 DK,NA	49	2.4		
	Total	663	32.4		
Total		2044	100.0		

Later in this book, when we focus on bivariate analysis, you may want to explore the causes of people's attitudes about ideal family size. For instance, it may be interesting to consider whether the number of siblings that respondents have (SIBS) impacts their conception of ideal family size (CHLDIDEL).

---

[1]Define the values –1 and 9 as "missing" for CHLDIDEL.

Writing Box 9.1

Almost half the respondents (49%) say two children would be ideal, while a fourth (25%) say three. Eleven percent offer an ambiguous answer, saying that "as many as you want" is the ideal number. Just under 4% of the sample say they want fewer than two children, and the rest want four or more. Clearly, the two-child family has become the norm. Both the median and mode are two children, while the mean is 3.2. (It should be noted that the actual mean may be higher since 7 represented those people who chose seven *or more* children.)

Child Training

What do you think are the most important qualities for children to develop as they grow up? Respondents to the General Social Survey (GSS) were asked that question also.

To frame the question more specifically, they were presented with several of the answers people commonly give and were asked how important each was. If you had to choose, which thing on this list would you pick as the most important for a child to learn to prepare him or her for life? The interviewer read the following list (which we've annotated with the GSS variable names):

OBEY	To obey
POPULAR	To be well liked or popular
THNKSELF	To think for himself or herself
WORKHARD	To work hard
HELPOTH	To help others when they need help

Once the respondents indicated which of these they felt was most important, they were asked,

Which comes next in importance?

Which comes third?

Which comes fourth?

This set of responses allowed the researchers to code the final responses as "Least important."

Demonstration 9.2: Important Qualities for Children

As with earlier topics, take a moment to notice how you feel about such matters. Then see if you can anticipate what public opinion is on these qualities of children.

It will be useful, by the way, to observe your reasoning process as you attempt to anticipate public opinion. What observations, clues, or cues prompt you to think that OBEY is the most important, or POPULAR, or whichever one you picked as the one most people would choose? Then you can see how people actually responded to the questions. You can use either of the univariate techniques reviewed in earlier chapters (Frequencies or Graphs) to examine how respondents to the 2010 GSS answered these questions. In this case, let's go ahead and run appropriate graphs and charts for these variables.[2] Once you've done that, you should review your earlier reasoning, either to confirm your predictions or to figure out where you went wrong. What can you infer from the differences among your opinions, your predictions, and the actual results?

---

[2]Define the values 0, 8, and 9 as "missing" for the variables OBEY, POPULAR, THNKSELF, WORKHARD, and HELPOTH.

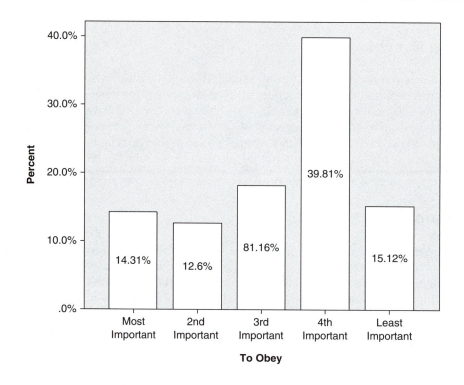

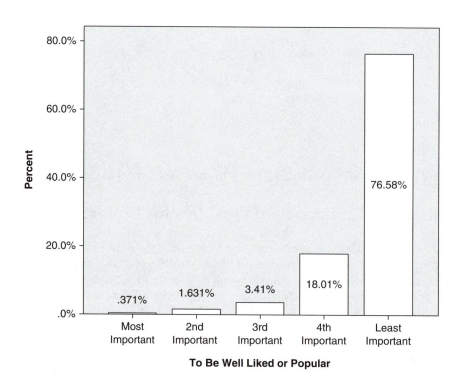

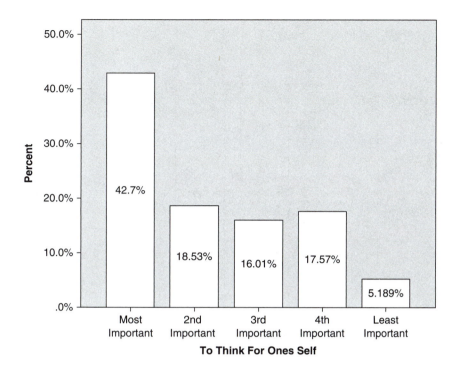

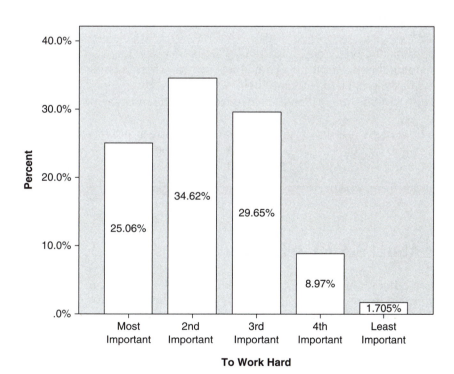

To Help Others

Writing Box 9.2

These graphs paint a pretty clear picture of what people say they value in their children.

Thinking for themselves is clearly the most prized, whereas being well liked or popular is in the basement. Hard work and helping others are both valued, and they evidence a normal distribution—minorities at the two extremes and most saying they are somewhat important.

A majority of the respondents ranked "to obey" as the fourth most important, or least important, value for children, an attitude that reflects the high value of children thinking for themselves.

Attitudes About Sexual Behavior

As with most other things, Americans differ in their feelings about sexual behavior. We thought you might be interested in this area of public opinion, so we've included three GSS variables dealing with two different kinds of sexual behavior.

HOMOSEX asks about homosexual sex relations, while PREMARSX focuses on premarital sex. Both variables measure respondents' attitudes toward the behavior of others. Either might be taken as an indication of overall orientation, but it is important always to remember exactly what variables represent. One question this raises, however, is whether self-reports of behavior can be considered generally reliable. A second question you may want to think about is whether this method of data collection (surveys) should be used primarily to measure opinions and attitudes rather than behavior, or doesn't it matter? We realize that you may very well have strong opinions about each of these issues. Your job as a social science researcher, however, is to find out what Americans as a whole think and do. Which do you think people tolerate more: premarital sex or homosexuality? Give it some thought and then check it out using one of the techniques we covered earlier.

It's not too early to begin asking yourself what would cause people to be more tolerant or less so in these regards. When we turn to bivariate analysis later on, you'll have a chance to test some of your expectations.

Demonstration 9.3: Index of Sexual Permissiveness

Here's an idea that could take you deeper into this general topic. See if you can use the **Transform → Count Values within Cases** command to create a composite measure of sexual permissiveness, combining the two items so the higher the index score, the more likely respondents will be sexually permissive.

Go ahead and construct an index called SEXPERM. As always, you will need to make sure of the nonresponses for HOMOSEX and PREMARSX. In this instance, there are four codes that should be defined as "missing" (0, 5, 8, and 9). As a result, you cannot use the "Discrete missing values" option discussed earlier (SPSS Statistics Command 5.4), because it allows you to list only three missing values. In cases where you want to designate four or more values as "missing," use the "Range plus one discrete missing value" option as described in SPSS Statistics Command 9.1.

- Access **Variable View** tab.
- Open the Missing Variables dialog box by double-clicking on the right side of the cell that corresponds with the appropriate variable and column labeled "Missing."
- Select the **Range plus one discrete missing value** option.
- Type 5 in the space next to "Low," 9 in the space next to "High," and 0 in the box next to the "Discrete value" option. This tells SPSS Statistics to designate any value from 5 to 9 and 0 as "missing" for this variable.
- Click **OK**.

SPSS Statistics Command 9.1: Setting Values and Labels as "Missing" Using "Range Plus One" Option

---

Access **Variable View** tab →

Double-click on right side of cell that corresponds with appropriate variable and column "Missing" →

Click on circle next to "Range plus one optional discrete missing value" →

Insert appropriate low, high, and discrete values →

Click **OK**

Don't forget to repeat that process for both HOMOSEX and PREMARSX. Once you are done, you will be ready to begin constructing your index by selecting **Transform → Count Occurrences of Values within Cases**.

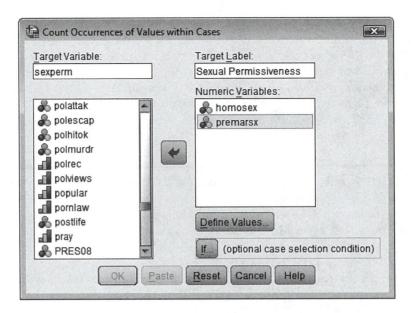

Since Response Code 4, "Not Wrong," indicates permissiveness, you will need to tell Count to count the instances of 4 for both variables.

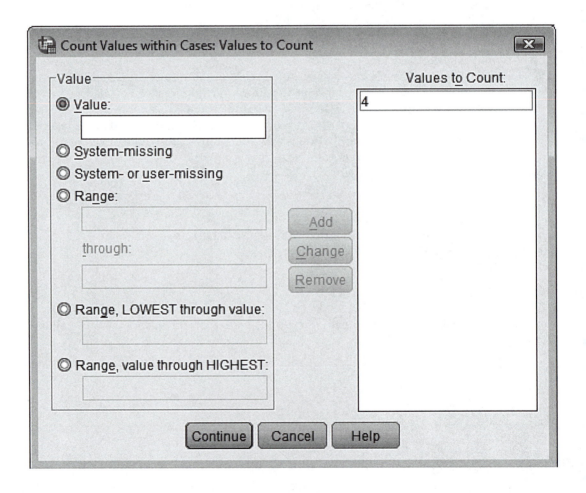

Use the NMISS function to ensure that only those cases that have responses for both questions are used to construct the index.

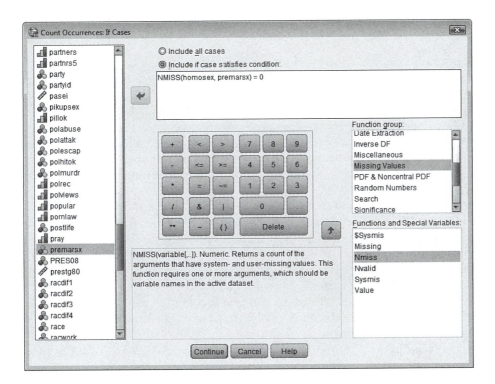

After you have created SEXPERM, run frequencies to check your index. What conclusions would you draw about the sexual attitudes and behavior of Americans? Are they generally permissive or not? You can check your findings against ours shown below.

sexperm Sexual Permissiveness

		Frequency	Percent	Valid Percent	Cumulative Percent
Valid	0 Non-permissive	245	12.0	39.3	39.3
	1 Conditionally permissive	157	7.7	25.2	64.5
	2 Permissive	221	10.8	35.5	100.0
	Total	623	30.5	100.0	
Missing	System	1421	69.5		
Total		2044	100.0		

Writing Box 9.3   Description of SEXPERM Index

SEXPERM is an index of sexual permissiveness, based on two indicators. HOMOSEX asks respondents whether "sexual relations between two adults of the same sex is always wrong, almost always wrong, wrong only sometimes, or not wrong at all." Those who chose the last response received a score of 1 on the index. The other item in the index, PREMARSX, asked a similar question about a man and a woman having "sex relations before marriage" and was scored the same as HOMOSEX. Thus, respondents who approved of both homosexuality and premarital sex received a score of 2 on the index; those who disapproved (even a little) of both were scored 0. Those who approved of only one (usually premarital sex), and disapproved of the other (even a little) were scored 1. The sexual permissiveness index has a range of 0 to 2. A considerable proportion of respondents (39%) had a score of 0, meaning they disapproved of both premarital sex and homosexuality. A quarter (25%) scored 1, indicating that they were permissive toward one of the two behaviors. Another 36% had a score of 2, indicating that they approved of both premarital sex and homosexuality.

Prejudice

Prejudice is a topic that has concerned social scientists for a long time, and the persistence of the problem keeps it a topic of interest and research. Your data file includes two items from the GSS that deal with aspects of anti-Black prejudice: MARBLK and RACDIF4. Both items deal with different components of prejudice and racial attitudes. MARBLK measures whether respondents would favor or oppose a close relative marrying a Black person, whereas RACDIF4 deals with the causes of differences in socioeconomic status. You may want to look at Appendix A for the exact wording of these questions. Remember, it's always important to know exactly how survey questions were asked in order to understand what the responses really mean.

These items present you with an interesting picture of racial attitudes and prejudice in the United States. Where, for instance, do you think the majority of adult Americans stand on the issue of interracial marriage when it comes to a close relative? Do you think most respondents would support or oppose interracial marriage under these circumstances? You may be able to guess that opposition to interracial marriage in general has decreased over the years. Before you look at this variable, however, take a moment to try to guess what the level of public support for interracial marriage is when it comes to a close family member. Then use one of the techniques introduced in earlier chapters to see how well you've been able to anticipate opinion on this issue.

Next, try to estimate overall opinion on the other variable, RACDIF4. It may be somewhat more difficult for you to estimate opinion on this issue, but after you have come up with an estimate, check to see whether you were correct by running frequencies.

You may want to consider creating a composite measure of prejudice called PREJIND from these two variables: MARBLK and RACDIF4. Remember, as in the last demonstration, you need to do a few things before you start constructing your index, beginning with defining the appropriate values for each variable as "missing" and then recoding as noted below.

Define appropriate values for each variable as missing. Designate 0, 8, and 9 as "missing" for both MARBLK and RACDIF4. Notice that the values/labels for these variables differ. Consequently, in order to create an index using Count, you should recode MARBLK to create RECMAR as follows.

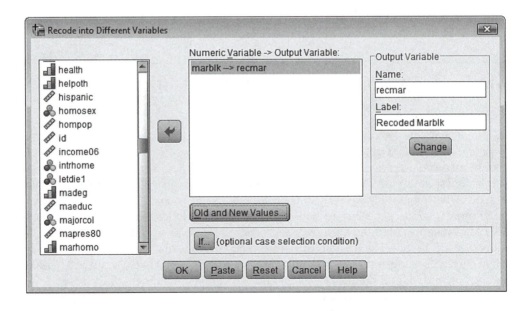

Pay particular attention to the new values and labels before proceeding:

Old Values/Labels	New Values/Labels
MARBLK	RECMAR
1 Strongly favor	2 No/Not oppose marriage
2 Favor	
4 Oppose	1 Yes/Oppose marriage
5 Strongly oppose	

Note: When a different variable is created, old values that are not assigned new values for recoding are set to system "missing" in the new variable unless the "Copy old value(s)" option is selected.

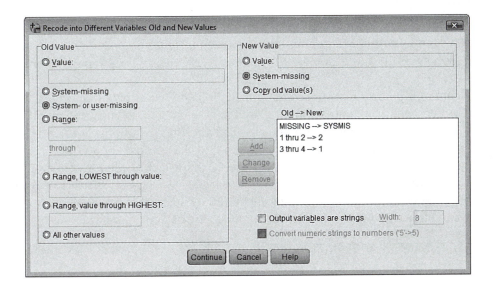

Now use the **Transform → Count Values within Cases** command to create a composite measure of prejudice. Combine the two items (your new, recoded variable RECMAR and RACDIF4) so the higher the index score, the less likely respondents will be prejudiced.

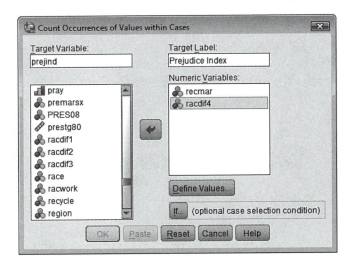

Afterward, be sure to define your new variable and then run frequencies.

prejind Prejudice Index

		Frequency	Percent	Valid Percent	Cumulative Percent
Valid	0 Prejudiced	356	17.4	28.8	28.8
	1 Conditionally prejudiced	621	30.4	50.2	78.9
	2 Not prejudiced	261	12.8	21.1	100.0
	Total	1238	60.6	100.0	
Missing	System	806	39.4		
Total		2044	100.0		

Writing Box 9.4

PREJIND is an index intended to measure respondents' levels of prejudice. Respondents could get between 0 and 2 points on the index, with lower scores representing higher degrees of prejudice.

Twenty-nine percent of the respondents had PREJIND scores of 0, indicating that they both opposed interracial marriage for a close family member and answered affirmatively to RACDIF4. More than half of respondents (50%) responded positively to at least one item, while the remainder of the sample (21%) said they would support the interracial marriage of a close family member and do not feel that the lack of willpower or motivation explains differences in jobs, income, and housing levels between Blacks and Whites. Interestingly, similar numbers of persons anchor both ends of the index with the majority (just over half) occupying the middle, more moderate position.

Conclusion

There are several other variables in the data set. As you may recall from our discussion in the previous chapters, you can get an overview of the whole thing by the command **Utilities →**
Variables. Once you see the list of variables on the left side of the window, you can click any of the variable names to get short descriptions and the codes used for categorical responses.

By the time you finish this chapter, you should be feeling fairly comfortable with SPSS Statistics and the GSS data set. Now you can add more strength to the facility you are developing. In Part III of this book, you are going to try your hand at bivariate analysis, which lets you start to search for the reasons people are the way they are.

Main Points

- Adventures.SAV contains a variety of variables that open a number of different research possibilities.
- In addition to religion, politics, and abortion, the data set also contains variables that deal with issues such as desired family size, child training, attitudes toward sexual behavior, and prejudice.
- Overpopulation is an issue of grave concern that can be dealt with in two primary ways.
- Because most people prefer a solution that deals with decreasing birth rates as opposed to increasing death rates, public opinion regarding ideal family size is an issue of great interest to many researchers.

- If you are interested in child development and child rearing, public attitudes regarding child training may be of particular interest to you.
- Attitudes toward sexual behavior can be combined to create an interesting measure of sexual permissiveness.
- It is important to note that while two of the variables dealing with sexual behavior measure public attitudes toward the behavior of others, the third variable focuses on the respondents' own behavior.
- Two variables deal with different aspects of prejudice, specifically anti-Black prejudice.
- You can use several of the techniques reviewed in earlier chapters to examine these and other topics in more depth.
- Some of the key techniques we have covered so far include frequencies, measures of central tendency and dispersion, recoding, graphs and charts, and index construction.

Key Terms

Doubling time

SPSS Statistics Commands Introduced in This Chapter

9.1   Setting Values and Labels as "Missing" Using "Range Plus One" Option

Review Questions

1. What topics besides religion, politics, and abortion are represented by the variables contained in your Adventures.SAV file?

2. List the variables on your Adventures.SAV file that pertain to ideal family size.

3. List two SPSS Statistics techniques reviewed previously that you could use to determine how respondents to the 2010 General Social Survey feel about this issue.

4. List the variables on your Adventures.SAV file that pertain to child training.

5. List two SPSS Statistics techniques reviewed earlier that you could use to determine which qualities respondents value in children.

6. List the variables on your Adventures.SAV file that pertain to attitudes about sexual behavior.

7. Which of the variable(s) listed in response to Question 6 measure attitudes, and which measure behavior?

8. Why might some researchers argue that a measure of self-reported behavior is potentially unreliable?

9. List the variables on your Adventures.SAV file that pertain to prejudice.

10. List two SPSS Statistics techniques reviewed previously that you could use to examine either the topic of sexual attitudes/behavior OR prejudice in more depth.

11. Of the four new topics introduced in this chapter, which would you be most interested in examining further and why?

NAME _____

CLASS _____

INSTRUCTOR _____

DATE _____

Your AdventuresPLUS.SAV data file contains a number of additional topics and research possibilities for you to consider, including gender roles, police, the environment, mass media use, national government spending priorities, teen sex, and affirmative action.

In this exercise, you will be asked to choose and examine one of these topics in more depth. Our goal is to allow you to use and apply several of the SPSS Statistics techniques introduced in earlier chapters. In addition, we want you to gain more practice not only using SPSS Statistics but interpreting your findings and writing up research results. Throughout the past few chapters, we have periodically supplied you with "Writing Boxes," which illustrate how a social scientist might communicate findings in writing. Remember, the ability to interpret and communicate your findings in writing is crucial. While this skill is as difficult to teach as it is to master, our experience has been that the more you read research reports in the media, scholarly journals, and the like, and the more you practice creating your own research reports, the better your chances of success.

Reminder: All the techniques we ask you to use in this exercise were discussed in Chapters 5 through 9. If you have any trouble recalling a particular command, please consult the appropriate chapter. There is a list of SPSS Statistics Commands at the end of each chapter and on the *Adventures in Social Research* student study website (http://www.sagepub.com/babbie8estudy/).

1. From the list provided above, choose a topic or issue you are interested in investigating further.

2. Use the **Utilities → Variables** command to identify the variables in your AdventuresPLUS. SAV file that pertain to this issue/topic. Then list the abbreviated variable name and variable label for each item in the space provided below.

   Abbreviated Variable Name   *Variable Label/Description*

3. Using the World Wide Web, the library, or another research tool, investigate the issue/topic you have chosen. Then write a few short paragraphs (similar to those in Chapter 9) that accomplish the following:

   ■ Provide a general overview or introduction to this issue/topic
   ■ State why the issue/topic is of interest to you (as well as others)
   ■ Highlight at least one of the central questions and controversies that make this issue/topic ripe for social research and further investigation

4. In the space provided below, list the abbreviated variable names, values, and labels for each variable listed in response to Question 2. Then indicate which of the values for each variable should be defined as "missing." Once you have done that, use SPSS Statistics to define the appropriate values for each variable as "missing."

5. Run frequencies for each variable listed in response to Question 2. Then summarize your findings in a paragraph or two in the space provided below. Be sure to print your findings and attach them to this sheet.

6. Run the appropriate graphs/charts for each variable listed in response to Question 2. Then summarize your findings in a few short paragraphs in the space below. Be sure to include a title on each of your charts, print out your findings, and attach them to this sheet.

7. Choose two of the following techniques:

   ■ Running measures of central tendency and dispersion
   ■ Recoding
   ■ Index construction

   Then use the techniques you chose to investigate the issue/topic you are considering in more depth. After examining the issue/topic:

   ■ Summarize your findings below.
   ■ Print and attach your output to this sheet.

Part III   Bivariate Analysis

This set of chapters adds another dimension to your analyses. By moving from the analysis of one variable (univariate analysis) to the analysis of two variables (bivariate analysis) we open the possibility of exploring matters of cause and effect.

Thus, in Chapter 10, we'll begin to examine what factors cause some people to be more religious than others. In this analysis, we'll be guided by an earlier analysis, which put forward a "deprivation theory" of church involvement.

In Chapters 11 and 12, our new bivariate techniques will be used to help us explore two important political and social issues. We will begin by trying to discover the roots of political orientation: Why are some people more liberal than others? Why do some people belong to the Democratic Party while others belong to the Republican Party and still others declare themselves Independents? What are some of the consequences of political orientations? What differences do they make in terms of other attitudes and orientations?

Chapter 12 takes us deeper inside the hotly controversial issue of abortion. You probably already have a few ideas about why some people are supportive of a woman's right to choose an abortion and others are opposed. Now you are going to have an opportunity to test your expectations and learn something about the roots of different points of view in this national debate.

In these chapters (10–12), we limit our analyses to percentage tables, a basic format for investigations in social research. However, there are many other methods for measuring the extent to which variables are related to one another. We'll examine some of these in Chapters 13 and 14, where we focus on some common measures of association such as lambda, gamma, Pearson's r, and linear regression. You'll learn the logic that lies behind these measures, as well as how they help us assess the strength (and in some cases, the direction) of association between variables.

Chapter 15 adds another set of techniques for your use in assessing the associations you discover among variables. Whenever samples have been chosen from a larger population, as is the case with the General Social Survey data, there is always some danger that the associations we discover among variables in our sample are merely results of sampling error and do not represent a genuine pattern in the larger population. Chapter 15 demonstrates several techniques social researchers use to guard against being misled in that fashion.

Finally, as we did in Part II, we'll conclude our examination of bivariate analysis in Chapter 16 with some suggestions for other avenues of inquiry.

Chapter 10   Examining the Sources of Religiosity

Up to this point, we have largely limited our discussion to ***univariate analysis***, the analysis of one variable. Now, however, we are going to turn our attention to ***bivariate analysis***, the analysis of two variables at a time. This will allow us to shift our focus from description (mainly the province of univariate analysis) to the more exciting world of explanation.

It may be useful for you to think about the differences between univariate and bivariate analyses in terms of the number of variables, major questions, and primary goals.

	Univariate Analysis	*Bivariate Analysis*
Number of variables	1	2
Major question	What?	Why?
Primary goal	Description	Explanation

Consequently, whereas in the previous section we looked at the extent of people's religiosity, we now turn our attention to trying to explain "why": *Why* are some people more religious than others? What *causes* some people to be more religious than others? How can we *explain* the fact that some people are more religious than others?

The Deprivation Theory of Religiosity

The reading titled "A Theory of Involvement," by Charles Y. Glock, Benjamin R. Ringer, and Earl R. Babbie, on the study website that accompanies this text (http://www.sagepub.com/babbie8estudy/) presents one explanation for differing degrees of religiosity. In it, the authors explain their social deprivation theory of religiosity. Simply put, they say that people who are denied gratification within the secular society will be more likely to turn to the church as an alternative source of gratification.

In their analysis, they looked for variables that distinguished those who were getting more gratification in the secular society from those who were getting less. For example, they reasoned that the United States is still a male-dominated society, meaning that women are denied the level of gratification enjoyed by men. Women often earn less for the same work, are denied equal access to prestigious occupations, are underrepresented in politics, and so on. According to the deprivation theory, therefore, women should be more religious than are men.

The data analyses done by Glock et al., based on a sample of Episcopalian church members, confirmed their hypothesis. The question we might now ask is whether the same is true of the general U.S. population. Our data allow us to test that hypothesis.

Testing Our Hypothesis: Correlating Religiosity and Gender

You may recall that at the end of Chapter 7, we asked you to save your recoded variables in a file named AdventuresPLUS.SAV. Since we want to use some of our recoded variables in this chapter, go ahead and open the AdventuresPLUS.SAV file. If you did not save your recoded variables, don't worry. You can easily recode the necessary variables (ATTEND → CHATT; AGE → AGECAT) by following the recode commands in Chapter 7.

As we noted earlier, this analysis requires us to advance our analytic procedures to bivariate analysis, involving two variables: a cause and an effect. If you think back to our discussion of theory and hypotheses in Chapter 1, you may recall that in this case, religiosity would be the effect, or *dependent variable*, and gender the cause, or *independent variable*. This means that your gender causes—to some extent—your degree of religiosity. Based on this notion, we might construct a *hypothesis* that states:

> You are more likely to be religious [dependent variable] if you are a woman [category of independent variable—gender] than if you are a man [category of independent variable—gender].

It is often easy to confuse the *categories* of the variable with the variable itself. Keep in mind that in this case, "woman" and "man" are categories of the independent variable "gender." To test this hypothesis, we need measures of both the independent and dependent variables. The independent variable (gender) is easy: The variable SEX handles that nicely. But what about our dependent variable (religiosity)? As you'll recall from our earlier discussion, there are several measures available to us. For the time being, let's use church attendance as our measure, even though we've noted that it is not a perfect indicator of religiosity in its most general meaning.

You will remember that when we looked at ATTEND in Chapter 7, it had nine categories. If we were to crosstabulate ATTEND and SEX without recoding ATTEND, we would expect a table with 18 (9 × 2) cells. To make our table more manageable, we are going to use the recoded CHATT as our measure of church attendance.[1]

Once we have identified the independent variable (SEX) and dependent variable (CHATT), we are going to ask SPSS Statistics to construct a crosstab or *crosstabulation*.[2] A crosstab is a table or matrix that shows the distribution of one variable for each category of a second variable. Crosstabs are powerful tools because they allow us to determine whether the two variables under consideration are associated. In this case, for instance, is gender (SEX) related to or associated with church attendance (CHATT)?

Demonstration 10.1: Running Crosstabs to Test Our Hypothesis

To access the Crosstabs dialog box, click on the following:

Analyze → Descriptive Statistics → Crosstabs

---

[1]If you did not save the recoded variable CHATT, simply go back to Chapter 7 and follow the recode commands to create CHATT before moving ahead with the demonstrations in this chapter.

[2]Some texts and statistical packages refer to crosstabulations as *bivariate distributions, contingency tables,* or *two-way frequency distributions,* but SPSS Statistics just refers to them as crosstabs.

Once you work your way through these menu selections, you should reach a window that looks like the following.

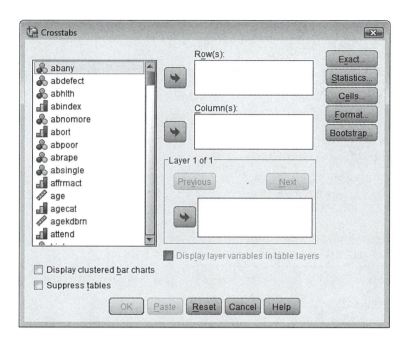

Because the logic of a crosstab will be clearer when we have an example to look at, just follow these steps on faith and we'll justify your faith in a moment.

When running Crosstabs, it is customary to specify the dependent variable as the row variable and the independent variable as the column variable. We will follow this convention by specifying CHATT (dependent variable) as the row variable and SEX (independent variable) as the column variable. (See Table 10.1 for help in determining independent and dependent variables.)

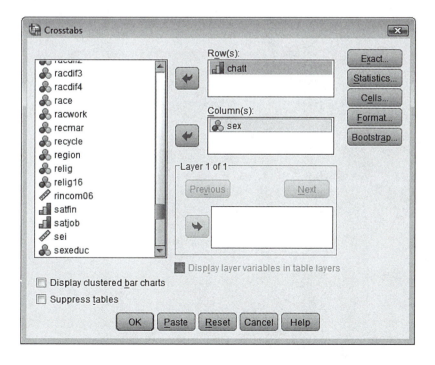

Table 10.1   Tips for Identifying Independent and Dependent Variables[a]

When running crosstabs on SPSS Statistics, it is important that you first determine which variable is the independent and which is the dependent. Then specify the independent as the column variable and the dependent as the row variable. Identifying independent and dependent variables can sometimes be tricky. If you are having difficulty differentiating between the independent and dependent variables, try using one of the following "hints":[b]
1. Restate the hypothesis as an "If _____ [independent variable], then _____ [dependent variable]" statement.
Example: If you are female [independent variable—gender], then you are more likely to be religious [dependent variable—religiosity].
2. The independent variable "influences" the dependent variable.
Example: Ask yourself if religiosity can "influence" gender. Once you have reasoned that the answer is "no," switch it around. Can gender "influence" religiosity? Since in this case the answer is "maybe," you know that gender is the independent variable and religiosity is the dependent variable.
3. The independent variable is the one that comes first in time (i.e., "before" the dependent variable).
Example: Ask yourself which came first, gender or religiosity? Since gender is determined before religiosity, you know that gender is the independent variable.

a. For further discussion, see Russell K. Schutt, *Investigating the Social World: The Process and Practice of Research,* 3rd edition (Thousand Oaks, CA: Pine Forge Press, 2001), pp. 39–41; Chava Frankfort-Nachmias and Ana Leon-Guerrero, *Social Statistics for a Diverse Society,* 2nd edition (Thousand Oaks, CA: Pine Forge Press, 2000), pp. 6–8.

b. If it isn't clear which variable is the independent or dependent, it may be that you are dealing with an ambiguous relationship or a case in which either variable could be treated as the independent. We examine this possibility in the next chapter.

Now we want to tell SPSS Statistics to compute column percentages. To do this, simply choose **Cells** and then select **Column** in the "Percentages" box; then click **Continue**.

Now hit the **OK** button to execute this command, and in a moment you should get the following result.

chatt Recoded Church Attendance * sex RESPONDENTS SEX Crosstabulation

| | | | sex RESPONDENTS SEX | | |
			1 MALE	2 FEMALE	Total
chatt Recoded Church Attendance	1 About weekly	Count	218	389	607
		% within sex RESPONDENTS SEX	24.5%	33.9%	29.8%
	2 About monthly	Count	133	185	318
		% within sex RESPONDENTS SEX	15.0%	16.1%	15.6%
	3 Seldom	Count	303	338	641
		% within sex RESPONDENTS SEX	34.1%	29.4%	31.5%
	4 Never	Count	234	236	470
		% within sex RESPONDENTS SEX	26.4%	20.6%	23.1%
Total		Count	888	1148	2036
		% within sex RESPONDENTS SEX	100.0%	100.0%	100.0%

Examining Your Output

Take a moment to examine the crosstab you created. You should notice that as requested, the categories of the independent variable (Male, Female—SEX) are displayed across the top in the columns. The categories of the dependent variable (About weekly, About monthly, Seldom, Never—CHATT) are displayed down the left side of the table in the rows. Now look at the numbers in each of the squares, or *cells*. You can see that there are two sets of numbers in each cell: the *frequency* (or count) and *column percentage*.

If you look in the first cell in the upper left corner of the table, you see there are 218 males who attend church about weekly. Below the frequency or count, you find the column percentage. In this instance, we know that 24.5% of male respondents reported attending church about weekly. The column percentage is calculated by dividing the cell frequency (218) by the column total (888) and then multiplying by 100.

In contrast, 389 or 33.9% of female respondents reported attending church about weekly. Once again, the column percentage (33.9) was calculated by dividing the cell frequency (389) by the column total (1,148) and then multiplying by 100. Luckily, you do not have to go through the process of calculating the column percentages by hand because you have SPSS Statistics to do the work for you.

If we want, we can also note that 39.5% of the men attend church either about weekly or about monthly (24.5% + 15.0%; "About weekly" + "About monthly"). Note also that 50% of the women attend either about weekly or about monthly (33.9% + 16.1%; "About weekly" + "About monthly").

In addition to the cell frequencies and column percentages, the table also contains *row totals* and *column totals* (sometimes called *marginals*). These are the frequencies for each variable. In this case, the column totals or marginals tell us that 888 of the respondents are male and 1,148 are female, while the row totals or marginals tell us that 607 respondents attend church about weekly, 318 about monthly, 641 seldom, and 470 never.

The column and row totals are followed by the column and row marginal percentages. The column marginal percentages are 100% for each column; the row marginal percentages tell us what percentage of all respondents attend church about weekly (29.8%), about monthly (15.6%), seldom (31.5%), and never (23.1%).

The *Total* **(N)** on the bottom right side tells us that a total of 2,036 of our 2,044 respondents gave valid responses to both the SEX and CHATT (recoded ATTEND) questions.

SPSS Statistics Command 10.1: Running Crosstabs— Specifying the Dependent and Independent Variables

Click **Analyze** → **Descriptive Statistics** → **Crosstabs**

　　→ Highlight dependent variable → Click arrow pointing to "Row(s):" box

　　→ Highlight independent variable → Click arrow pointing to "Column(s):" box

　　→ Click **Cells** → Select **Column** in the "Percentages" box

　　→ Click **Continue** → **OK**

Interpreting Crosstabs

The demonstration asked you to run crosstabs to test a hypothesis with two variables, one that is ordinal (CHATT—dependent variable) and one that is nominal (SEX—independent variable). While we can run crosstabs with either nominal or ordinal variables, as a practical matter, the larger the number of categories, the more tricky it becomes to read or interpret your table.

To start, we'll keep it simple by focusing on a table with two nominal variables and limited to two categories each. Then we will look at a few crosstabs with variables that have more than two categories.

Interpreting Crosstabs: Association, Strength, and Direction

As we noted earlier, we run crosstabs to determine whether there is an *association* between two variables. In addition, crosstabs may tell us other important things about the relationship between the two variables—namely, the *strength of association* and, in some cases, the direction of association. We should stress that you can determine the direction of association only when both the variables in your table are greater than nominal—that is, capable of "greater than, less than" relationships. If your table contains one or more nominal variables, it is *not* possible to determine the direction.

Once you have created your crosstab, you should ask yourself the following questions:

1. Is there an *association* between the two variables?

   IF THE ANSWER TO QUESTION 1 IS YES (OR MAYBE) . . .

2. What is the *strength of association* between the two variables?

   IF BOTH VARIABLES ARE ORDINAL . . .

3. What is the direction of association?

Demonstration 10.2: Interpreting a Crosstab With Limited Categories

Since we have been focusing on the relationship between religiosity and gender, let's run a crosstab with an indicator of religiosity (POSTLIFE) and gender (SEX). POSTLIFE measures respondents' belief in life after death and is measured at the nominal level. In accordance with the deprivation theory, we may hypothesize that women are more likely than men to believe in life after death. Consequently, POSTLIFE is the dependent variable and SEX is once again the independent variable.

Before running crosstabs, make sure to define the values 0, 8, and 9 as "missing" for the variable POSTLIFE. Now go ahead and request crosstabs, specifying POSTLIFE (dependent variable) as the row variable, SEX (independent variable) as the column variable, and cells to be percentaged by column. If you are doing this all in one session, remember to click **Reset** or move CHATT back to the variable list before proceeding.

First Question: Is There an Association?

As we noted earlier, the first question you want to ask is whether there is an association between the variables. In examining a crosstab to determine if there is an *association* between two variables, what we are really trying to determine is whether knowing the value of one variable helps us predict the value of the other variable. In other words, if gender is associated with belief in the afterlife, knowing the independent variable (SEX) should help us predict the value of the dependent variable (POSTLIFE). If gender and belief in the afterlife are *not* related or associated, knowing the value of SEX will *not* help us predict the value of POSTLIFE.

postlife BELIEF IN LIFE AFTER DEATH * sex RESPONDENTS SEX Crosstabulation

| | | | sex RESPONDENTS SEX | | |
			1 MALE	2 FEMALE	Total
postlife BELIEF IN LIFE AFTER DEATH	1 YES	Count	611	851	1462
		% within sex RESPONDENTS SEX	76.7%	84.6%	81.1%
	2 NO	Count	186	155	341
		% within sex RESPONDENTS SEX	23.3%	15.4%	18.9%
Total		Count	797	1006	1803
		% within sex RESPONDENTS SEX	100.0%	100.0%	100.0%

In order to determine whether SEX and POSTLIFE are associated, we read across the rows of the dependent variable to see if there are differences in the column percentages.[3] Let's go ahead and do that for the table we just created. By reading across the first row, we see that women (85%) are more likely than men (77%) to believe in life after death.

You can also examine the second category of the dependent variable, or the second row (those who do not believe in life after death). Here we see that a slightly higher percentage of men (23%) are likely to report that they do not believe in life after death than are women (15%).

The crosstab suggests that women are a little more likely to believe in the afterlife than are men, but knowing respondents' gender helps us very little in predicting whether or not they believe in life after death.

Second Question: How Strong Is the Association?

If there seems to be an association between the variables, the next logical question is, How strong is the association? In Chapter 13, we provide you with statistical techniques that provide measures of strength for associations. For now, we'll provide a crude method of assessing the strength of associations by examining the size of the differences in the percentages. One general rule to keep in mind is that the larger the percentage differences across the categories, the stronger the association. Conversely, the smaller the percentage differences across categories, the weaker the association between the variables.

Some researchers use a rough *10-percentage-point rule.* That is, if the percentage point difference is 10 or more, the relationship between the variables is probably worth examining further. Of course, the larger the percentage point difference, the stronger the association.

---

[3]One useful rule to keep in mind is that when a table is presented with column percentages (as it is in this case and will be for the remainder of the text) the table should be read across the columns. Conversely, when a table shows row percentages, it should be read across the rows.

Keep in mind that this "rule of thumb" is just a rough indicator. Whereas some researchers may see a 6- or 8-percentage-point difference as an indication of a potentially noteworthy relationship, others may not. (Percentage point "rules" also tend to fluctuate depending on how many columns and/or rows make up the table. For example, a greater number of columns often means that a smaller percentage-point difference can be noteworthy.)

In the case of our crosstab between POSTLIFE and SEX, we have an 8-percentage-point difference (23% – 15%; 85% – 77%). Based on the 10-percentage-point rule, the relationship would be judged to be weak. However, bear in mind that a number of researchers may still feel that an 8-percentage-point difference is enough to warrant some further investigation.

Finally, because POSTLIFE and SEX are nominal variables, we cannot ask the third question regarding the direction of association. That question applies only when *both* the variables in your table are measured at the ordinal level or higher.

Demonstration 10.3: Correlating Another Measure of Religiosity and Gender

Your file contains a third measure of religiosity we can use to continue our examination: PRAY. As the name implies, this variable measures how often respondents pray: several times a day, once a day, several times a week, etc.

Take a moment to examine the response categories for the variable PRAY. Remember, you can do this in a number of ways (accessing the **Variable View** tab and then clicking on **Values** or using the **Utilities** → **Variables** commands).

Based on your examination, it should be clear that PRAY is measured at the ordinal level. It should also be apparent that before running crosstabs, you need to define the values 0, 8, and 9 for this variable as "missing." Once you have done that, go ahead and run crosstabs, specifying PRAY as the row variable and SEX as the column variable, with cells percentaged by column. Once again, if you are doing this all in one session, you should click **Reset** or move POSTLIFE back to the variable list before you begin.

pray HOW OFTEN DOES R PRAY * sex RESPONDENTS SEX Crosstabulation

| | | | sex RESPONDENTS SEX | | |
			1 MALE	2 FEMALE	Total
pray HOW OFTEN DOES R PRAY	1 SEVERAL TIMES A DAY	Count	180	407	587
		% within sex RESPONDENTS SEX	20.4%	35.7%	29.0%
	2 ONCE A DAY	Count	228	364	592
		% within sex RESPONDENTS SEX	25.8%	31.9%	29.3%
	3 SEVERAL TIMES A WEEK	Count	115	125	240
		% within sex RESPONDENTS SEX	13.0%	11.0%	11.9%
	4 ONCE A WEEK	Count	61	57	118
		% within sex RESPONDENTS SEX	6.9%	5.0%	5.8%
	5 LT ONCE A WEEK	Count	130	92	222
		% within sex RESPONDENTS SEX	14.7%	8.1%	11.0%
	6 NEVER	Count	169	95	264
		% within sex RESPONDENTS SEX	19.1%	8.3%	13.0%
Total		Count	883	1140	2023
		% within sex RESPONDENTS SEX	100.0%	100.0%	100.0%

Now take a moment to examine your table. Remember, because one of your variables (SEX) is measured at the nominal level, it is not possible to determine the direction of association between the variables. It is possible, however, to draw some tentative conclusions regarding whether there is a relationship between the variables and, if so, the strength of the association.

Once you have examined your table, compare your answer to the description that follows. Keep in mind that because crosstabs give us only a rough indication of the strength of association, your interpretation may be somewhat different from the one below.

Writing Box 10.1

As we see in the table, women pray more often than men. For example, 68% of the women in the sample say they pray at least once a day, compared with 46% of the men. Or, looking at the other end of the table, we see that men are more likely (54%) to say they pray less than once a day than are women (32%).

Drawing Conclusions Carefully: Reassessing Our Original Hypothesis

Now that we have had a chance to examine three crosstabs involving gender and religiosity, it is important to go back to our original hypothesis. Would you say that we have found support for the thesis put forth by Glock and his coauthors? Remember, we began with a hypothesis based on the deprivation theory, which suggested that people who were deprived of gratification in the secular society would be more likely to turn to religion as an alternative source of gratification.

Based on our analysis, would you say that we have found evidence that supports the thesis with regard to gender differences—that women are still deprived of gratification in American society in comparison with men? Why or why not? It is important to note that while we may have found relationships between some of the religiosity variables and gender that are worth investigating further, we still have not *proven* our theory. For instance, while there seems to be a fairly strong association between PRAY and SEX, this association is probably best looked at as *evidence of*, not *proof of*, a causal relationship. Because two variables can be associated without it necessarily being a causal relationship, we need to proceed with care when interpreting our findings.

Demonstration 10.4: Interpreting a Crosstab With Ordinal Variables—Religiosity and Age

Now that you've had a chance to practice interpreting crosstabs with variables measured at the nominal level, we will consider a case in which both variables are measured at the ordinal level. As we noted earlier, the greater the number of categories, the more difficult it is to interpret your output.

In addition to focusing on gender, Glock et al. argue that the United States is a youth-oriented society, with gratification being denied to elderly people. Whereas some traditional societies tend to revere their elders, this is not the case in the United States. The deprivation thesis, then, would predict that older respondents are more religious than younger ones. The researchers confirmed this expectation in their data from Episcopalian church members.

Let's check out the relationship between age and religiosity. To make the table manageable, we'll use the recoded AGECAT variable created earlier in Chapter 7.[4] So request a Crosstab using[5]

- CHATT (our dependent variable) as a row variable,
- AGECAT (our independent variable) as a column variable, and
- cells to be percentaged by column.

---

[4]If you did not save the recoded variable AGECAT, simply go back to Chapter 7 and follow the Recode command to create AGECAT before moving ahead with the demonstrations in this chapter.

[5]If you are doing this all in one session, remember to click **Reset** before specifying your row and column variables for this demonstration.

The resulting table should look like this:

chatt Recoded Church Attendance * agecat Recoded Age Crosstabulation

			agecat Recoded Age				Total
			1 Under 21	2 21-39 years old	3 40-64 years old	4 65 and older	
chatt Recoded Church Attendance	1 About weekly	Count	18	137	274	177	606
		% within agecat Recoded Age	31.0%	20.2%	30.2%	45.3%	29.8%
	2 About monthly	Count	14	129	130	44	317
		% within agecat Recoded Age	24.1%	19.0%	14.3%	11.3%	15.6%
	3 Seldom	Count	18	232	300	90	640
		% within agecat Recoded Age	31.0%	34.2%	33.1%	23.0%	31.5%
	4 Never	Count	8	180	202	80	470
		% within agecat Recoded Age	13.8%	26.5%	22.3%	20.5%	23.1%
Total		Count	58	678	906	391	2033
		% within agecat Recoded Age	100.0%	100.0%	100.0%	100.0%	100.0%

Interpreting Crosstabs With Ordinal Variables

Because our table contains two ordinal variables, our method of interpretation varies slightly from what it was when we were looking at a table with at least one nominal variable.

Consider the hypothetical example in Table 10.2 of a crosstab with two ordinal variables (Variables A and B). Notice that we have identified the highest percentage in each row of our hypothetical table. Because the highest percentages occur in the three cells that form a diagonal from the upper left to the lower right, this is an indication of a *positive association* between Variables A and B.

Table 10.2 Hypothetical Example of a Positive Association

Variable B (Dependent)	Variable A (Independent)		
	Low	Medium	High
Low	50%	25%	25%
Medium	25%	50%	25%
High	25%	25%	50%
Total	100%	100%	100%

Positive associations are ones in which increases in one variable are related to increases in the other variable. You may be able to think of some variables that you suspect are positively related. One common example is level of education and income. The more education you have, the more money you are likely to earn. If this is truly the case, these variables are positively related or associated. In terms of religiosity, you may hypothesize that church attendance and the amount you pray are positively related—meaning that the more you attend church, the more likely you are to pray.

Now consider another example (Table 10.3) depicting a negative relationship between two hypothetical variables. Once again, you will notice that we have indicated the highest percentage in each row. However, in this case, the higher percentages occur in the three cells that form a diagonal line from the bottom left to the upper right side of the table. This indicates a *negative association* between the variables—an increase in one variable is associated with a decrease in the other variable.

Table 10.3 Hypothetical Example of a Negative Association

Variable B (Dependent)	Variable A (Independent)		
	Low	Medium	High
Low	25%	25%	50%
Medium	25%	50%	25%
High	50%	25%	25%
Total	100%	100%	100%

Can you think of two variables that you suspect may be negatively related? How about levels of education and prejudice? Students' alcohol consumption and GPA? If the variables are negatively associated, you will find that as education increases, the level of prejudice decreases. Similarly, as alcohol consumption increases, GPA decreases.

Examining Your Output

Of course, as we move from hypothetical tables to tables with "real social data," we seldom, if ever, find such clear positive or negative relationships.

Look back at your crosstab of AGECAT and CHATT. If you take a moment to identify the highest percentage in each row, you will quickly notice that it does not form a diagonal line depicting a positive or negative association such as the ones we mentioned earlier. Instead, as is typical when you are dealing with real-life data, the relationship is a little more "messy." Nevertheless, with some slight variations, we can see that older respondents are more likely to report that they attend church than are younger respondents. This would tend to support the deprivation thesis in terms of age, at least to some extent. You should be aware that some variations such as this become more common as the number of groups being compared increases.

Once again, while the table gives us an indication of association and strength, other statistics, such as the measures of association we discuss in Chapter 13, are a much better indication of the strength and direction of the association.

Demonstration 10.5: Correlating Other Measures of Religiosity and Age

We can request more than one table in our Crosstab window. You will notice that the commands for this procedure are the same as those listed in SPSS Statistics Command 10.1, except that in this case, we are going to click on two dependent variables and place them both in the "Row(s):" field.

Once you open your Crosstabs window, highlight **POSTLIFE** and move it to the "Row(s):" field. Then do the same with **PRAY**, placing it in the "Row(s):" field directly below **POSTLIFE**. Now go ahead and move **AGECAT** to the "Column(s):" field, request that your cells be percentaged by column, and then click **OK**.[6]

---

[6]Don't forget to click **Reset** or move CHATT back to the variable list before proceeding with this demonstration.

postlife BELIEF IN LIFE AFTER DEATH * agecat Recoded Age Crosstabulation

			agecat Recoded Age				
			1 Under 21	2 21-39 years old	3 40-64 years old	4 65 and older	Total
postlife BELIEF IN LIFE AFTER DEATH	1 YES	Count	39	482	660	278	1459
		% within agecat Recoded Age	81.2%	79.4%	81.8%	82.2%	81.1%
	2 NO	Count	9	125	147	60	341
		% within agecat Recoded Age	18.8%	20.6%	18.2%	17.8%	18.9%
Total		Count	48	607	807	338	1800
		% within agecat Recoded Age	100.0%	100.0%	100.0%	100.0%	100.0%

Now take a moment to examine your tables. As you do this, keep in mind that one table has an ordinal and a nominal variable, while the other contains two ordinal variables.

pray HOW OFTEN DOES R PRAY * agecat Recoded Age Crosstabulation

			agecat Recoded Age				
			1 Under 21	2 21-39 years old	3 40-64 years old	4 65 and older	Total
pray HOW OFTEN DOES R PRAY	1 SEVERAL TIMES A DAY	Count	12	156	278	138	584
		% within agecat Recoded Age	21.1%	23.1%	30.9%	35.7%	28.9%
	2 ONCE A DAY	Count	19	183	265	125	592
		% within agecat Recoded Age	33.3%	27.1%	29.4%	32.3%	29.3%
	3 SEVERAL TIMES A WEEK	Count	3	91	117	29	240
		% within agecat Recoded Age	5.3%	13.5%	13.0%	7.5%	11.9%
	4 ONCE A WEEK	Count	7	38	53	20	118
		% within agecat Recoded Age	12.3%	5.6%	5.9%	5.2%	5.8%
	5 LT ONCE A WEEK	Count	8	89	88	37	222
		% within agecat Recoded Age	14.0%	13.2%	9.8%	9.6%	11.0%
	6 NEVER	Count	8	118	100	38	264
		% within agecat Recoded Age	14.0%	17.5%	11.1%	9.8%	13.1%
Total		Count	57	675	901	387	2020
		% within agecat Recoded Age	100.0%	100.0%	100.0%	100.0%	100.0%

While examining the tables, you may want to consider the following types of questions: Do age and belief in the afterlife appear to have an association? How about age and prayer? Are older people more likely to believe in the afterlife than are younger people? Do younger people report praying less often than do elderly people, more often, or about the same? Based on your interpretation, what conclusions might you draw in regard to our original hypothesis about the relationship between age and religiosity? Is there support for the notion that the older you are, the more religious you are likely to be? Why or why not? Once you have taken a few minutes to examine these tables, compare your interpretation to that in Writing Box 10.2.

Writing Box 10.2

Age appears to have no impact on belief in life after death. More than four fifths hold this belief in each of the four age groups.

The relationship between age and prayer is more complex than other relationships we've examined. If we look only at praying several times a day, we find that as age increases, so does the likelihood of frequent prayer. Why might this be true? Is it that older people are more concerned about that which they may be approaching as they age? Or is age a marker for the time when people of that age were born and the state of religiosity in general at that time? Or is there another explanation?

A more complicated pattern emerges, however, if we combine the first two response categories—representing prayer at least once a day. Fifty-four percent of the youngest group say that they pray at least once a day. This decreases to 50% among the 21-to-39 group, rises further to 60% among the 40-to-64 group, and reaches a high of 68% in the oldest group. Could this signal the reversal of a trend of secularization in the United States, since the youngest group is more likely to pray at least once a day than is the next-youngest group?

Epsilon

Before concluding our initial examination of crosstabs, we want to mention one simple statistic you may find useful. *Epsilon* is a statistic often used to summarize percentage differences such as those in the tables above. Epsilon is calculated by identifying the largest and smallest percentages in either row and then subtracting the smallest from the largest.

For example, look back at the first crosstab we ran in this chapter (Demonstration 10.1) comparing gender and church attendance. In comparing men and women in terms of "about weekly" church attendance, the percentage difference (epsilon) is 9 points (33.9 − 24.5). This simple statistic is useful because it gives us a tool for comparing sex differences on other measures of religiosity.

When discussing epsilon, it is important to note that technically, tables that have more than two columns have several epsilons, one for each pair of cells being compared. In these cases, researchers will often use epsilon to refer to the largest difference in any row of cell percentages. As mentioned earlier in the chapter, it is also important to consider the number of columns that make up the table. A crosstabulation with more columns is more likely to exhibit a relationship with a smaller epsilon than is one with fewer columns. Consider that two columns would have 50% each if split evenly. A table with five columns would leave 20% in each cell if divided evenly. An epsilon of 5 for the former may not seem impressive but would be noteworthy in the latter case.

These data, then, seem to show that women are somewhat more likely than men to attend church frequently. This would seem to support the deprivation theory of religiosity to a limited extent. Take a moment to determine epsilon for the other tables we created in this chapter, and see if they produce a similar result.

This completes our initial foray into the world of bivariate analysis. We hope you've gotten a good sense of the potential for detective work in social research.

Conclusion

In this chapter, we made a critical logical advance in the analysis of social scientific data. Up to now, we have focused our attention on description. With this examination of religiosity, we've crossed over into explanation. We've moved from asking *what* to asking *why*.

Much of the excitement in social research revolves around discovering why people think and act as they do. You've now had an initial exposure to the logic and computer techniques that make such inquiries possible.

Let's apply your new capabilities to other subject matter. In the next chapters, we're going to examine the sources of different political orientations and why people feel as they do about abortion.

Main Points

- In this chapter, we shifted our focus from univariate analysis to bivariate analysis.
- Univariate analysis is the analysis of one variable at a time.
- Bivariate analysis is the analysis of two variables at a time.
- You can think of the differences between univariate and bivariate analysis in terms of the number of variables, major questions, and primary goals.
- We began our bivariate analysis by considering why some people are more religious than others.
- In this analysis, we were guided by the social deprivation theory of religiosity.
- We used this theory to develop a hypothesis stating that women are more likely to be religious than are men.
- This hypothesis contains two variables: gender (independent variable/cause) and religiosity (dependent variable/effect).
- It is important not to confuse the categories of a variable with the variable itself.
- We tested this hypothesis by running crosstabs with column percentages.

- When running crosstabs, it is customary to specify the dependent variable as the row variable and the independent variable as the column variable.
- Throughout the chapter, we ran several crosstabs correlating religiosity (as measured by CHATT, PRAY, and POSTLIFE) with gender (as measured by SEX) and then age (as measured by AGECAT).
- While you can run crosstabs with both nominal and ordinal variables, as a general rule, the more categories you are dealing with, the more difficult it becomes to interpret your table.
- When interpreting crosstabs, we are generally looking to see whether an association exists between two variables.
- We do this by reading across categories of the dependent variable.
- A percentage point difference of 10 or more is sometimes taken as an indication that the association between the variables is worth investigating further. This is just a general rule of thumb, however, and additional details about number of columns and so forth are discussed in this chapter.
- In addition to looking for an association between the variables, crosstabs also give us a rough indication of the strength and, in some cases, the direction of association.
- Bear in mind, however, that the measures of association we focus on in Chapter 13 give us a much better basis for drawing conclusions about the nature, strength, and direction of association between variables.
- Finding an association between variables is best looked at as evidence of, not proof of, a causal relationship.
- Epsilon is a simple statistic used to summarize percentage differences.

Key Terms

Univariate analysis	Categories
10-percentage-point rule	Strength of association
Crosstabulation	Dependent variable
Cells	Epsilon
Independent variable	Total (N)
Hypothesis	Frequency
Column percentage	Positive association
Bivariate analysis	Negative association
Column total	Row total
Marginals	Association

SPSS Statistics Commands Introduced in This Chapter

10.1   Running Crosstabs—Specifying the Dependent and Independent Variables

Review Questions

1. What is univariate analysis?

2. What is bivariate analysis?

3. What are the major differences between univariate and bivariate analysis in terms of the number of variables, major questions, and primary goals?

4. In a hypothesis, the variable that is said to "cause" (to some extent) variation in another variable is referred to as what type of variable?

5. What three general questions might you ask yourself when examining a crosstab with two ordinal variables?

Identify the independent and dependent variables in the following hypotheses (Questions 6 and 7):

6. Those employed by companies with more than 20 employees are more likely to have some form of managed-choice health care than are those employed by companies with fewer employees.

7. In the United States, women are more likely to vote Democratic than men are.

8. What are the categories of the independent variable in the hypothesis in Question 7?

9. When running crosstabs, is it customary to specify the dependent variable as the row or column variable?

10. If you were running crosstabs to test the relationship between the variables in the hypothesis in Question 6, which variable would you specify as the row variable and which would you specify as the column variable?

11. If you produce a crosstab for the variables SEX and PARTYID (with the following categories: Democrat, Republican, Independent, other), is it possible to determine the direction of association between these two variables? Why or why not?

12. A researcher produces a crosstab for the variables AGECAT and level of happiness (with the following categories: low, medium, and high) and finds a negative relationship between these two variables. Does this mean that the older you are the happier you are likely to be, or the older you are the less happy you are likely to be?

13. A researcher produces a crosstab for the variables class (with the following categories: lower, working, middle, and upper) and level of contentment (with the following categories: low, medium, and high) and finds that the upper class is more likely to be content than are the middle, working, and lower classes. Based on this hypothetical example, how would you describe the direction of association between these variables?

14. What is epsilon?

15. How is epsilon calculated?

16. Can there be more than one epsilon in a table that has more than two columns?

17. If you run crosstabs and find a strong relationship between the independent and dependent variables in your hypothesis, are you better off looking at this association as *evidence of* or *proof of* a causal relationship?

NAME _____

CLASS _____

INSTRUCTOR _____

DATE _____

To complete the following exercises, you should load the data file AdventuresPLUS.SAV.

A number of studies have addressed the relationship between race and attitudes toward sex roles. In a 1992 study, Jill Grisby argued that Whites are more likely than Blacks to believe that a woman's working has a detrimental impact on her children.

We are going to test this hypothesis using the variables RACE (as a measure of race) and FECHLD (as a measure of opinions regarding the impact of working women on their children). Simply follow the steps listed below and supply the information requested in the spaces provided (Questions 1–10).

1. Restate the hypothesis linking RACE and FECHLD.

2. Identify the independent and dependent variables in the hypothesis.
   Race = Inde   Fe = depend

3. When running crosstabs, which variable should you specify as the row variable?
   Fechld

4. When running crosstabs, which variable should you specify as the column variable?
   Race

5. What is the level of measurement for RACE?
   Nominal

6. What is the level of measurement for FECHLD?
   ordinal

7. Now run crosstabs with column percentages to test the hypothesis (do not forget to define 0, 8, and 9 as missing values for FECHLD before you begin). When you have produced your table, present your results by filling in the following information:

   a. List the categories of the independent variable in the spaces provided on Line A.

   b. In the spaces provided on Line B, list the percentage of respondents who "strongly agree" and "agree" with the statement that a woman's working does not hurt children (i.e., sum of those who "strongly agree" + "agree" on FECHLD).

      i. FECHLD by RACE
      ii. LINE A ____white____   ____black____   ____other____
      iii. LINE B ____75.7____   ____76.4____   ____70.8____

8. Are the results consistent with your hypothesis as stated in response to Question 1? Explain.

9. Compare Blacks and Whites in terms of agreement ("strongly agree" + "agree") with the statement that a woman's working does not hurt children, and give the percentage difference (epsilon) below.

10. Do your findings show evidence of a causal relationship between RACE and FECHLD? Explain.

Continue to research the causes of differing attitudes toward sex roles by selecting one independent and one dependent variable from the following lists:

Independent variables: SEX, RACE, CLASS [select one]

Dependent variables: FEFAM, FEHIRE, FEPRESCH [select one]

As before, run crosstabs with column percentages to test your hypothesis and then fill in the information requested in the spaces provided (Questions 11–18).

11. State and explain your hypothesis involving the [one] independent variable and [one] dependent variable you chose from the lists above. *Preschool Kids are more likely to suffer if the mother works. compared with father*

12. Identify the independent and dependent variables in your hypothesis. *Sex = ind   Fepr = dep*

13. When running crosstabs, which variable should you specify as the column variable and which should you specify as the row variable? *Sex = column   Fepr = row*

14. Identify the level of measurement for each of the variables in your hypothesis. *Sex = nominal   Fepr = ordi*

15. Now run crosstabs with column percentages to test your hypothesis. Remember to define the appropriate values as "missing" for both of your variables. Then complete the exercise below:

    a. List the abbreviated variable names of your independent and dependent variables on Line A.

    b. List the categories of the independent variable in the spaces provided on Line B (use only as many blank spaces as necessary).

    c. List the percentage of respondents who "strongly agree" and "agree" for either FEFAM, FEHIRE, or FEPRESCH on Line C (i.e., sum of those who "strongly agree" + "agree" for the variable you have chosen); use only as many blank spaces as necessary.

SPSS STATISTICS LAB EXERCISE 10.1 (CONTINUED)

LINE A ___Fepresch___ by ___sex___

[Dependent Variable]   [Independent Variable]

LINE B ___male, female___

LINE C ___42.1   29.7___

16. Are these results consistent with your hypothesis as stated in response to Question 11 above? Explain.

17. Compute epsilon.

18. Do your findings show evidence of a causal relationship between your independent and dependent variables? Explain.

Continue to research the causes of differing attitudes toward sex roles by selecting one independent and one dependent variable from the following lists:

Independent variables: AGE, EDUC [select one]

Dependent variables: FEFAM, FEHIRE, FEPRESCH [select one]

This time, however, recode the independent variable before proceeding. Then run crosstabs with column percentages to test your hypothesis and fill in the information requested in the spaces provided (Questions 19–23).

19. State and explain your hypothesis involving the [one recoded] independent variable and [one] dependent variable you chose from the lists above.

20. Identify the independent and dependent variables in your hypothesis.

Indep = class
dep = fehire

21. When running crosstabs, which variable should you specify as the column variable and which should you specify as the row variable?

Column = class
row = fehire

22. Identify the level of measurement for each of the variables in your hypothesis.

ordinal

23. Now run crosstabs with column percentages to test your hypothesis. Then print and attach a copy of your table to this sheet. (Remember to define the appropriate values as "missing" before proceeding.) Analyze and then write a short description of your findings below (similar to those in Writing Box 10.1). In particular, you may want to consider whether the results are consistent with your hypothesis and whether there appears to be an association between the variables. If so, how strong is the association? If applicable, what is the direction of association? Based on your examination, is this association worth investigating further? Why or why not? Include relevant percentages.

24. Access the SPSS Statistics Help feature Tutorial and run through the section titled "Crosstabulation Tables."

Hint: Click **Help → Tutorial → Crosstabulation Tables**

Chapter 11   **Political Orientations as Cause and Effect**

In looking for the sources of religiosity, we worked with a coherent theory (the deprivation theory). As we noted earlier, this process is called *deductive research*, and it is usually the preferred approach to data analysis. Sometimes, however, it's appropriate to take a less structured route beginning with data and then proceeding to theory. As you may recall, this process is known as *inductive research*. As we turn our attention to politics in this chapter, we're going to be more inductive than deductive so you can become familiar with this approach as well.

In Chapter 6, we examined two 2010 General Social Survey (GSS) variables: POLVIEWS and PARTYID. In the analyses to follow, we'll look at the relationship between these variables. You can do that now that you understand the Crosstabs command. Next, we'll explore some of the variables that cause differences in political philosophies and party identification, such as age, religion, gender, race, education, class, and marital status. Finally, we'll look at POLVIEWS and PARTYID as independent variables to determine what impact they have on other variables.

The Relationship Between POLVIEWS and PARTYID

Let's begin with the recoded forms of our two key political variables, POLVIEWS and PARTYID. You may recall that when we recoded these items, we named them POLREC and PARTY. We then recommended that you save these variables on a file named AdventuresPLUS .SAV. Go ahead and open your AdventuresPLUS.SAV file now. If you did not save the recoded variables, don't worry. Simply follow the recode commands listed in Chapter 7 before continuing.

As we indicated earlier, the consensus is that Democrats are more liberal than Republicans and Republicans are more conservative than Democrats, although everyone recognizes the existence of liberal Republicans and conservative Democrats. The GSS data allow us to see what the relationship between these two variables actually is. Because neither variable is logically prior to the other, we could treat either as the independent variable. For our present purposes, it is probably useful to explore both possibilities: (1) Political philosophy causes party identification (political philosophy as the independent variable), and (2) party identification causes political philosophy (party identification as the independent variable).

Demonstration 11.1: POLREC by PARTY

To begin, then, let's see if Democrats are more liberal or more conservative than Republicans. Before beginning your examination, make sure you define the value 4 ("Other") for the variable PARTY as "missing." Now go ahead and run Crosstabs, specifying

- POLREC as the row variable,
- PARTY as the column variable, and
- cells to be percentaged by column.

polrec Recoded Political Views * party Recoded PartyID Crosstabulation

			party Recoded PartyID				Total
			1 Democrat	2 Independent	3 Republican	4 Other	
polrec Recoded Political Views	1 Liberal	Count	321	194	38	12	565
		% within party Recoded PartyID	47.7%	24.7%	8.3%	25.5%	28.8%
	2 Moderate	Count	248	366	110	16	740
		% within party Recoded PartyID	36.8%	46.6%	24.1%	34.0%	37.7%
	3 Conservative	Count	104	225	308	19	656
		% within party Recoded PartyID	15.5%	28.7%	67.5%	40.4%	33.5%
Total		Count	673	785	456	47	1961
		% within party Recoded PartyID	100.0%	100.0%	100.0%	100.0%	100.0%

The data in this table confirm the general expectation. Of the Democrats in the GSS sample, 48% describe themselves as liberals in contrast to 8% of the Republicans. The Independents fall between the two parties, with 25% saying they are liberals. The relationship can also be seen by reading across the bottom row of percentages: 16% of the Democrats, versus 68% of the Republicans, call themselves conservatives.

Demonstration 11.2: PARTY by POLREC

We can also turn the table around logically and ask whether liberals or conservatives are more likely to identify with the Democratic Party (or which are more likely to say they are Republicans). You can get this table by simply reversing the location of the two variable names in the earlier command. Run the Crosstabs procedure again, only this time, make PARTY the row variable and POLREC the column variable. The results are shown opposite.

party Recoded PartyID * polrec Recoded Political Views Crosstabulation

			polrec Recoded Political Views			Total
			1 Liberal	2 Moderate	3 Conservative	
party Recoded PartyID	1 Democrat	Count	321	248	104	673
		% within polrec Recoded Political Views	56.8%	33.5%	15.9%	34.3%
	2 Independent	Count	194	366	225	785
		% within polrec Recoded Political Views	34.3%	49.5%	34.3%	40.0%
	3 Republican	Count	38	110	308	456
		% within polrec Recoded Political Views	6.7%	14.9%	47.0%	23.3%
	4 Other	Count	12	16	19	47
		% within polrec Recoded Political Views	2.1%	2.2%	2.9%	2.4%
Total		Count	565	740	656	1961
		% within polrec Recoded Political Views	100.0%	100.0%	100.0%	100.0%

Again, the relationship between the two variables is evident. Liberals are more likely (57%) to say they are Democrats than are moderates (34%) or conservatives (only 16%). Now, why don't you state the relationship between these two variables in terms of the likelihood that they will support the Republican Party? Either way of stating the relationship is appropriate.

In summary, then, an affinity exists between liberalism and Democrats and between conservatism and Republicans. At the same time, it is not a perfect relationship, and you can find plenty of liberal Republicans and conservative Democrats in the tables.

Now, let's switch gears and see if we can begin to explain why people are liberals or conservatives, Democrats or Republicans. Whereas in the previous chapter, when we began our discussion of bivariate analyses, we examined why some people are more religious than others, in this chapter we are going to ask similar questions regarding political orientation and party identification: Why are some people more liberal (or conservative) than others? Why do some people identify themselves as Democrats, while others identify themselves as Republicans, Independents, or "other"? What causes people to be liberals or conservatives, Democrats or Republicans?

Age and Politics

Often the search for causal variables involves the examination of *demographic* (or background) *variables* such as age, religion, sex, race, education, class, and marital status. Such variables often have a powerful impact on attitudes and behaviors. Let's begin with age.

Common belief is that young people are more liberal than are older people—that people get more conservative as they get older. As you can imagine, liberals tend to see this as a trend toward stodginess, whereas conservatives tend to explain it as a matter of increased wisdom. Regardless of the explanation you might prefer, let's see if it's even true that older people are more conservative than are young people.

Demonstration 11.3: POLREC by AGECAT

To find out, run Crosstabs. In this case, age (as measured by AGECAT) would be the independent variable and political views (as measured by POLREC) would be the dependent variable.[1] Consequently, you should specify POLREC as the row variable and AGECAT as the column variable. Here's what you should get:

polrec Recoded Political Views * agecat Recoded Age Crosstabulation

			agecat Recoded Age				
			1 Under 21	2 21-39 years old	3 40-64 years old	4 65 and older	Total
polrec Recoded Political Views	1 Liberal	Count	17	207	240	101	565
		% within agecat Recoded Age	32.1%	31.3%	27.2%	26.9%	28.7%
	2 Moderate	Count	29	259	334	124	746
		% within agecat Recoded Age	54.7%	39.2%	37.9%	33.1%	37.8%
	3 Conservative	Count	7	195	308	150	660
		% within agecat Recoded Age	13.2%	29.5%	34.9%	40.0%	33.5%
Total		Count	53	661	882	375	1971
		% within agecat Recoded Age	100.0%	100.0%	100.0%	100.0%	100.0%

How would you interpret the table above? Do older age groups appear to be more conservative than younger age groups? Are older age groups more liberal than younger age groups, or would you interpret the table in another way? One thing you probably notice is

---

[1] If you did not save the recoded variable AGECAT, go back to Chapter 7 and follow the recode commands before continuing.

that the relationship between age and ideology is not as strong as you may have thought initially, though there is certainly a relationship—what is it?

Demonstration 11.4: PARTY by AGECAT

What would you expect to find in terms of political party identification (as measured by PARTY)? Do you expect to find growing strength for Republicans as people grow older? Are young people more likely to identify themselves as Democrats? Here's an opportunity to test these ideas. Why don't you try it yourself and see what you get? Then compare your table to the following one.

party Recoded PartyID * agecat Recoded Age Crosstabulation

| | | | agecat Recoded Age | | | | |
			1 Under 21	2 21-39 years old	3 40-64 years old	4 65 and older	Total
party Recoded PartyID	1 Democrat	Count	16	227	295	156	694
		% within agecat Recoded Age	28.1%	33.6%	32.6%	40.2%	34.3%
	2 Independent	Count	29	300	373	120	822
		% within agecat Recoded Age	50.9%	44.4%	41.2%	30.9%	40.6%
	3 Republican	Count	11	129	216	105	461
		% within agecat Recoded Age	19.3%	19.1%	23.9%	27.1%	22.8%
	4 Other	Count	1	20	21	7	49
		% within agecat Recoded Age	1.8%	3.0%	2.3%	1.8%	2.4%
Total		Count	57	676	905	388	2026
		% within agecat Recoded Age	100.0%	100.0%	100.0%	100.0%	100.0%

Interpreting Your Table: The Relationship Between Age and Party Identification

How would you interpret this table? What's the relationship between age and party identification? See if you can interpret this table yourself before moving on.

As you can see, the relationship between AGECAT and PARTY is somewhat similar to the relationship between AGECAT and POLREC. Younger people are somewhat less likely to call themselves Democrats than are older people. When we examine the bottom row, we discover that the older people are, the more likely they are to describe themselves as Republicans. Thus, younger people are less likely to ascribe a political party to their own status.

As such, the middle row, those identifying themselves as Independents, is perhaps most interesting of all. This identification appears to be much more common among the young than among the old. The clearest relationship in this table is that the likelihood of identifying with some political party increases dramatically with age.

Realize that the observed pattern is amenable to more than one explanation. It could be that people become more likely to identify with the major parties as they grow older. On the other hand, the relationship might reflect a trend phenomenon: a disenchantment with the major parties in recent years, primarily among young people. To test these competing explanations, you would need to analyze *longitudinal data*, those representing the state of affairs at different points in time. Because the GSS has been conducting surveys since 1972, we have such data extending back more than a quarter century, making further determinations possible.

Religion and Politics

In the United States, the relationship between religion and politics is somewhat complex, especially with regard to Roman Catholics. Let's begin with political philosophies. We will ask SPSS Statistics to run Crosstabs connecting RELIG with POLREC.

Demonstration 11.5: POLREC by RELIG

Before we run Crosstabs, however, we need to make our measure of religious affiliation (RELIG) more manageable. You may recall from our earlier discussion that RELIG contains 16 categories. Because we are primarily interested in the four largest categories (Protestant, Catholic, Jewish, agnostic/atheist) as opposed to the smaller categories labeled "Other" (Mormon, Buddhist, Moslem), we want to instruct SPSS to define the latter as "missing."

You can review the labels and values for this item by accessing the Variable View tab and then double-clicking in the rectangle that corresponds with RELIG and Values (or by looking at the codebook in Appendix A). Do that now and you will see that we want to define the values 0, 5 to 13, 98, and 99 as "missing." Now go ahead and define these values as "missing" by using the "Range plus one discrete missing value" option. In this case, we want to list the range of values as 5 through 99 and the discrete value as 0.

Now we are ready to run your Crosstabs connecting RELIG (independent/column variable) and POLREC (dependent/row variable).

polrec Recoded Political Views * relig RS RELIGIOUS PREFERENCE Crosstabulation

			relig RS RELIGIOUS PREFERENCE				
			1 PROTESTANT	2 CATHOLIC	3 JEWISH	4 NONE	Total
polrec Recoded Political Views	1 Liberal	Count	203	131	20	161	515
		% within relig RS RELIGIOUS PREFERENCE	21.5%	28.4%	55.6%	46.0%	28.7%
	2 Moderate	Count	361	180	9	120	670
		% within relig RS RELIGIOUS PREFERENCE	38.2%	39.0%	25.0%	34.3%	37.3%
	3 Conservative	Count	382	151	7	69	609
		% within relig RS RELIGIOUS PREFERENCE	40.4%	32.7%	19.4%	19.7%	33.9%
Total		Count	946	462	36	350	1794
		% within relig RS RELIGIOUS PREFERENCE	100.0%	100.0%	100.0%	100.0%	100.0%

We did not define "None" as "missing," because, as you can see, it is a meaningful category. Notice that agnostics and atheists ("None," 46%) are not too far away from the likelihood of Jews (56%) to identify themselves as politically liberal. It is not surprising that of the three religious groups, Jews are the least conservative (19%). Protestants (40%) and Catholics (33%), on the other hand, are much more conservative among these three groups.

If you were to make a gross generalization about the relationship between religious affiliation and political philosophy, it would place Protestants and Catholics on the right end of the political spectrum and Jews and "Nones" on the left.

Demonstration 11.6: PARTY by RELIG

Political party identification, however, is a somewhat different matter. Like the Jews, Roman Catholics have been an ethnic minority throughout much of U.S. history, and the Democratic Party, in the past century at least, has focused more on minority rights than has the Republican Party. That would explain the relationship between religion and political party. Why don't you run that table now? Make PARTY the row variable and RELIG the column variable.

party Recoded PartyID * relig RS RELIGIOUS PREFERENCE Crosstabulation

| | | | relig RS RELIGIOUS PREFERENCE | | | | |
			1 PROTESTANT	2 CATHOLIC	3 JEWISH	4 NONE	Total
party Recoded PartyID	1 Democrat	Count	335	158	17	125	635
		% within relig RS RELIGIOUS PREFERENCE	34.6%	33.1%	45.9%	34.6%	34.4%
	2 Independent	Count	330	207	12	191	740
		% within relig RS RELIGIOUS PREFERENCE	34.1%	43.3%	32.4%	52.9%	40.1%
	3 Republican	Count	282	111	7	33	433
		% within relig RS RELIGIOUS PREFERENCE	29.1%	23.2%	18.9%	9.1%	23.5%
	4 Other	Count	22	2	1	12	37
		% within relig RS RELIGIOUS PREFERENCE	2.3%	0.4%	2.7%	3.3%	2.0%
Total		Count	969	478	37	361	1845
		% within relig RS RELIGIOUS PREFERENCE	100.0%	100.0%	100.0%	100.0%	100.0%

As we see in the table, Jews are the most likely to identify themselves with the Democratic Party, with 46% reporting such affiliation. While Protestants are slightly more likely to identify themselves as Democrats than are Catholics, there is not a meaningful difference between the two groups in this regard. Protestants are more likely than any other of these groups to report affiliation with the Republican Party.

If you are interested in these two variables, you might want to explore the relationship between politics and the other religious variables we've examined: POSTLIFE and PRAY. On the other hand, you could look for other consequences of RELIG. What else do you suppose might be affected by differences of religious affiliation?

Gender and Politics

Gender is a demographic variable associated with a great many attitudes and behaviors. Take a minute to think about the reasons women might be more liberal or more conservative than men.

Demonstration 11.7: PARTY and POLREC by SEX

Once you've developed expectations regarding the relationship between gender and political views, use SPSS Statistics to examine the actual association between these variables.

polrec Recoded Political Views * sex RESPONDENTS SEX Crosstabulation

| | | | sex RESPONDENTS SEX | | |
			1 MALE	2 FEMALE	Total
polrec Recoded Political Views	1 Liberal	Count	228	339	567
		% within sex RESPONDENTS SEX	26.5%	30.5%	28.7%
	2 Moderate	Count	317	429	746
		% within sex RESPONDENTS SEX	36.8%	38.6%	37.8%
	3 Conservative	Count	316	344	660
		% within sex RESPONDENTS SEX	36.7%	30.9%	33.5%
Total		Count	861	1112	1973
		% within sex RESPONDENTS SEX	100.0%	100.0%	100.0%

A larger percentage of men (37%, versus 31% of women) identify themselves as conservative. Furthermore, a larger percentage of women (31%) identify themselves as liberal compared with men (27%). With epsilon equal to 6, the relationship between gender and political ideology may warrant further investigation. Still, the relationship appears not to be a particularly strong one.

party Recoded PartyID * sex RESPONDENTS SEX Crosstabulation

| | | | sex RESPONDENTS SEX | | |
			1 MALE	2 FEMALE	Total
party Recoded PartyID	1 Democrat	Count	244	452	696
		% within sex RESPONDENTS SEX	27.6%	39.5%	34.3%
	2 Independent	Count	388	434	822
		% within sex RESPONDENTS SEX	43.8%	38.0%	40.5%
	3 Republican	Count	226	235	461
		% within sex RESPONDENTS SEX	25.5%	20.6%	22.7%
	4 Other	Count	27	22	49
		% within sex RESPONDENTS SEX	3.1%	1.9%	2.4%
Total		Count	885	1143	2028
		% within sex RESPONDENTS SEX	100.0%	100.0%	100.0%

The association between party identification and gender is indeed noteworthy. Women are more likely to identify with the Democratic Party (40%) than are men (28%). This may reflect the fact that the Democratic Party has been more explicit in its support for women's issues in recent years than has the Republican Party.

Race and Politics

Given our brief discussion above about politics and ethnic minority groups such as Jews and Roman Catholics, what relationship do you expect to find between politics and race? The variable available to you for analysis (RACE) codes only "White," "Black," and "other," so it's not possible to examine this relationship in great depth, but you should be able to make some educated guesses about how Caucasians and African Americans might differ politically.

Demonstration 11.8: POLREC by RACE

After you've thought about the likely relationship between race and politics, why don't you run the tables and test your ability to predict such matters?

polrec Recoded Political Views * race RACE OF RESPONDENT Crosstabulation

| | | | race RACE OF RESPONDENT | | | |
			1 WHITE	2 BLACK	3 OTHER	Total
polrec Recoded Political Views	1 Liberal	Count	404	102	61	567
		% within race RACE OF RESPONDENT	26.8%	34.8%	35.7%	28.7%
	2 Moderate	Count	565	112	69	746
		% within race RACE OF RESPONDENT	37.4%	38.2%	40.4%	37.8%
	3 Conservative	Count	540	79	41	660
		% within race RACE OF RESPONDENT	35.8%	27.0%	24.0%	33.5%
Total		Count	1509	293	171	1973
		% within race RACE OF RESPONDENT	100.0%	100.0%	100.0%	100.0%

The table shows that Whites are notably more likely to be conservative (36%) than are Blacks (27%). There is a smaller but also noteworthy difference when it comes to liberalism. Blacks are more likely than Whites (35% and 27%, respectively) to identify themselves as liberal.

Demonstration 11.9: PARTY by RACE

While the correlation between political views and race may not have been as pronounced as you might have expected, the relationship between race and political party identification is very strong, reflecting the Democratic Party's orientation toward minority groups. Blacks are far more likely to identify as Democrats (64%) than are Whites (28%). While about the same percentage of Whites claim the Republican label (27%), just 6% of Blacks identify as Republicans.

party Recoded PartyID * race RACE OF RESPONDENT Crosstabulation

			race RACE OF RESPONDENT			
			1 WHITE	2 BLACK	3 OTHER	Total
party Recoded PartyID	1 Democrat	Count	435	199	62	696
		% within race RACE OF RESPONDENT	28.3%	64.2%	34.6%	34.3%
	2 Independent	Count	645	90	87	822
		% within race RACE OF RESPONDENT	41.9%	29.0%	48.6%	40.5%
	3 Republican	Count	418	19	24	461
		% within race RACE OF RESPONDENT	27.2%	6.1%	13.4%	22.7%
	4 Other	Count	41	2	6	49
		% within race RACE OF RESPONDENT	2.7%	0.6%	3.4%	2.4%
Total		Count	1539	310	179	2028
		% within race RACE OF RESPONDENT	100.0%	100.0%	100.0%	100.0%

Education and Politics

Education, a common component of social class, is likely to be of interest to you, especially if you are currently a college student. From your own experience, what would you expect to be the relationship between education and political philosophy?

Demonstration 11.10: Recoding EDUC → EDCAT

Before we test the relationship between these variables, we need to recode our measure of education (EDUC) to make it more manageable. We will create a new variable called EDCAT and save it on our AdventuresPLUS.SAV file.

Before you begin recoding, make sure to define the values 97, 98, and 99 for EDUC as "missing." Then follow the instructions listed below.

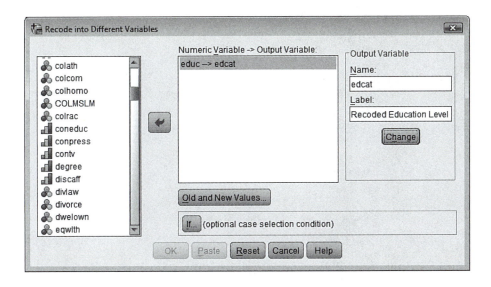

Old Values	New Values	Labels
Lowest to 11	1	Less than high school
12	2	High school graduate
13 to 15	3	Some college
16	4	College graduate
17 to 20	5	Graduate studies (beyond college)

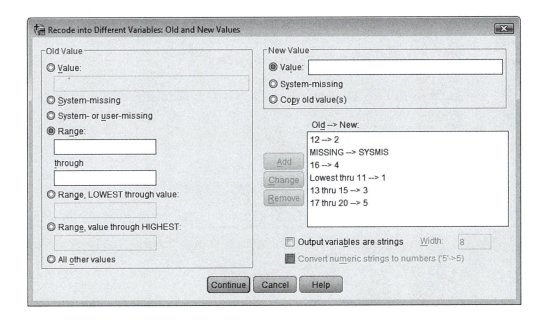

Don't forget to go back and add the labels for EDCAT in the Variable View.

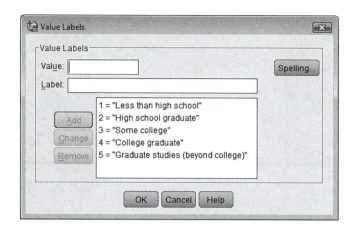

After you have successfully recoded and labeled EDCAT, check your results by running Frequencies on the new variable. Your table should look like the one below.

edcat Recoded Education Level

		Frequency	Percent	Valid Percent	Cumulative Percent
Valid	1 Less than high school	346	16.9	17.0	17.0
	2 High school graduate	558	27.3	27.4	44.3
	3 Some college	524	25.6	25.7	70.0
	4 College graduate	334	16.3	16.4	86.4
	5 Graduate studies (beyond college)	277	13.6	13.6	100.0
	Total	2039	99.8	100.0	
Missing	System	5	.2		
Total		2044	100.0		

Demonstration 11.11: POLREC by EDCAT

Now we are ready to run the Crosstabs for EDCAT (independent/column variable) and POLREC (dependent/row variable) so you can find out whether your expectations regarding the relationship between these variables are accurate.

polrec Recoded Political Views * edcat Recoded Education Level Crosstabulation

			edcat Recoded Education Level					
			1 Less than high school	2 High school graduate	3 Some college	4 College graduate	5 Graduate studies (beyond college)	Total
polrec Recoded Political Views	1 Liberal	Count	92	111	143	109	111	566
		% within edcat Recoded Education Level	29.6%	20.6%	27.9%	32.8%	40.5%	28.7%
	2 Moderate	Count	120	248	207	96	73	744
		% within edcat Recoded Education Level	38.6%	45.9%	40.4%	28.9%	26.6%	37.8%
	3 Conservative	Count	99	181	163	127	90	660
		% within edcat Recoded Education Level	31.8%	33.5%	31.8%	38.3%	32.8%	33.5%
Total		Count	311	540	513	332	274	1970
		% within edcat Recoded Education Level	100.0%	100.0%	100.0%	100.0%	100.0%	100.0%

As you can see, while liberalism generally increases with education (except for the least-educated group), this does not mean that conservatism declines with increasing education. Perhaps the most important row on the table is the middle one, which shows that those with less than a college degree tend to be more moderate than those with bachelor's and advanced degrees.

Demonstration 11.12: PARTY by EDCAT

But how about political party? You decide how to structure your Crosstabs SPSS Statistics command to obtain the following table.

party Recoded PartyID * edcat Recoded Education Level Crosstabulation

| | | | edcat Recoded Education Level | | | | | |
			1 Less than high school	2 High school graduate	3 Some college	4 College graduate	5 Graduate studies (beyond college)	Total
party Recoded PartyID	1 Democrat	Count	120	188	176	114	97	695
		% within edcat Recoded Education Level	35.2%	33.9%	33.8%	34.3%	35.1%	34.3%
	2 Independent	Count	168	229	211	108	103	819
		% within edcat Recoded Education Level	49.3%	41.3%	40.6%	32.5%	37.3%	40.5%
	3 Republican	Count	49	126	123	101	62	461
		% within edcat Recoded Education Level	14.4%	22.7%	23.7%	30.4%	22.5%	22.8%
	4 Other	Count	4	12	10	9	14	49
		% within edcat Recoded Education Level	1.2%	2.2%	1.9%	2.7%	5.1%	2.4%
Total		Count	341	555	520	332	276	2024
		% within edcat Recoded Education Level	100.0%	100.0%	100.0%	100.0%	100.0%	100.0%

The relationship here is consistent in that the less educated tend to be Independents. Those with a college degree are most likely to identify as Republican. Those with graduate studies beyond college, however, tend to be Democrats. One thing is for certain: The trend is not monotonically moving toward Democrat or Republican as years of education increase. This may have to do with other interdependent factors, such as the major one chooses in college or family income. Another issue we have neglected to discuss in detail is that our EDCAT variable is an estimate of degree level. Perhaps actual degree earned produces a different picture. All three of those variables (major, income, and degree), by the way, are in the Adventures.SAV data file, and so the adventure of finding out awaits you!

Some Surprises: Class, Marital Status, and Politics

Sometimes, the inductive method of analysis produces some surprises. As an example, you might take a look at the relationship between our political variables and the demographic variables CLASS and MARITAL.[2]

Social Class

You might expect that social class is related to political philosophy and party identification because the Democratic Party has traditionally been strong among the working class, whereas the well-to-do have seemed more comfortable as Republicans. If you are so inclined,

---

[2]Define 0 and 5 to 9 as "missing" for CLASS and 9 as "missing" for MARITAL.

why don't you check to see if this relationship still holds true? You can use the variable CLASS, which is a measure of subjective social class, asking respondents how they view themselves in this regard.

After you have done that, try to interpret your tables in much the same way we did with the tables above. What do your tables show? Is there a relationship between subjective social class and political philosophy or party identification? Are the results consistent with your expectations, or are they surprising in some way?

Marital Status

In addition, you may also want to take a look at the relationship between marital status and political orientations. If even the suggestion that there is a relationship between marital status and political orientations sounds far-fetched to you, the results of this analysis may be surprising.

Once you've run the tables, try to interpret them and think of any good reasons for the observed differences. Here's a clue: Try to think of other variables that might account for the patterns you've observed. Later, when we engage in multivariate analysis, you'll have a chance to check out some of your explanations.

The Impact of Party and Political Philosophy

Let's shift gears now and consider politics as an independent variable. What impact do you suppose political philosophy and/or political party might have in determining people's attitudes on some of the political issues we looked at earlier? Ask yourself where liberals and conservatives would stand on the following issues. Then run the tables to find out if your hunches are correct: GUNLAW—registration of firearms; CAPPUN—capital punishment. Remember, when POLREC is the independent variable, you need to alter its location in the Crosstabs command, making it the column variable.

After you've examined the relationship between political philosophies and these more specific political issues, consider the impact of political party. In forming your expectations in this latter regard, you might want to review recent political platforms of the two major parties or the speeches of political candidates from the two parties. Then see if the political party identification of the American public falls along those same lines.

Saving Recoded Variable: EDCAT

Before concluding your SPSS Statistics session, be sure to save the data set, including the newly recoded variable EDCAT under the name "AdventuresPLUS.SAV" so we can use it later on.

Conclusion

We hope this chapter has given you a good look at the excitement possible in the detective work called social science research. We're willing to bet that some of the results you've uncovered in this chapter pretty much squared with your understanding of American politics, whereas other findings came as a surprise.

The skills you are learning in this book, along with your access to SPSS Statistics and the GSS data, make it possible for you to conduct your own investigations into the nature of American politics and other issues that may interest you. In the chapter that follows, we're going to return to our examination of attitudes toward abortion. This time, we want to learn what causes differences in attitudes on this controversial topic.

Main Points

- While in the previous chapter we began with a theory of religiosity, in this chapter we took a more inductive approach.
- We examined the causes (and effects) of political orientations by focusing primarily on the recoded forms of our two main political variables: POLVIEWS and PARTYID.
- We examined what causes some people to be liberals or conservatives, Democrats or Republicans.
- In searching for causal variables, we focused primarily on demographic items such as age, religion, gender, race, education, class, and marital status.
- The bulk of the chapter was devoted to explaining what causes people to hold various political views or to identify with one political party or another. In these analyses, we identified POLREC and PARTY as dependent variables.
- It is also possible to look at political views and party identification as independent variables to see what role they might have in determining people's attitudes on political issues such as gun control and capital punishment.

Key Terms

Deductive research
Demographic variables

Inductive research
Longitudinal data

SPSS Statistics Commands Introduced in This Chapter

No new commands were introduced in this chapter.

Review Questions

1. What is the difference between deductive and inductive data analysis?

2. Which approach did we use in Chapter 10 to examine religiosity?

3. Which approach did we use to examine political orientations in this chapter (Chapter 11)?

4. What did we find when we crosstabulated POLREC and PARTY?

5. Do our findings suggest that all conservatives are Republican and all liberals are Democrats?

6. Name three demographic variables in your Adventures.SAV data file.

7. Is the variable CAPPUN a demographic variable? How about INCOME?

8. Of the three religious categories we examined, which is the most liberal? Which is the most conservative?

9. Summarize the relationship between gender and our political variables.

10. There is a strong relationship between race and party identification. Describe that relationship. Can we conclude from this that there is an equally strong relationship between race and political views?

11. Describe the relationship between education levels and party identification.

12. Summarize the relationship between marital status and political orientation. How can we explain these results?

13. Does either political philosophy or party identification determine attitudes about issues such as gun control or capital punishment?

NAME _____

CLASS _____

INSTRUCTOR _____

DATE _____

To complete the following exercises, you need to load the AdventuresPLUS.SAV data file.

1. What causes some people to feel that the national government is spending too little on improving and protecting Americans' health? We will use the variables NATHEAL and HEALTH to begin to address this question.

   ■ Define the values 0, 8, and 9 as "missing" for both NATHEAL and HEALTH.
   ■ Run Crosstabs with NATHEAL as your dependent variable, HEALTH as your independent variable, and cells to be percentaged by column.
   ■ Print your output.
   ■ On Line A below, write the names of the categories of HEALTH using as many spaces as necessary. On Line B, fill in the spaces noting the percentage who answered "Too little" on NATHEAL for each category of HEALTH.

NATHEAL by HEALTH

Line A _____

Line B _____

Do the column percentages in the summary table above change, suggesting that there is a relationship between the variables?

2. Choose one demographic variable from your AdventuresPLUS.SAV data file and state your expectations regarding the relationship between that variable (as the independent variable) and NATHEAL (as the dependent variable). (Depending on which variable you chose, you may need to recode it to make it more manageable.)

   a. Was it necessary for you to recode the demographic variable you chose as your independent variable? If so, explain how you recoded the variable by listing the name of the variable you recoded, the name of the new (recoded) variable, the old value labels, and the new value labels.

   b. Run Crosstabs with NATHEAL as your dependent variable and the variable you chose as the independent variable and print your table.

   c. Then on Line A below, write the name of your independent variable. On Line B, write the names of the categories of your independent variable using as many spaces as necessary. On Line C, fill in the spaces noting the percentage who answered "Too little" on NATHEAL for each category of the independent variable.

Line A NATHEAL by _____

Line B _____

Line C _____

    d. Do the column percentages in the summary table you created above change, suggesting that there is a relationship between the variables?

    e. Summarize and explain your findings below. You may want to note, for instance, whether the findings match your expectations and whether you were surprised by the findings.

  3. Choose one of the following variables as your dependent variable: NATEDUC, NATRACE, or NATFARE, and note the name of the variable you chose and what it measures below.

    a. Now choose a second variable from your data set as your independent variable. Note the name of the variable and what it measures below. If it is necessary to recode your variable, note how you did that below.

    b. What are your expectations regarding the relationship between your independent and dependent variables?

    c. Run Crosstabs and print your table. Then summarize and explain your findings below. You may want to note, for instance, whether there is a relationship between your independent and dependent variables. Do the findings match your expectations? Were you at all surprised by the results? If so, why? If not, why not?

In this chapter, we examined the relationship between PARTY and POLREC in two ways: (1) PARTY as the "cause" or independent variable and (2) POLREC as the "cause" or independent variable. In the following exercise, we are going to consider the relationship between two variables on your AdventuresPLUS.SAV file having to do with opinions about the environment (GRNSOL and GRNPRICE), both of which can be technically examined as the "cause" and as the "effect."

  4. What do the variables GRNSOL and GRNPRICE measure?

    a. Define the values 0, 8, and 9 as "missing" for both variables.

    b. Recode both variables to create RECSOL and RECPRICE as follows:

Old Value	New Value	New Label
1 to 2	1	Willing
3	2	Neither willing nor unwilling
4 to 5	3	Unwilling

c. Construct a hypothesis in which RECPRICE is the independent variable and RECSOL is the dependent variable.

d. Using your recoded variables, test your hypothesis by running Crosstabs. Print and attach your output to this sheet. Then in the space provided below, analyze your findings. You may want to consider questions such as, How does the dependent variable change with changes in the independent variable? How strong is the relationship between the variables?

5. Now construct a hypothesis in which RECPRICE is the dependent variable and RECSOL is the independent variable.

a. Use your recoded variables and test the hypothesis by running Crosstabs. Print and attach your output to this sheet. Then in the space provided below, analyze your findings. You may want to consider questions such as, How does the dependent variable change with changes in the independent variable? How strong is the relationship between the variables?

6. Now choose two other variables on your AdventuresPLUS.SAV file that can be treated as the "cause" and as the "effect."

■ Once you have identified the variables, make sure you define the appropriate values for each variable as "missing," and recode if necessary.
■ Then run Crosstabs to examine the relationship between the variables. Remember, you should examine each variable as the "cause" (or independent variable).
■ Once you have run Crosstabs, print your output and attach it to this sheet. Then write a short description of your findings below. Your primary goal is to identify and describe the general pattern (and any exceptions to that pattern) in the tables. In addition, you may want to ask yourself: What is the nature of association between the variables in each table? How does the dependent variable change with changes in the independent variable? How strong is the association between the variables? What general pattern(s) is/ are evident in each table? Are there any exceptions to this pattern that are worth noting? Is one variable "better" viewed as the "cause" or independent variable?

Chapter 12 What Causes Different Attitudes Toward Abortion?

One of the most controversial issues of recent years has concerned whether a woman has the right to have an abortion. Partisans on both sides of this issue are often extremely vocal and demonstrative.

In this chapter, we're going to use your new analytic skills to begin exploring the causes of different points of view on abortion. Whereas in the previous two chapters we focused on running and interpreting Crosstabs, in this chapter we want to use our output to create summary tables that help us understand the causes of different attitudes toward abortion. Why are some people permissive and others not? As we pursue this question, we can profit from an excellent review of the research on abortion attitudes: Cook, Jelen, and Wilcox's *Between Two Absolutes: Public Opinion and the Politics of Abortion* (1992). Chapter 2 from that book is available on the *Adventures in Social Research* student study website (http://www.sagepub .com/babbie8estudy/) to serve as a background for the analyses we'll undertake in this chapter and to suggest additional directions of analysis if you would like to pursue this topic beyond these first steps.

Demonstration 12.1: Gender and Abortion

Go ahead and open your AdventuresPLUS.SAV file so we can use some of the variables we created earlier.

As you think about possible causes of different attitudes about abortion, the first one that probably comes to mind is gender, given that abortion affects women more directly than it does men. In a quote that has become a popular pro-choice bumper sticker, Florence Kennedy put it this way several years ago: "If men could get pregnant, abortion would be a sacrament." There is reason to believe, therefore, that women would be more supportive of abortion than would men. Let's see.

Here is a table that summarizes attitudes toward abortion by gender. It is not SPSS Statistics output, but we've created it from several SPSS Statistics tables. Your task is to figure out how to get the SPSS Statistics tables that would allow you to create this table.[1]

---

[1]If you are having trouble figuring out how to access the information used to create this table, don't worry, the commands are as follows: Run Crosstabs for the seven abortion variables (row variables) and SEX (column variable), cells to be percentaged by column. You should get seven crosstabs that can be used to create this table (i.e., the percentage of men and women who responded "Yes" to each item). Make sure you are clear as to how this table was created before moving ahead, because we will ask you to set up similar tables later in the chapter.

Percentage Approving of Abortion Under the Following Conditions		Men	Women
ABHLTH	Woman's health endangered	88	86
ABRAPE	Resulted from rape	84	77
ABDEFECT	Serious defect likely	77	74
ABNOMORE	Family wants no more	50	46
ABSINGLE	Woman is unmarried	44	42
ABANY	For any reason	46	42
ABPOOR	Can't afford more children	46	45

Contrary to what we expected, women are not more supportive of abortion than are men. Actually, men are more supportive. The differences between men and women range between 2 and 7 percentage points.

We used the individual items concerning abortion for this analysis because it was possible that men and women would differ on some items but not on others and/or differ by varying amounts. For instance, we might have expected women to be more supportive on the item concerning the woman's health, but this was not the case.

As an alternative strategy for examining the sources of attitudes toward abortion, let's make use of the index ABORT that we created by combining responses to ABDEFECT and ABSINGLE. To do this, we will use the ABORT index we created earlier. Based on the summary table above, we would expect men to exhibit slightly more support for abortion than would women. And our modified expectations are more accurate, as you'll discover when you create the following table.

abort Simple Abortion Index * sex RESPONDENTS SEX Crosstabulation

			sex RESPONDENTS SEX		
			1 MALE	2 FEMALE	Total
abort Simple Abortion Index	0 Yes/Approve	Count	242	268	510
		% within sex RESPONDENTS SEX	44.1%	41.9%	42.9%
	1 Conditional Support	Count	182	202	384
		% within sex RESPONDENTS SEX	33.2%	31.6%	32.3%
	2 No/Disapprove	Count	125	170	295
		% within sex RESPONDENTS SEX	22.8%	26.6%	24.8%
Total		Count	549	640	1189
		% within sex RESPONDENTS SEX	100.0%	100.0%	100.0%

Recall that a score of 0 on the index represents those who supported a woman's right to have an abortion in both circumstances: if there was a chance of a birth defect and if she was single. Overall, 43% of the sample took that position. As we suspected, this table indicates a small difference between men and women, with men a bit more likely (44%) to score 0 on the index than women (42%).

So far, then, we have learned that while men are slightly more likely to be supportive of abortion, gender is not the sole explanation for differences in attitudes toward abortion. Let's see if age has an impact.

Demonstration 12.2: Age and Abortion

As Cook and her colleagues (1992) point out, abortion is somewhat more relevant to young people because they are more likely to experience unwanted pregnancies than are older people. What would that lead you to expect in the way of a relationship between age and support for abortion? Think about that, and then use SPSS Statistics to run the tables that answer the question for you.

Recall that we recoded AGE into AGECAT and saved the variable on our AdventuresPLUS .SAV file. We'll want to use AGECAT for our examination of the relationship between age and abortion attitudes. Now you can request the SPSS Statistics tables that relate age to abortion attitudes.

Here's a summary of the tables you should have created.

Percentage Approving of Abortion Under the Following Conditions		*Under 21*	*21 to 39*	*40 to 64*	*Over 64*
ABHLTH	Woman's health endangered	90	86	89	85
ABRAPE	Resulted from rape	95	81	79	78
ABDEFECT	Serious defect likely	84	73	76	77
ABSINGLE	Woman is unmarried	53	42	46	33
ABANY	For any reason	50	46	46	32
ABPOOR	Can't afford more children	42	46	48	39
ABNOMORE	Family wants no more	58	48	52	37

Take a minute to look over the data presented in this summary table. How do the analytic results square with your expectations? These data suggest that age is not a clear predictor of attitude on abortion. Those 21 and under are more likely than any of the other groups to support abortion if a woman's health is at risk, if the pregnancy is the result of rape, if a serious defect is likely, if the woman is single, and if the woman doesn't want any more children. Those 40 to 64 are more likely than those in other age groups to support abortion in instances where the woman cannot afford any more children. For most of these conditions, support neither increases monotonically nor decreases monotonically.

As an aside, these data change each time the General Social Survey (GSS) is administered. Remember that we use the 2010 GSS for the analyses in this book. Just 6 years ago, for instance, in the 2004 GSS, those under 21 were less likely than those in any of the other age categories to support abortion under each of the conditions presented. In just 6 years, public opinion about abortion has indeed changed along lines of age.

Now let's run Crosstabs using our ABORT index and AGECAT to see what we find. This table indicates that the relationship between age and abortion as measured by our ABORT index continues to be murky. Whereas 50% of those 21 and under approve of abortion in both cases, this percentage is lower in all the other age categories (42%, 46%, 35%), though without monotonically increasing or decreasing. Likewise, the percentage who disapprove of abortion in both cases changes from 17% (21 and under) to 27% (21–39) to 24% (40–64) to 25% (65 and older.) Again, there isn't a clear continuous pattern. What is clear is that those 21 and under seem to have a higher likelihood of approval in both these instances.

abort Simple Abortion Index * agecat Recoded Age Crosstabulation

			agecat Recoded Age				
			1 Under 21	2 21-39 years old	3 40-64 years old	4 65 and older	Total
abort Simple Abortion Index	0 Yes/Approve	Count	9	177	250	72	508
		% within agecat Recoded Age	50.0%	41.9%	46.2%	35.0%	42.8%
	1 Conditional Support	Count	6	132	163	83	384
		% within agecat Recoded Age	33.3%	31.3%	30.1%	40.3%	32.4%
	2 No/Disapprove	Count	3	113	128	51	295
		% within agecat Recoded Age	16.7%	26.8%	23.7%	24.8%	24.9%
Total		Count	18	422	541	206	1187
		% within agecat Recoded Age	100.0%	100.0%	100.0%	100.0%	100.0%

Demonstration 12.3: Religion and Abortion

If we began this analysis by asking you what variable you thought might account for attitudes toward abortion, there is a good chance you would have guessed religion, given the unconditional and public opposition of the Roman Catholic Church. Your growing facility with SPSS Statistics and the 2010 GSS data make it possible for you to test that expectation.

Let's start with the possible impact of religious affiliation. Here's a summary of what you should discover if you run the several abortion items by RELIG. (Remember to define the values 0 and 5 to 99 as "missing" for the variable RELIG before running Crosstabs and creating your summary table.)

Percentage Approving of Abortion Under the Following Conditions		Protestant	Catholic	Jewish	None
ABHLTH	Woman's health endangered	85	82	100	97
ABRAPE	Resulted from rape	76	75	100	92
ABDEFECT	Serious defect likely	71	71	97	89
ABNOMORE	Family wants no more	40	42	83	67
ABANY	For any reason	37	36	75	63
ABSINGLE	Woman is unmarried	35	35	79	63
ABPOOR	Can't afford more children	38	37	79	64

There are several observations you might make about these data. To begin with, the expectation that Catholics would be the most opposed to abortion is not confirmed. In fact, Protestants are just as likely, if not slightly more likely, to disapprove of abortion under some of these seven conditions. Based on our summary table, we can conclude that upward of 71% of both Protestants and Catholics are likely to approve of abortion in cases of health, rape, and birth defect, while between only 35% and 42% are likely to support it in the other cases.

You might also note that the level of support for abortion among American Catholics is greatly at variance with the Roman Catholic Church's official position. Under the traumatic conditions summarized at the top of the table, 71% to 82% of the Catholics say they would approve of abortion. Even under the less traumatic conditions, 35% to 42% of the Catholics would support a woman's right to an abortion.

In contrast, Jews and those with no religious affiliation are consistently more supportive of a woman's right to an abortion. It is interesting to note that 100% of the Jewish respondents said they would support abortion in two of the three most traumatic cases (ABHLTH and ABRAPE; 97% for ABDEFECT). In the other cases, Jewish respondents' approval ranges from 79% to 83%.

To examine this relationship further, you could also use the index ABORT to see if this pattern is reflected in scores on a composite measure. Your table should look like the one below.

abort Simple Abortion Index * relig RS RELIGIOUS PREFERENCE Crosstabulation

| | | | relig RS RELIGIOUS PREFERENCE | | | | |
			1 PROTESTANT	2 CATHOLIC	3 JEWISH	4 NONE	Total
abort Simple Abortion Index	0 Yes/Approve	Count	200	96	23	136	455
		% within relig RS RELIGIOUS PREFERENCE	35.0%	35.8%	79.3%	63.6%	42.1%
	1 Conditional Support	Count	203	94	5	54	356
		% within relig RS RELIGIOUS PREFERENCE	35.6%	35.1%	17.2%	25.2%	32.9%
	2 No/Disapprove	Count	168	78	1	24	271
		% within relig RS RELIGIOUS PREFERENCE	29.4%	29.1%	3.4%	11.2%	25.0%
Total		Count	571	268	29	214	1082
		% within relig RS RELIGIOUS PREFERENCE	100.0%	100.0%	100.0%	100.0%	100.0%

In our earlier examinations of religion, we've sometimes gone beyond affiliation to examine the measure of religiosity, or religiousness. How do you suppose church attendance would relate to abortion attitudes? To find out, let's use CHATT, the recoded variable created earlier. Run the appropriate tables. Here's a summary of what you should have learned.

Percentage Approving of Abortion Under the Following Conditions		Weekly	Monthly	Seldom	Never
ABHLTH	Woman's health endangered	74	85	94	95
ABDEFECT	Serious defect likely	53	73	89	85
ABRAPE	Resulted from rape	58	82	91	90
ABPOOR	Can't afford more children	22	45	56	61
ABSINGLE	Woman is unmarried	17	42	55	59
ABNOMORE	Family wants no more	23	49	60	62
ABANY	For any reason	20	43	54	58

What does this table tell you about religion and abortion attitudes? The overall relationship is pretty clear: Increased church attendance is related to decreased support for abortion. There is a substantial difference, however, between those who attend church about weekly and those who attend less often. Those who attend one to three times per month are considerably more likely to support unconditional abortion than are those who attend church weekly. Even in the case of the traumatic conditions at the top of the table, only those who attend church weekly stand out overwhelmingly in their comparatively low level of support.

The fact that the other three groups do not differ much from one another (particularly the "seldom" and "never" groups), by the way, is a result of what we call a *ceiling effect*.

Whenever the overall percentage of people agreeing with something approaches 100%, considerable variation among subgroups is not possible.

In the extreme case, if everyone agreed on something, there would be no way for men and women to differ, because 100% of both would have to agree. Similarly, there could be no differences among age groups, religions, and so on. When the overall percentage approaches zero, a similar situation occurs that we call a *floor effect.*

Now check the relationship between CHATT and the abortion index, ABORT, and then compare your results with ours below.

abort Simple Abortion Index * chatt Recoded Church Attendance Crosstabulation

			chatt Recoded Church Attendance				
			1 About weekly	2 About monthly	3 Seldom	4 Never	Total
abort Simple Abortion Index	0 Yes/Approve	Count	59	78	211	161	509
		% within chatt Recoded Church Attendance	17.4%	41.1%	55.5%	57.9%	42.9%
	1 Conditional Support	Count	119	60	126	79	384
		% within chatt Recoded Church Attendance	35.1%	31.6%	33.2%	28.4%	32.4%
	2 No/Disapprove	Count	161	52	43	38	294
		% within chatt Recoded Church Attendance	47.5%	27.4%	11.3%	13.7%	24.8%
Total		Count	339	190	380	278	1187
		% within chatt Recoded Church Attendance	100.0%	100.0%	100.0%	100.0%	100.0%

Are these results consistent with those in the summary table we created? Do you find evidence of a ceiling effect in this table as well? The GSS data we've provided for your use permit you to explore this general topic even further, if you wish. Why don't you check out the effect of POSTLIFE and PRAY on abortion attitudes? Before running the tables, however, take some time to reflect on what might logically be expected. How should beliefs about an afterlife affect support for or opposition to abortion? You may be surprised by what you learn. Then again, maybe you won't be surprised.

Writing Box 12.1

It might seem ironic that those who believe in a life after death are less likely to support abortion than are those who do not—and vice versa—but that's what the data indicate. Of those who believe in an afterlife, only 38% give unconditional support to a woman's right to have an abortion, contrasted with 59% of those who disbelieve in an afterlife.

Religiosity, as measured by prayer, also affects attitudes toward abortion. Among those who pray several times a day, only 28% give unconditional support for abortion, as do about 39% of those who pray once a day. The real difference shows up among those who pray less than once a week: 62% of respondents who pray less than once a week and 65% of those who never pray support a woman's right to an abortion unconditionally.

Demonstration 12.4: Politics and Abortion

A strong and consistent relationship exists between political philosophy and abortion attitudes. What do you suppose that relationship is? Who would you expect to be the more supportive of abortion: liberals or conservatives? To carry out this investigation, you'll probably want to use the recoded variable POLREC.

Now you can examine the relationship between political philosophy and abortion attitudes to see if your hunch is correct. Here's a summary of what you should discover.

Percentage Approving of Abortion Under the Following Conditions		Liberal	Moderate	Conservative
ABHLTH	Woman's health endangered	93	89	79
ABRAPE	Resulted from rape	90	85	66
ABDEFECT	Serious defect likely	86	79	62
ABSINGLE	Woman is unmarried	61	42	30
ABPOOR	Can't afford more children	63	45	33
ABNOMORE	Family wants no more	68	48	33
ABANY	For any reason	63	43	30

How would you describe these results? Before reading ahead, try your hand at writing a sentence or two to report on the impact of political philosophy on abortion attitudes.

One way to summarize the relationship between political philosophy and abortion attitudes is to say the following: Liberals are strongly and consistently more supportive of a woman's right to an abortion than are conservatives, with moderates falling in between.

To pursue this relationship further, you may want to use the ABORT index.

abort Simple Abortion Index * polrec Recoded Political Views Crosstabulation

			polrec Recoded Political Views			
			1 Liberal	2 Moderate	3 Conservative	Total
abort Simple Abortion Index	0 Yes/Approve	Count	204	183	114	501
		% within polrec Recoded Political Views	60.9%	41.8%	29.9%	43.4%
	1 Conditional Support	Count	83	165	122	370
		% within polrec Recoded Political Views	24.8%	37.7%	32.0%	32.1%
	2 No/Disapprove	Count	48	90	145	283
		% within polrec Recoded Political Views	14.3%	20.5%	38.1%	24.5%
Total		Count	335	438	381	1154
		% within polrec Recoded Political Views	100.0%	100.0%	100.0%	100.0%

Are these results consistent with those in the summary table above? Does the interpretation of the relationship between political philosophy and abortion attitudes as being strong and consistent hold up? Another direction you might want to follow in investigating the relationship between politics and abortion attitudes concerns political party identification. As you no doubt realize, the Democratic Party has been generally more supportive of a woman's right to have an abortion than has the Republican Party. This difference was dramatically portrayed by the U.S. presidential and vice-presidential candidates during the 2000 and 2004 elections. Economic recession and war took some of the spotlight away from the abortion issue during the 2008 presidential election.

How do you suppose the official differences separating the parties show up in the attitudes of the rank and file? Among the general public, who do you suppose are the most supportive of abortion: Democrats or Republicans? See if you can find out for yourself by creating a summary table similar to those above and then running Crosstabs with PARTY (our measure of party identification) and ABORT (our index of abortion attitudes). Once you have done that, see if the findings are consistent with your expectations.

Writing Box 12.2

The relationship between political party identification and support for abortion is what you may have anticipated, based on official party platforms and the pronouncements of party spokespersons. Of the Democrats, 51% give unconditional support to a woman's right to an abortion. Among Republicans, only 32% do so. Not surprisingly, Independents fall almost directly in the middle, with 45% giving unconditional support for abortion.

Demonstration 12.5: Sexual Attitudes and Abortion

Recalling that abortion attitudes are related to differences in political philosophy, it might occur to you that other philosophical differences might be relevant as well. As you may recall from our earlier discussion, the GSS data set contains three items dealing with sexual permissiveness/restrictiveness:[2]

PREMARSX—attitudes toward premarital sex

HOMOSEX—attitudes toward homosexual sex relations

XMOVIE—attendance at an X-rated movie during the year

PIKUPSEX—respondent had sex with a casual date within the past year

Here's the relationship between attitudes toward premarital sex (PREMARSX) and toward abortion. Notice how permissiveness on one is related to permissiveness on the other.

abort Simple Abortion Index * premarsx SEX BEFORE MARRIAGE Crosstabulation

			premarsx SEX BEFORE MARRIAGE				
			1 ALWAYS WRONG	2 ALMST ALWAYS WRG	3 SOMETIMES WRONG	4 NOT WRONG AT ALL	Total
abort Simple Abortion Index	0 Yes/Approve	Count	17	6	48	187	258
		% within premarsx SEX BEFORE MARRIAGE	12.7%	13.3%	45.3%	56.5%	41.9%
	1 Conditional Support	Count	42	22	43	90	197
		% within premarsx SEX BEFORE MARRIAGE	31.3%	48.9%	40.6%	27.2%	32.0%
	2 No/Disapprove	Count	75	17	15	54	161
		% within premarsx SEX BEFORE MARRIAGE	56.0%	37.8%	14.2%	16.3%	26.1%
Total		Count	134	45	106	331	616
		% within premarsx SEX BEFORE MARRIAGE	100.0%	100.0%	100.0%	100.0%	100.0%

---

[2]Use the "Range plus one optional discrete missing value" option to define the values 0, 5, 8, and 9 as "missing" for the variables PREMARSX and HOMOSEX. In addition, remember to define the values 0, 8, and 9 as "missing" for the variable XMOVIE and PIKUPSEX.

Now, why don't you check to see if the same pattern holds for the other sexual permissiveness items?

abort Simple Abortion Index * homosex HOMOSEXUAL SEX RELATIONS Crosstabulation

			homosex HOMOSEXUAL SEX RELATIONS				
			1 ALWAYS WRONG	2 ALMST ALWAYS WRG	3 SOMETIMES WRONG	4 NOT WRONG AT ALL	Total
abort Simple Abortion Index	0 Yes/Approve	Count	109	18	44	322	493
		% within homosex HOMOSEXUAL SEX RELATIONS	21.3%	36.7%	48.4%	64.7%	42.9%
	1 Conditional Support	Count	204	15	32	122	373
		% within homosex HOMOSEXUAL SEX RELATIONS	39.9%	30.6%	35.2%	24.5%	32.5%
	2 No/Disapprove	Count	198	16	15	54	283
		% within homosex HOMOSEXUAL SEX RELATIONS	38.7%	32.7%	16.5%	10.8%	24.6%
Total		Count	511	49	91	498	1149
		% within homosex HOMOSEXUAL SEX RELATIONS	100.0%	100.0%	100.0%	100.0%	100.0%

abort Simple Abortion Index * xmovie SEEN X-RATED MOVIE IN LAST YEAR Crosstabulation

			xmovie SEEN X-RATED MOVIE IN LAST YEAR		
			1 YES	2 NO	Total
abort Simple Abortion Index	0 Yes/Approve	Count	84	163	247
		% within xmovie SEEN X-RATED MOVIE IN LAST YEAR	55.6%	39.7%	44.0%
	1 Conditional Support	Count	45	139	184
		% within xmovie SEEN X-RATED MOVIE IN LAST YEAR	29.8%	33.8%	32.7%
	2 No/Disapprove	Count	22	109	131
		% within xmovie SEEN X-RATED MOVIE IN LAST YEAR	14.6%	26.5%	23.3%
Total		Count	151	411	562
		% within xmovie SEEN X-RATED MOVIE IN LAST YEAR	100.0%	100.0%	100.0%

abort Simple Abortion Index * pikupsex R HAD SEX WITH CASUAL DATE LAST YEAR Crosstabulation

			pikupsex R HAD SEX WITH CASUAL DATE LAST YEAR		
			1 HAD SEX WITH PICK-UP	2 NOT SELECTED	Total
abort Simple Abortion Index	0 Yes/Approve	Count	41	54	95
		% within pikupsex R HAD SEX WITH CASUAL DATE LAST YEAR	58.6%	47.4%	51.6%
	1 Conditional Support	Count	21	38	59
		% within pikupsex R HAD SEX WITH CASUAL DATE LAST YEAR	30.0%	33.3%	32.1%
	2 No/Disapprove	Count	8	22	30
		% within pikupsex R HAD SEX WITH CASUAL DATE LAST YEAR	11.4%	19.3%	16.3%
Total		Count	70	114	184
		% within pikupsex R HAD SEX WITH CASUAL DATE LAST YEAR	100.0%	100.0%	100.0%

As you probably expected, abortion attitudes are related to attitudes toward homosexuality, to attendance at an X-rated movie, and to whether the respondent had sex with a casual date in the past year. In other words, those who are sexually permissive tend to be permissive in the area of abortion as well.

Other Factors You Can Explore on Your Own

A number of other factors can affect attitudes toward abortion. We'll suggest a few more demographic variables for you to check out. You may want to review the excerpt from Cook et al. on the website (http://www.sagepub.com/babbie8estudy/) for further ideas.

You might suspect that education is related to abortion attitudes. If so, which direction do you suppose that relationship goes? Why do you suppose that is? Race is another standard demographic variable that might be related to abortion attitudes. You should examine the relationship between attitudes toward abortion in particular circumstances and race carefully, because race may have slightly different effects on different items.

Finally, you might want to explore the relationship between abortion attitudes and some family variables. How do you suppose abortion attitudes relate to respondents' views of the ideal number of children to have? The family variable that may surprise you in its relationship to abortion attitudes is MARITAL. You may recall that we also uncovered a surprising effect of marital status on political philosophy. Check this one out, and we'll take a more in-depth look once we begin our multivariate analyses.

Conclusion

In this chapter, you've had an opportunity to search for explanations for the vast differences in people's feelings about abortion. We've found that religion and politics, for example, are powerful influences. We've also just seen that permissiveness and restrictiveness regarding abortion are strongly related to permissiveness and restrictiveness on issues of sexual behavior.

Thus far, we've opened up the search only for explanations, limiting ourselves to bivariate analyses. In the analyses to come, we'll dig even deeper into the reasons for differences in the opinions people have. Ultimately, you should gain a well-rounded understanding of the logic of social scientific research, as well as mastering some of the fundamental techniques for acting on that logic through SPSS Statistics.

Main Points

- This chapter focused on the causes of differing attitudes toward abortion.
- Cook, Jelen, and Wilcox's book *Between Two Absolutes: Public Opinion and the Politics of Abortion* was used as a starting point for our analysis.
- Throughout the chapter, we examined the relationship between attitudes toward abortion and several potential independent or causal variables.
- In addition to examining SPSS Statistics output (primarily Crosstabs), we also created our own summary tables based on several SPSS Statistics tables. These summary tables made it easy to examine the relationship between attitudes toward abortion and several independent variables.
- Gender and age tend not to be related to attitudes toward abortion, whereas religion, politics, and sexual attitudes do.
- You may also want to try looking for relationships between attitudes toward abortion and other demographic variables such as race, education, and family.

Key Terms

Ceiling effect   Floor effect

SPSS Statistics Commands Introduced in This Chapter

No new commands were introduced in this chapter.

Review Questions

1. One way we explored the relationship between attitudes toward abortion and several potential independent variables was by constructing a summary table based on SPSS Statistics output. How were these tables constructed?

2. What are the commands used to instruct SPSS Statistics to construct a table looking at the relationship between our abortion index ABORT and gender?

3. Do our findings support the contention that women are more likely to support abortion than are men?

4. Why did we use the recoded variable AGECAT as opposed to the variable AGE to examine the relationship between age and attitudes toward abortion?

5. Summarize the relationship between abortion attitudes and religious affiliation.

6. Do the majority of American Catholics differ from the Roman Catholic Church on abortion in particular circumstances? If so, what circumstances?

7. What is the ceiling effect?

8. What is the floor effect?

9. We found that liberals are more likely than conservatives to support abortion. Does this mean that Democrats are more likely to support abortion than are Republicans?

10. Are those favoring small families more or less likely to support abortion?

11. Is there any difference between those who are separated and those who are divorced in terms of their unconditional support for abortion?

12. Is there a relationship between attitudes toward sexual behavior and attitudes toward abortion?

SPSS STATISTICS LAB EXERCISE 12.1

NAME _____

CLASS _____

INSTRUCTOR _____

DATE _____

To complete the following exercises, you need to retrieve the data file AdventuresPLUS.SAV.

We are going to begin by looking at what causes people to be permissive (or restrictive) in the area of teen sex. We will focus primarily on the three variables from your data file that deal with this issue: PILLOK, SEXEDUC, and TEENSEX.

1. What do each of the following variables measure?

    a. PILLOK

    b. SEXEDUC

    c. TEENSEX

2. List the values and labels of each of the following variables in the spaces provided below. Check to make sure all "missing" and "don't know" codes have been designated as missing values for each variable:

    ■ PILLOK—0, 8, 9
    ■ SEXEDUC—0, 3, 8, 9
    ■ TEENSEX—0, 5, 8, 9

Hint: If needed, use "Range plus one optional discrete missing value" for SEXEDUC and TEENSEX.

a. PILLOK

Value	*Label*
_____	_____
_____	_____
_____	_____
_____	_____
_____	_____

b. SEXEDUC

	Value	*Label*	

c. TEENSEX

	Value	*Label*	

3. Recode the variables PILLOK and TEENSEX as indicated below.

   a. PILLOK—recode to create PILLREC

      1 to 2 → 1 Permissive

      3 to 4 → 2 Restrictive

   b. TEENSEX—recode to create TEENREC

      3 to 4 → 1 "Permissive"

      1 to 2 → 2 "Restrictive"

4. Relabel SEXEDUC in the following manner:

   1—Permissive (favor)

   2—Restrictive (oppose)

Hint: Access the **Variable View** tab → Click the cell corresponding with SEXEDUC and the column labeled "Values" → Remove the label (1—Favor; 2—Oppose), and add labels listed above.

5. Use SEX and our three measures of permissiveness on teen sex (PILLREC, TEENREC, and SEXEDUC) to instruct SPSS Statistics to run tables that can be used to fill in the summary table below.

Percentage Permissive on Teen Sex Given the Following Situations	Men	Women
PILLREC		
SEXEDUC		
TEENREC		

6. What does this table tell you about the strength of the relationship between gender and permissiveness on teen sex? Do the column percentages change? If so, specify how the dependent variable changes with changes in the independent variable.

7. Now we will move away from demographic variables. Try using FEFAM (which, as you may recall, is one of our measures of attitudes toward family and sex roles) and our three measures of permissiveness on teen sex (PILLREC, TEENREC, and SEXEDUC) to instruct SPSS Statistics to run tables that can be used to fill in the summary table below. (Don't forget to define 0, 8, and 9 as "missing" for FEFAM.)

Percentage Permissive on Teen Sex Given the Following Situations	Strongly Agree	Agree	Disagree	Strongly Disagree
PILLREC				
SEXEDUC				
TEENREC				

8. What does this table tell you about the strength of the relationship between attitudes toward family/sex roles and permissiveness on teen sex? Do the column percentages change? If so, specify how the dependent variable changes with changes in the independent variable.

9. Use FEPRESCH and our three measures of permissiveness on teen sex (PILLREC, TEENREC, and SEXEDUC) to instruct SPSS Statistics to run tables that can be used to fill in the following summary table. (Don't forget to define 0, 8, and 9 as "missing" for FEPRESCH.)

Percentage Permissive on Teen Sex Given the Following Situations	*Strongly Agree*	*Agree*	*Disagree*	*Strongly Disagree*
PILLREC				
SEXEDUC				
TEENREC				

10. What does this table tell you about the strength of the relationship between attitudes toward family/sex roles and permissiveness on teen sex? Do the column percentages change? If so, specify how the dependent variable changes with changes in the independent variable.

11. Now choose another variable from AdventuresPLUS.SAV that you think may be related to permissiveness on teen sex. Write the name of the variable and explain why you chose it (i.e., how it might be related to permissiveness on teen sex).

12. List the values and labels of the item you chose and indicate whether it is necessary to recode/relabel. If so, explain how you did that below. Remember, you may also need to define values as "missing" for your variable.

13. Examine the relationship between the variable you chose and our three measures of permissiveness on teen sex (PILLREC, SEXEDUC, and TEENREC). Then fill in the following summary table below with the results of your analysis. If you need more space, use a separate sheet of paper. (Don't forget, depending on which variable you choose, you may have to designate "DK" as "missing.")

Percentage Permissive on Teen Sex

LABELS OF VARIABLE CHOSEN _____

PILLREC

SEXEDUC

TEENREC

14. What does this table tell you about the strength of the relationship between _____ [variable you chose] and permissiveness on teen sex? Do the column percentages change? If so, specify how the dependent variable changes with changes in the independent variable.

15. In previous SPSS Statistics lab exercises, we examined several variables on your data file that ask respondents under which circumstances, if any, they would support a police officer striking an adult male (POLABUSE, POLATTAK, POLESCAP, POLHITOK, POLMURDR). In Lab Exercise 8.1, we created an index, named POLIN, based on two of

these items. In the following exercises, we want you to explore these variables further, specifically by asking why some people are more supportive of the use of violence by police officers than are others.

a. To begin, choose two of the demographic variables from your file that you think may be associated with attitudes toward the use of force. List the abbreviated variable names and explain why you think these variables may be related to attitudes toward the use of force by police officers.

b. Is it necessary to recode either of the demographic variables to make them more amenable to examination? If so, recode the variable(s) and show how you did that in the space below. If necessary, define appropriate values for each variable as "missing."

c. Use SPSS Statistics to explore the association between these demographic variables and the five variables (POLABUSE, POLATTAK, POLESCAP, POLHITOK, POLMURDR).

Then create two summary tables similar to those we focused on in the demonstrations in Chapter 12. Your tables should be titled

"Percentage Approving of Police Officer Striking a Citizen Under the Following Conditions"

Demographic/Independent Variable 1 _____

Categories/Labels of Demographic/Independent Variable

_____ _____ _____ _____

POLABUSE

POLATTAK

POLESCAP

POLHITOK

POLMURDR

You should use a separate sheet to create two summary tables, one for each demographic/independent variable with which you are working.

d. After creating your summary tables, write a few sentences showing what each table tells us about the relationship between the independent/demographic variables you are working with and POLABUSE, POLATTAK, POLESCAP, POLHITOK, and POLMURDR.

e. Now, examine the relationship between the two demographic variables you chose and POLIN. If you did not save POLIN on your AdventuresPLUS.SAV file, simply follow the instructions in SPSS Statistics Lab Exercise 8.1 to create the index before moving ahead. Once you have run your crosstabs, print and attach your output to this sheet. Then summarize your findings in a few sentences. What do your findings tell us about the relationship between the variables? Are the variables associated? If so, how strongly? Be sure to cite percentages.

Reminder: Before ending this session, be sure to save your new recoded variables on your AdventuresPLUS.SAV file.

Chapter 13   Measures of Association for Nominal and Ordinal Variables

In the preceding analyses, we depended on percentage tables as our format for examining the relationships among variables. While crosstabs are a useful way to examine the relationship between two variables, it is often difficult to get a clear sense of how strong the association between the variables is. While a relationship appears to exist between POLREC and PARTY, it can be difficult to say how strong the association actually is. In this chapter, we are going to explore some measures that allow us to determine more precisely the strength of a relationship. By and large, these techniques, called *measures of association*, summarize relationships (strength and, in some cases, direction) in contrast to the way percentage tables lay out the details.

Another way you might think of measures of association is in contrast to the statistics (measures of central tendency and dispersion) we introduced in Chapter 5. Both are considered *descriptive statistics*, but measures of central tendency and dispersion summarize the distribution of categories of a single variable, whereas measures of association summarize the strength of association between two variables. Additionally, when two variables are at either the ordinal or interval/ratio level of measure, measures of association also indicate the direction of association. These capabilities enable us to use measures of association to answer two important questions:

1. How strong is the association or relationship between two variables?

2. For ordinal and interval/ratio variables, what is the direction of association between two variables?

We begin our discussion with a thought experiment designed to introduce you to the logic of statistical association. We will then focus on two of the most commonly used measures of association for nominal variables, each of which is appropriate for variables at different levels of measurement: lambda (λ) and gamma (γ). In Chapter 14, we will cover measures of association for interval and ratio variables: Pearson's r, the coefficient of determination (r^2), and linear regression. In each case, we will begin by introducing the logic of the measure and then show you how it can be calculated using SPSS Statistics. Finally, at the end of Chapter 14, we will briefly discuss measures appropriate for mixed types of variables.

The Logic of Statistical Association: Proportionate Reduction of Error

To introduce the logic of statistical association, we would like you to take a minute for a "thought experiment." Imagine that there is a group of 100 people in a lecture hall and you are standing in the hallway outside the room. The people will come out of the room one at a time, and your task will be to guess the gender of each before he or she comes into view. Take a moment to consider your best strategy for making these guesses.

If you know nothing about the people in the room, there really is no useful strategy for guessing—no way to make educated guesses. But now suppose you know that 60 of the people in the room are women. This would make educated guesses possible: You should guess "woman" every time. By doing this, you would be right 60 times and wrong 40 times.

Now suppose that every time a person prepares to emerge from the room, his or her first name is announced. This would probably improve your guessing substantially. You'd guess "woman" for every Nancy or Joanne and "man" for every Joseph or Wendell. Even so, you probably wouldn't be totally accurate, given the ambiguity of names such as Pat, Chris, Taylor, and Leslie.

It is useful to notice that we could actually calculate how much knowing first names improved your guessing. Let's say you would have made 40 errors out of 100 guesses without knowing names and only 10 errors when you knew the names. You would have made 30 fewer mistakes. Out of an original 40 mistakes, that's a 75% improvement. Statisticians refer to this as a *proportionate reduction of error (PRE)*. The measures of association we are going to focus on are largely based on this logic.

Lambda (λ): A Measure Appropriate for Nominal Variables

Lambda (λ) is a measure of association appropriate for use with two nominal variables, and it operates on the PRE logic. Essentially, this means that the two variables are related to each other to the extent that knowing a person's attribute on one will help you guess his or her attribute on the other (i.e., the extent to which one variable is "associated" with, affects, or has an impact on another variable).

An Indication of Strength of Association

The value of lambda, which can vary between 0.00 and 1.00, indicates the strength of association or relationship between two nominal variables. The closer the value of lambda is to 1.00, the stronger the relationship between the variables. Conversely, the closer the value of lambda is to 0.00, the weaker the relationship between the variables.

You'll remember that nominal variables are just sets of categories with no greater-than or less-than relationships between them. Based on that, you have probably already correctly surmised that lambda provides *no* indication of the direction of association. In the absence of an order between categories, there can be no direction of relationship. Even if there were a strong relationship between eye color and hair color, it would make no sense to say it was positive or negative.

Measure of association	Lambda
Type of variables	Nominal × Nominal
Values (strength)	0.00 to +1.00
	0 = no association
	1.00 = perfect association
Direction	Not applicable

Example 1: The Logic of Lambda (λ)[1]

Here's a very simple example of lambda. Suppose we have data on the employment status of 1,000 people. Half are employed; half are unemployed. If we were to begin presenting you with person after person, asking you to guess whether each was employed or not, you'd get about half wrong and half right by guessing blindly. So the logic of lambda begins with the assumption that you'd make 500 errors in this case. Let's call these your "uneducated errors." Now, take a look at the table below, which gives you additional information: the ages of the subjects.

	Young	Old	Total
Employed	0	500	500
Unemployed	500	0	500
Total	500	500	1,000

Now suppose we were to repeat the guessing exercise. This time, however, you would know whether each person is young or old. What would be your strategy for guessing employment status? Clearly, you should make an educated guess of "unemployed" for every young person and "employed" for every old person. Do that and you'll make no errors. You will have reduced your errors by 500, in comparison with the first attempt. Given that you will have eliminated all your former errors, we could also say that you have reduced your errors by 100%.

Here's the simple equation for lambda that allows you to calculate the reduction of errors:

$$\frac{(uneducated\,errors) - (educated\,errors)}{(uneducated\,errors)} = \frac{500 - 0}{500} = 1.00$$

Notice that the calculation results in 1.00, which we treat as 100% in the context of lambda. So, in this hypothetical exercise, age is a perfect predictor of employment status: 100% of the variation in employment status can be accounted for by the variation in age.

Example 2: The Logic of Lambda (λ)

To be sure the logic of lambda is clear to you, let's consider another hypothetical example, similar to the previous example.

	Young	Old	Total
Employed	250	250	500
Unemployed	250	250	500
Total	500	500	1,000

In this new example, we still have half young and half old, and we also have half employed and half unemployed. Notice the difference in the relationship between the two variables, however. Just by inspection, you should be able to see that they are independent of each other. In this case, age has no impact on employment status.

---

[1]Throughout this chapter, we review how measures of association are computed in an effort to provide the rationale for matching the measure to the level of data. Instructors and students who would rather focus on interpretation are encouraged to skip these sections and proceed directly to the demonstrations.

The lack of a relationship between age and employment status here is reflected in the "educated" guesses you would make about employment status if you knew a person's age. It wouldn't help you at all, and you would get half the young people wrong and half the old people wrong. You would have made 500 errors in uneducated guesses, and you wouldn't have improved by knowing their ages.

Lambda reflects this new situation:

$$\frac{(uneducated\,errors) - (educated\,errors)}{(uneducated\,errors)} = \frac{500 - 500}{500} = 0.00$$

Knowing age would have reduced your errors by 0%. Moreover, none (0%) of the variation in employment status can be accounted for by the variation in age.

Demonstration 13.1: Instructing SPSS Statistics to Calculate Lambda (λ)

The real relationships between variables are seldom this simple, of course, so let's look at a real example using SPSS Statistics and the General Social Survey (GSS) data. You'll be pleased to discover that you won't have to calculate the errors or the proportion of reduction, because SPSS Statistics does it for you.

Go ahead and open your AdventuresPLUS.SAV file. Then set up a Crosstabs request using ABANY as the row variable and RELIG as the column variable.[2]

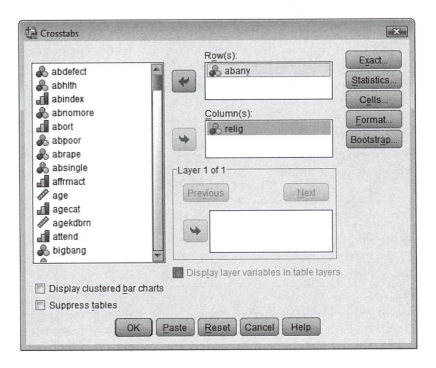

---

[2]Define the values 0, 8, and 9 as "missing" for the variable ABANY. Define the values 0 and 5 to 99 as "missing" for the bivariate variable RELIG.

For the time being, we will not request percentages. To double-check that this option is *not* selected, you can click on **Cells** and turn off percentages by "Column."[3] Then return to the Crosstabs window by clicking **Continue**. Before executing the Crosstabs command, however, click the **Statistics** button. Here's what you should see:

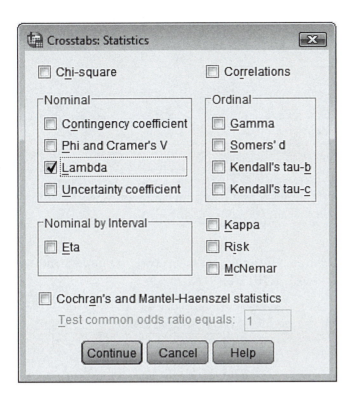

You will notice that SPSS Statistics gives you a variety of options, including some of the measures of association we are going to introduce in this chapter (as well as some we will not introduce but which you may want to explore on your own). You can see that lambda is listed in the box on the left-hand side under "Nominal," because, as we know, lambda is a measure of association used for two nominal variables.

Click **Lambda** and a check mark will appear in the small box to the left. Once you have done that, you can leave the Statistics window by clicking **Continue** and then execute the Crosstabs command by selecting **OK**. Here's the result that should show up in your Output window:

abany ABORTION IF WOMAN WANTS FOR ANY REASON * relig RS RELIGIOUS PREFERENCE Crosstabulation

Count

		relig RS RELIGIOUS PREFERENCE				Total
		1 PROTESTANT	2 CATHOLIC	3 JEWISH	4 NONE	
abany ABORTION IF WOMAN WANTS FOR ANY REASON	1 YES	216	103	21	137	477
	2 NO	372	182	7	81	642
Total		588	285	28	218	1119

We've omitted the request for percentages in this table because it will be useful to see the actual number of cases in each cell of the table; however, keep in mind that in most cases you

---

[3]Usually, percentages are requested in order to see how the column percentages change or move. We omitted them here only to make the table somewhat easier to read.

want to request percentages to see how the column percentages change or move. At the far right of the table, notice that 477 people supported the idea of a woman being able to get an abortion just because she wanted one, while 642 were opposed. If we were to make uneducated guesses about people's opinions on this issue, we'd do best always to guess "opposed." But by doing that, we would make 477 errors.

If we knew each person's religion, however, we would improve our record somewhat. Here's what would happen.

Religion	Guess	Errors
Protestant	No	216
Catholic	No	103
Jewish	Yes	7
None	Yes	81
Total		407

Compare this table to your output to ensure that you understand how we created it. To calculate lambda, then,

$$\lambda = \frac{(uneducated\ errors) - (educated\ errors)}{(uneducated\ errors)} = \frac{477 - 407}{477} = .147$$

This indicates, therefore, that we have improved our guessing of abortion attitudes by 15% as a result of knowing religious affiliation. How SPSS Statistics reports the result is shown below. SPSS Statistics reports more information than we need right now, so let's focus our attention on the second row of numbers. Because we have been testing whether we could predict abortion attitudes (ABANY) by knowing religion, that makes ABANY the dependent variable. As you can see, the value of lambda in that instance is .147, the value we got by calculating it for ourselves.

Directional Measures

			Value	Asymp. Std. Error[a]	Approx. T[b]	Approx. Sig.
Nominal by Nominal	Lambda	Symmetric	.069	.015	4.503	.000
		abany ABORTION IF WOMAN WANTS FOR ANY REASON Dependent	.147	.030	4.503	.000
		relig RS RELIGIOUS PREFERENCE Dependent	.000	.000	[c]	[c]
	Goodman and Kruskal tau	abany ABORTION IF WOMAN WANTS FOR ANY REASON Dependent	.055	.013		.000[d]
		relig RS RELIGIOUS PREFERENCE Dependent	.019	.005		.000[d]

a. Not assuming the null hypothesis.

b. Using the asymptotic standard error assuming the null hypothesis.

c. Cannot be computed because the asymptotic standard error equals zero.

d. Based on chi-square approximation

Knowing a person's religious affiliation, then, allows us to predict his or her attitude on abortion 15% more accurately. The implicit assumption in this analysis is that religious affiliation is associated with or to some extent "causes" attitudes toward abortion. We use the value of lambda as an indication of how strong the causal link is.

However, it is important that you bear in mind that *association alone does not prove causation*. When we observe a statistical relationship between two variables, that relationship strengthens the probability that one causes the other, but it is not sufficient proof. To be satisfied that a causal relationship exists, social scientists also want the link to make sense logically (in this case, the role of churches and clergy in the abortion debate offers that reasoning). And, finally, we want to be sure that the observed relationship is not an artifact produced by the effects of a third variable. This latter possibility will be examined later when we focus on multivariate analysis.

For curiosity's sake, notice the third row, which treats RELIG as the dependent variable. This deals with the possibility that we might be able to guess people's religions by knowing where they stand on abortion. Take a moment to look at the crosstabulation above.

If we were to make uneducated guesses about people's religions, we'd always guess Protestant, because Protestants are by far the largest group. Knowing attitudes toward abortion wouldn't help matters, however. In either case, we'd still guess Protestant, even among those who were in favor of abortion rights. If RELIG were the dependent variable, then, knowing ABANY would improve our guessing by 0%, which is the calculation presented by SPSS Statistics.

Interpreting Lambda and Other Measures

After all this discussion about lambda, you may still be wondering what a value of .140, for instance, actually means in terms of the strength of the relationship between these two variables. Does it signify a strong, relatively strong, moderate, or weak association? Is it something worth noting, or does it indicate such a weak relationship that it is not even worth paying attention to? Unfortunately, there are no easy answers to these questions. Because we are going to say more about *statistical significance* in the next chapter, we do want to give you some (albeit general) basis for interpreting these values.

We can say for certain that not only for lambda but also for all the PRE measures of association we are considering in this chapter, a value of 1.00 (both positive and negative in the case of gamma and Pearson's *r*) indicates a perfect association or the strongest possible relationship between two variables. Conversely, with the exception noted above, a value of 0.00 indicates no relationship between the variables. We can also say that the closer the value is to 1.00 (both positive and negative in the case of gamma and Pearson's *r*), the stronger the relationship. And the closer the value is to 0.00, the weaker the relationship.

Determining whether or not a relationship is noteworthy depends on what other researchers have found. If other researchers had not discovered any variables that related to abortion attitudes, then our lambda of .140 would be important. As you will see later, there are variables that have much stronger relationships to abortion than does religion—so strong that they make our .140 quite unremarkable. In short, all relationships must be interpreted in the context of other findings in the same general area of inquiry.

That said, in order to give you a sense of one way you can approach these statistics, we include some general guidelines in Table 13.1 for interpreting the strength of association between variables. As the title of the table clearly indicates, these are merely loose guidelines with arbitrary cutoff points that must be understood within the context of our discussion above regarding the absence of any absolute "rules" of interpretation.

You will note that Table 13.1 lists the possible values as positive or negative. While the value of lambda is, as we noted, always a positive value between 0.00 and 1.00, the values of the other measures we will discuss later in this chapter (see discussions of gamma and Pearson's *r* next) run from −1.00 to +1.00 and, thus, include both positive and negative values.

Table 13.1 Some General Guidelines for Interpreting Strength of Association
(Lambda [λ], Gamma [γ], etc.)

Strength of Association	Value of Measures of Association (Lambda (λ), Gamma (γ), etc.)
None	0.00
Weak/uninteresting association	±.01 to .09
Moderate/worth noting	±.10 to .29
Evidence of strong association/extremely interesting	±.30 to .99
Perfect/strongest possible association	±1.00

Note: This table is adapted from a more in-depth discussion in Healey et al. (1999). In the text, the authors note that whereas "this scale may strike you as too low . . . remember that, in the social sciences, we deal with probabilistic causal relationships . . . expecting measures of association to approach 1 is unreasonable. Given the complexity of the social world the (admittedly arbitrary) guideline presented is serviceable in most instances" (p. 84). In comparison with similar guides to interpretation, the scale presented here is fairly low. Frankfort-Nachmias et al. (2000), for instance, include a table that gives the following guide to interpretation: .00 (no relationship); ±.20 (weak relationship); ±.40 (moderate relationship); ±.60 (strong relationship); ±.80 (very strong relationship); ±1.00 (perfect relationship). They go on to note that "these are only rough guidelines" (p. 259).

Caveat: Interpreting Lambdas of 0.00

Now that we have told you quite firmly that lambdas of 0.00 always indicate that there is no association between the variables, we need to revise our statement slightly. Lambdas of 0.00 must be treated with great caution. When one of the totals for the dependent variable is much larger than the rest, lambda can take on the value zero even when an inspection of the percentages indicates a strong relationship. Moreover, lambda is typically not useful when crosstabulating two *dichotomous variables*.

To be safe, lambda should be used only when the marginal totals are relatively equal in magnitude and when at least one of the variables is not a dichotomy (i.e., when at least one of the variables has at least three categories). If those requirements are not met, a chi-square-based measure of association, such as Cramer's *V*, should be used.

Based on these general guidelines, what might you say about the strength of the relationship between ABANY and RELIG? Take a moment to develop your answer, and then compare your response to Writing Box 13.1 below.[4]

Writing Box 13.1

The relationship between ABANY and RELIG can be summarized by a value of .147 on lambda. This means that if we wanted to predict how someone felt about abortion, we'd make only about 15% fewer errors if we knew their religious affiliation than we'd make without knowing that.

Now, why don't you choose two nominal variables and then experiment with instructing SPSS Statistics to calculate lambda? If you need help recalling how to run lambda on SPSS Statistics, refer to SPSS Statistics Command 13.1 below.

---

[4]A more complete discussion may be found in Hubert M. Blalock's *Social Statistics* (1960, pp. 310–311).

SPSS Statistics Command 13.1: Running Crosstabs With Lambda (λ)

Click **Analyze** → **Descriptive Statistics** → **Crosstabs**
→ Specify dependent variable as the "Row(s):" variable
→ Specify independent variable as the "Column(s):" variable

   → Click **Cells**
   → Make sure "Column" under "Percentages" is not selected [there should not be a check mark next to "Column"][5]
   → Click **Continue** →

→ Click **Statistics**
   → **Lambda** → **Continue** → **OK**

Gamma (γ): A Measure Appropriate for Ordinal Variables

Whereas lambda is used to examine the association between two nominal variables, *gamma* (γ) is a measure of association based on the logic of proportionate reduction of error appropriate for two ordinal variables (or one ordinal variable and one nominal dichotomous variable.)

An Indication of Strength and Direction (With a Caveat) of Association

Unlike lambda, gamma not only indicates the *strength* of association, but it also indicates the direction of association between two ordinal variables. With each variable having a greater-than/less-than relationship between its categories, we can observe whether or not high values of one variable are associated with high values of the other (a positive association) or if high values of one variable are associated with low values of the other (a negative association). The values of gamma range from −1.00 to +1.00.

In terms of the *strength of association*, the closer to −1.00 or +1.00, the stronger the relationship between the two variables, whereas the closer to 0.00, the weaker the association between the variables.

In terms of the *direction of association*, a negative sign indicates a *negative association*; as one variable increases, the other decreases (the items change in opposite directions). Conversely, a positive sign indicates a *positive association*; both items change in the same direction (they both either increase or decrease).

A negative association between social class and prejudice, for instance, indicates that the variables change in opposite directions: As one increases, the other decreases. Conversely, a positive association between social class and prejudice indicates that the variables change in the same direction: They both either increase or decrease (see Table 13.2).

We want to underscore the fact that you can determine the direction of association between two variables only if they are both measured on scales that express greater-than/less-than relationships between their points. Furthermore, if an ordinal variable's values have been arbitrarily assigned in such a way that high numbers or values indicate an absence of what is being measured and low numbers or values indicate a presence of the phenomenon, the sign (− or +) of the relationship will be reversed.

---

[5]Usually, percentages are requested in order to see how the column percentages change or move. We purposefully omitted the percentages in Demonstration 13.1 in order to make the table easier to read.

Table 13.2   Direction of Association: Using Class and Prejudice as an Example

Negative Association	Value of Gamma Is Negative (–)
	As social class increases, prejudice decreases.
	↑Social class, ↓Prejudice
	OR
	As social class decreases, prejudice increases.
	↓Social class, ↑Prejudice
Positive Association	Value of Gamma Is Positive (+)
	As social class increases, so, too, does prejudice.
	↑Social class, ↑Prejudice
	OR
	As social class decreases, so, too, does prejudice.
	↓Social class, ↓Prejudice

For an example of what we mean by *arbitrary* values, think back to when we reduced the number of ordinal categories on the variable ATTEND to create CHATT. In that case, we chose the value of 1 to represent "About weekly," 2 to represent "About monthly," and so on. These are arbitrary values in the sense that we could just as easily have arranged the categories in the opposite direction so that 1 represented "Never," 2 represented "Seldom," 3 represented "About monthly," and 4 represented "About weekly." With reversed codes, you need to keep in mind that a negative value for gamma in this case would not correctly indicate a direction of association; it would merely be a result of the way we arbitrarily arranged the categories of an item.

Measure of association	Gamma
Type of variables	Ordinal × Ordinal
Values (strength)	–1.00 to +1.00
	0 = no association
	–1.00 = perfect (negative) association
	+1.00 = perfect (positive) association
Direction	+ indicates positive association (variables change or move in same direction, i.e., both ↓↓ OR both ↑↑)
	– indicates negative association (variables change or move in opposite directions, i.e., as one goes ↑, other goes ↓ OR as one goes ↓, other goes ↑)
	Caveat: Have the values been arbitrarily assigned? If so, ± sign may be reversed.

Example 1: The Logic of Gamma (γ)

We judge two variables to be related to each other to the extent that knowing what a person is like in terms of one variable helps us guess what he or she is like in the other. Whereas the application of this logic in the case of lambda lets us make predictions for individuals (e.g., if a person is Protestant, we guess he or she is also Republican), the logic is applied to pairs of people in the case of gamma.

To see the logic of gamma, let's consider the following nine people, placed in a matrix that indicates their social class standing and their level of prejudice: two ordinal variables.

Prejudice	Lower Class	Middle Class	Upper Class
Low	Jim	Tim	Kim
Medium	Mary	Harry	Carrie
High	Nan	Jan	Fran

Our purpose in this analysis is to determine which of the following best describes the relationship between social class and prejudice:

1. The higher your social class, the more prejudiced you are.

2. The higher your social class, the less prejudiced you are.

3. Your social class has no effect on your level of prejudice.

To begin our analysis, we should note that the only pairs who are appropriate to our question are those who differ in both social class and prejudice. Jim and Harry are an example; they differ in both social class and level of prejudice. Here are the 18 pairs that qualify for analysis:

Jim-Harry	Kim-Mary	Harry-Nan
Jim-Carrie	Kim-Harry	Harry-Fran
Jim-Jan	Kim-Nan	
Jim-Fran	Kim-Jan	
Tim-Mary	Mary-Jan	Carrie-Nan
Tim-Nan	Mary-Fran	Carrie-Jan
Tim-Carrie		
Tim-Fran		

Take a minute to assure yourself that no other pair of people satisfies the criterion that they differ in both social class and prejudice.

If you study the table, you should be able to identify pairs of people who would support Conclusions 1 and 2; we'll come back to Conclusion 3 a little later.

Suppose now that you have been given the list of pairs but you've never seen the original table. Your task is to guess which member of each pair is the more prejudiced. Given that you will simply be guessing blindly, chances are that you'll get about half right and half wrong: nine correct answers and nine errors. Gamma helps us determine whether knowing how two people differ on social class would reduce the number of errors we'd make in guessing how they differ on prejudice.

Let's consider Jim-Harry for a moment. If they were the only two people you could study, and if you had to reach a conclusion about the relationship between social class and prejudice, what would you conclude? Notice that Harry is higher in social class than Jim (middle class vs. lower class), and Harry is also higher in prejudice (medium vs. low). If you were to generalize from this single pair of observations, there is only one conclusion you could reach: "The higher your social class, the more prejudiced you are." As we noted earlier, this is referred to as a positive association: the higher on one variable, the higher on the other. In the more specific language of gamma, we'll refer to this as a *same pair*: The direction of the difference between Jim and Harry on one variable is the same as the direction of their difference on the other. Harry is higher than Jim on both.

Suppose you had to base your conclusion on the Jim-Jan pair. What would you conclude? Look at the table and you'll see that Jan, like Harry, is higher than Jim on both social class and prejudice. This pair would also lead you to conclude that the higher your social class, the more prejudiced you are. Jim-Jan, then, is another same pair in the language of gamma.

Suppose, on the other hand, we observed only Tim and Mary. They would lead us to a very different conclusion. Mary is lower than Tim on social class, but she is higher on prejudice. If this were the only pair you could observe, you'd have to conclude that the higher your social class, the lower your prejudice. In the language of gamma, Tim-Mary is an *opposite pair*: The direction of their difference on one variable is the opposite of their difference on the other.

Now, we hope you've been feeling uncomfortable about the idea of generalizing from only one pair of observations, although that's what many people often do in everyday life. In social research, however, we would never do that.

Moving a little bit in the direction of normal social research, let's assume that you have observed all nine of the individuals in the table. What conclusion would you draw about the association between social class and prejudice? Gamma helps you answer this question.

Let's see how well each of the alternative conclusions might assist you in guessing people's prejudice based on knowing about their social class. If you operated on the basis of the conclusion that prejudice increases with social class, for example, and I told you Fran is of a higher social class than Harry, you would correctly guess that Fran is more prejudiced. If, on the other hand, I told you that Harry is higher in social class than Nan, you would incorrectly guess that Harry is more prejudiced.

Take a minute to go through the list of pairs above and make notations of which ones are same pairs and which are opposite pairs. Once you've done that, count the numbers of same and opposite pairs.

You should get nine of each type of pair. This means if you assume that prejudice increases with social class, you will get the nine opposite pairs wrong; if you assume prejudice decreases with social class, you will get the nine same pairs wrong. In other words, neither strategy for guessing levels of prejudice based on knowing social class will do you any good in this case. In either case, we make as many errors as we would have made if we didn't know the social class differences in the pairs. Gamma gives us a method for calculating that result.

The formula for gamma is as follows:

$$\gamma = \frac{same - opposite}{same + opposite}$$

To calculate gamma, you must first count the number of same pairs and the number of opposite pairs. Once you've done that, the mathematics is pretty simple.

Now, you can complete the formula as follows:

$$\gamma = \frac{9-9}{9+9} = \frac{0}{18} = 0$$

In gamma, this result is interpreted as 0%, meaning that knowing how two people differ on social class would improve your guesses as to how they differ on prejudice by 0—or not at all.

Consider the following modified table, however. Suppose for the moment that there are only three people to be studied:

Prejudice	Lower Class	Middle Class	Upper Class
Low	Jim		
Medium		Harry	
High			Fran

Just by inspection, you can see how perfectly these three people fit the pattern of a positive association between social class and prejudice. Each of the three pairs—Jim-Harry, Harry-Fran, and Jim-Fran—is a same pair. There are no opposite pairs. If we were to give you each of these pairs, telling you who is higher in social class, the assumption of a positive association between the two variables would let you guess who is higher in social class with perfect accuracy.

Let's see how this situation would look in terms of gamma.

$$\gamma = \frac{same - opposite}{same + opposite} = \frac{3-0}{3+0} \; 1.00 \; or \; 100\%$$

In this case, we would say that gamma equals 1.00, meaning you have reduced the number of errors by 100%. To understand this meaning of gamma, we need to go back to the idea of guessing differences in prejudice without knowing social class.

Recall that if you were guessing blindly, you'd be right about half the time and wrong about half the time. In this hypothetical case, you'd be wrong 1.5 times out of 3 (that would be your average if you repeated the exercise hundreds of times). As we've seen, however, knowing social class in this instance lets us reduce the number of errors by 1.5—down to zero. It is in this sense that we say we have reduced our errors by 100%.

Now, let's consider a slightly different table.

Prejudice	Lower Class	Middle Class	Upper Class
Low			Nan
Medium		Harry	
High	Kim		

Notice that in this case we could also have a perfect record if we use the assumption of a negative association between social class and prejudice: the higher your social class, the lower your prejudice. The negative association shows up in gamma as follows:

$$\gamma = \frac{same - opposite}{same + opposite} = \frac{3-0}{3+0} = -1.00 \; or \; 100\%$$

Once again, the gamma indicates that we have reduced our errors by 100%. The minus sign in this result simply signals that the relationship is negative.

Example 2: The Logic of Gamma (γ)

We are finally ready for a more realistic example. Just as you would not want to base a generalization on as few cases as we've been considering so far, neither would it make sense to calculate gamma in such situations. Notice how gamma helps you assess the relationship between two variables when the results are not as obvious to those inexperienced in statistics.

Prejudice	Lower Class	Middle Class	Upper Class
Low	200	400	700
Medium	500	900	400
High	800	300	100

In this table, the names of individuals have been replaced with the number of people having a particular social class and level of prejudice. There are 200 lower-class people in the table, for example, who are low on prejudice. On the other hand, there are 100 upper-class people who are high on prejudice.

Perhaps you can get a sense of the relationship in this table by simple observation. The largest cells are those lying along the diagonal running from lower left to upper right. This would suggest a negative association between the two variables (direction of association). Gamma lets us determine with more confidence whether that's the case and gives us a yardstick for measuring how strong the relationship is (strength of association).

In the simpler examples, every pair of cells represented one pair because there was only one person in each cell. Now, it's a little more complex. Imagine for a moment just one of the people in the upper left cell (lower class, low prejudice). If we matched that person up with the 900 people in the center cell (middle class, medium prejudice), we'd have 900 pairs. The same would result from matching each of the people in the first cell with all those in the second. We can calculate the total number of pairs produced by the two cells by simple multiplication: 200 × 900 gives us 180,000 pairs. Notice, by the way, that these are same pairs.

As a further simplification, notice that there are 900 + 400 + 300 + 100 people who will match with the upper left cell to form same pairs. That makes a total of 1,700 × 200 = 340,000. Here's an overview of all the same pairs in the table:

200 × (900 + 300 + 400 + 100)	= 340,000
500 × (300 + 100)	= 200,000
400 × (400 + 100)	= 200,000
900 × 100	= 90,000
Total same pairs	= 830,000

Following the same procedure, here are all the opposite pairs:

700 × (500 + 800 + 900 + 300)	= 1,750,000
400 × (800 + 300)	= 440,000
400 × (500 + 800)	= 520,000
900 × 800	= 720,000
Total opposite pairs	= 3,430,000

Even though this procedure produces quite a few more pairs than we've been dealing with, the formula for gamma still works the same way:

$$\gamma = \frac{same - opposite}{same + opposite} = \frac{830,000 - 3,430,000}{830,000 + 3,430,000} = \frac{-2,600,000}{4,260,000} = -0.61$$

The minus sign in this result confirms that the relationship between the two variables is a negative one (direction). The numerical result indicates that knowing the social class ranking in each pair reduces our errors in predicting their ranking in terms of prejudice by 61% (strength).

Suppose, for the moment, that you had tried to blindly predict differences in prejudice for each of the 4,260,000 pairs. You would have been wrong about 2,130,000 times. By assuming that the person with higher social class is less prejudiced, you would have made only 830,000 errors, or 2,130,000 − 830,000 = 1,300,000 fewer errors. Dividing the 1,300,000 improvement by the 2,130,000 baseline gives us .61, indicating that you have reduced your errors by 61%.

Demonstration 13.2: Instructing SPSS Statistics to Calculate Gamma (γ)—Example 1

Now, here's the good news. Although it's important for you to understand the logic of gamma, it is no longer necessary for you to do the calculations by hand. Whenever you run Crosstabs in SPSS Statistics, you can request that gamma be calculated by making that request when you set up the table.

Go to Crosstabs. Make EDCAT the column variable and CLASS the row variable.[6] Then click on **Statistics**.

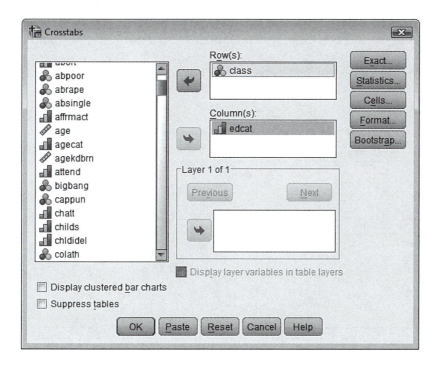

---

[6]We walked through the steps of recoding EDUC to create the new variable EDCAT previously. If you did not save EDCAT on your AdventuresPLUS.SAV file, simply refer back to the recode commands in the earlier chapter before continuing with this demonstration. For the variable CLASS, the values 0, 5, 8, and 9 should be defined as "missing."

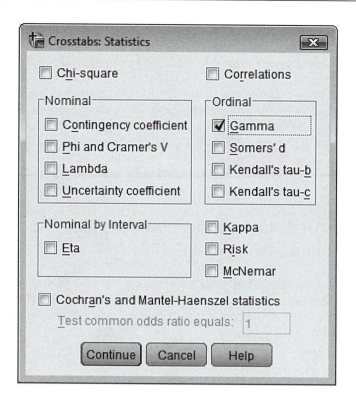

Notice that "Gamma" is appropriately listed as a choice under "Ordinal" data. Click it. If "Lambda" still has a check mark next to it, turn it off before selecting **Continue** to return to the main window. Once you have returned to the Crosstabs window, you can instruct SPSS Statistics to run the procedure by selecting **OK**. You should get the following crosstab table as well as the table reporting on gamma (titled "Symmetric Measures").

class SUBJECTIVE CLASS IDENTIFICATION * edcat Recoded Education Level Crosstabulation

Count

		edcat Recoded Education Level					
		1 Less than high school	2 High school graduate	3 Some college	4 College graduate	5 Graduate studies (beyond college)	Total
class SUBJECTIVE CLASS IDENTIFICATION	1 LOWER CLASS	70	52	37	7	6	172
	2 WORKING CLASS	170	316	300	101	57	944
	3 MIDDLE CLASS	95	181	170	205	199	850
	4 UPPER CLASS	7	7	12	17	15	58
Total		342	556	519	330	277	2024

Symmetric Measures

		Value	Asymp. Std. Error[a]	Approx. T[b]	Approx. Sig.
Ordinal by Ordinal	Gamma	.419	.025	15.944	.000
N of Valid Cases		2024			

a. Not assuming the null hypothesis.

b. Using the asymptotic standard error assuming the null hypothesis.

Notice that gamma is reported as .419. This means that knowing a person's level of education would improve our estimate of his or her class by 42%. Based on our "general guidelines" for interpreting measures of association (Table 13.1), you can see that this value of gamma is fairly high and is evidence of a strong, extremely interesting association between these two variables.

Remember that in addition to indicating the strength of the association between two variables, which in this case is fairly high, gamma also tells us the direction of association. In this instance, the positive sign indicates a positive association between the variables. That is, as level of education rises, so does class.

Writing Box 13.2

Education is a powerful predictor of social class, as these data illustrate. The gamma value of +.419 indicates that our ability to predict accurately which of two people has the higher social class is improved by more than 42% if we know their respective educational levels.

Demonstration 13.3: Running Gamma (γ)—
Example 2 (Reverse Scoring Case)

Now that you are fairly comfortable requesting gamma, let's look at an example that makes the reverse scoring case. An instance in which the arbitrary arrangement of the categories means that you cannot rely on the positive or negative sign is an indication of the direction of association between the variables.

For this example, let's look at the association between CHATT and AGECAT. Go ahead and run Crosstabs, then make CHATT the row variable and AGECAT the column variable. Once you have done that, click on **Statistics** and request "Gamma."

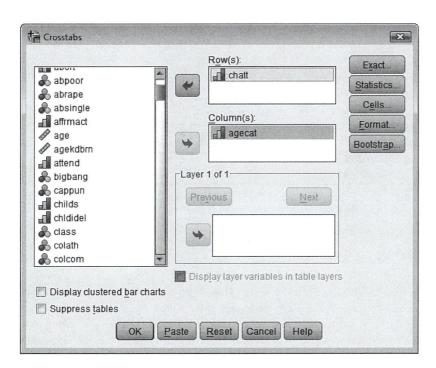

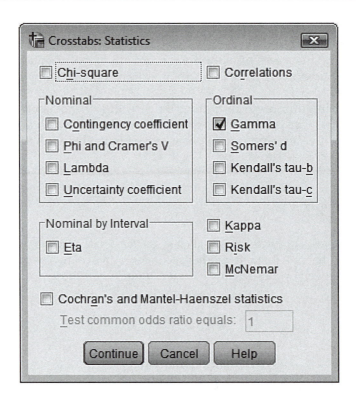

Your output should look like the tables below.

chatt Recoded Church Attendance * agecat Recoded Age Crosstabulation

Count

		agecat Recoded Age				
		1 Under 21	2 21-39 years old	3 40-64 years old	4 65 and older	Total
chatt Recoded Church Attendance	1 About weekly	18	137	274	177	606
	2 About monthly	14	129	130	44	317
	3 Seldom	18	232	300	90	640
	4 Never	8	180	202	80	470
Total		58	678	906	391	2033

Symmetric Measures

		Value	Asymp. Std. Error[a]	Approx. T[b]	Approx. Sig.
Ordinal by Ordinal	Gamma	-.154	.028	-5.538	.000
N of Valid Cases		2033			

a. Not assuming the null hypothesis.

b. Using the asymptotic standard error assuming the null hypothesis.

Notice that gamma is reported as –.154. This means that knowing a person's age would improve our estimate of his or her church attendance by 15%.

Although you might think that the negative sign means there is a negative relationship between age and church attendance, we know now that this is not the case. In this example, the minus sign results from our choosing to arrange attendance categories arbitrarily from the most frequent at the top to the least frequent at the bottom. If we had arranged the categories in the opposite direction, the gamma would have been positive.

Whenever you ask SPSS to calculate gamma, it is important that you note how the variable categories are arranged. If coding is not consistent—that is, if low codes don't indicate low amounts of what is being measured and vice versa—gamma's signs will be reversed. You can also determine the direction of the association by inspecting the table. Looking at the first row in the table above, it is fairly clear that church attendance tends to increase as age increases; hence, the relationship between the two variables is a positive one.

Now take a moment to compare the two tables we just created: CLASS-EDCAT and CHATT-AGECAT. You can see that in the first case, the values of CLASS run down the left-hand side from lowest (top) to highest (bottom), whereas the values of EDCAT run across the top from the lowest education level (left side) to the highest (far right side). Because the categories are arranged in this way, you can be sure that the sign (positive in this case) is meaningful.

Example 1: CLASS × EDCAT					
	EDCAT				
	Less Than High School	*High School*	*Some College*	*College Graduate*	*Graduate Students+*
CLASS	LOWEST education level → HIGHEST education level				
Lower	LOWEST				
Working	Class				
Middle	↓				
Upper	HIGHEST				
	Class				

To the contrary, if you examine the second example (CHATT-AGECAT), you can see a major difference. In this case, CHATT is arranged down the left side from the most frequent attendance (top) to the least frequent attendance (bottom), whereas AGECAT is arrayed across the top from the youngest (left side) to the oldest (right side).

As a result, in this case, the sign is reversed. If we rearranged the categories for CHATT, we would get the opposite sign. However, instead of recoding the variable, it is easier just to make sure you are careful when reading the table and pay attention to cases such as this in which the sign is reversed.

Example 2: CHATT × AGECAT				
	AGECAT			
	Under 21	*21 to 39*	*40 to 64*	*65 and older*
CHATT	YOUNGEST age category → OLDEST age category			
About weekly	Attend			
About monthly	MOST often			
Seldom	↓			
Never	Attend LEAST often			

To gain some more experience with gamma, why don't you select two ordinal variables that interest you and examine their relationships with each other by using gamma. Remember to pay attention to cases such as the one we just reviewed in which the scoring is reversed.

SPSS Statistics Command 13.2: Running Crosstabs With Gamma (γ)

Click **Analyze** → **Descriptive Statistics** → **Crosstabs**

→ Specify dependent variable as the "Row(s):" variable

→ Specify independent variable as the "Column(s):" variable

→ Click **Cells**

→ Make sure "Column" under "Percentages" is not selected [there should not be a check mark next to "Column"][7]

→ Click **Continue** →

Click **Statistics** → **Gamma** → **Continue** → **OK**

Additional Measures of Association

The measures of association we have focused on are each appropriate for variables at a particular level of measurement. For instance, lambda is used to examine the association between two nominal variables, whereas gamma is used to explore the relationship between two ordinal variables; and later in Chapter 14, we will explore Pearson's r, the coefficient of determination (r^2), and regression, which are appropriate for two interval/ratio items. Table 13.3 lists the measures we have reviewed in this chapter, along with their appropriate level of measurement.

Table 13.3 Measures of Association Reviewed in This Chapter

Level of Measurement	Measure of Association (Range of Values) of Each Variable
Nominal	Lambda [0 to +1]
Ordinal	Gamma [−1 to +1]

Keep in mind, however, that the measures we reviewed are just a few of the most popular measures available for nominal and ordinal variables. In Table 13.4 below, we list some other measures applicable for variables at particular levels of measurement. You can access all of these (and other measures) using SPSS Statistics. However, we want to stress that you should use these measures with caution, because like the others we reviewed in the chapter, each has its own peculiar benefits and limitations. For instance, unlike gamma (γ), which is a symmetrical measure of association, Somers' d is an asymmetrical measure of association. So, with Somers' d, if you switched the positions (columns vs. rows) of the variables, the value would be different. With gamma (γ), the value would remain the same.

---

[7]Usually, percentages are requested in order to see how the column percentages change or move. We purposefully omitted the percentages in Demonstrations 13.2 and 13.3 in order to make the tables easier to read.

Table 13.4   Additional Measures of Association

Level of Measurement	
Nominal	Goodman and Kruskal's tau
	Phi
	Cramer's *V*
Ordinal	Somers' *d*
	Kendall's tau-*b*
	Kendall's tau-*c*

Analyzing the Association Between Variables at Different Levels of Measurement

In some cases, you may discover that you want to analyze the association between variables at different levels of measurement: for instance, nominal by ordinal, nominal by interval/ratio, and so on. While there are specialized bivariate measures of association for use when two variables have different levels of measurement, the safest course of action for the novice is to use a measure of association that meets the assumptions for the variable with the lowest level of measurement.

This often means that one variable will have to be recoded to reduce its level of measurement. For instance, when we examined the relationship between church attendance (ordinal) and age (interval/ratio), we recoded people's ages into ordinal categories to carry out the analysis in a meaningful way.

While it would not be appropriate to use Pearson's *r* or *r*² (measures of association for interval/ratio variables) to correlate our CHATT and AGE, it is perfectly legitimate to use gamma to correlate CHATT with AGECAT.

While it is beyond the scope of this chapter to review all the statistical measures appropriate for each particular situation, those of you who are interested in pursuing the relationship between mixed variables may want to consult a basic statistics or research methods text for a discussion of appropriate measures.[8]

Conclusion

In this chapter, we've seen a number of statistical techniques that can be used to summarize the degree of relationship or association between two variables. We've seen that the appropriate technique depends on the level of measurement represented by the variables involved. Lambda (λ) is appropriate for nominal variables, while gamma (γ) is appropriate for ordinal and dichotomous nominal variables.

We realize that you may have had trouble knowing how to respond to the results of these calculations. How do you decide, for example, if a gamma of .25 is high or low? Should you

---

[8]Among the many basic statistics texts you may want to consult are Freeman's *Elementary Applied Statistics* (1968), Siegel and Castellan's *Nonparametric Statistics for the Behavioral Sciences* (1998), and Frankfort-Nachmias et al.'s *Social Statistics for a Diverse Society* (2006). Alreck and Settle's *The Survey Research Handbook* (1985, pp. 287–362) contains a table titled "Statistical Measures of Association" (Table 10-5, p. 303), which, depending on the type of independent and dependent variables you are working with (i.e., categorical or continuous), specifies appropriate measures of association. The table is accompanied by a discussion of statistical analysis and interpretation.

get excited about it or yawn and continue looking? The following chapter on statistical significance offers one basis for making such decisions.

Main Points

- In this chapter, we looked at new ways to examine the relationships among variables.
- Measures of association summarize the strength (and in some cases, the direction) of association between two variables in contrast to the way percentage tables lay out the details.
- This chapter focuses on only two of the many measures available: lambda (λ) and gamma (γ).
- The measures are largely based on the logic of proportionate reduction of error.
- Lambda (λ) is appropriate for two nominal variables; its values range from 0 to 1.
- Lambda (λ) is an asymmetrical measure of association.
- Gamma (γ) is appropriate for two ordinal variables (as well as dichotomous nominal variables); its values range from -1 to $+1$.
- Gamma (γ) is a symmetrical measure of association.
- The closer the value is to positive or negative 1, the stronger the relationship between the items (with $+1$ or -1 indicating a perfect association). The closer the value is to 0, the weaker the relationship between the items (with 0 indicating no association between the items).
- There are no absolute rules for interpreting strength of association.
- Pearson's r, r^2, and linear regression analysis are appropriate for analyzing relationships between interval and ratio variables (see Chapter 14).
- There are a variety of appropriate statistics for examining the relationship between mixed types of variables.

Key Terms

Measures of association	Strength of association
Proportionate reduction of error (PRE)	Dichotomous variables
Lambda (λ)	Direction of association
Gamma (γ)	Same pair
Positive association	Opposite pair
Negative association	Statistical significance
Descriptive statistics	

SPSS Statistics Commands Introduced in This Chapter

13.1   Running Crosstabs With Lambda (λ)

13.2   Running Crosstabs With Gamma (γ)

Review Questions

1. If two variables are strongly associated, does that mean they are necessarily causally related?

2. What is PRE?

3. List two measures of association.

4. Measures of association give us an indication of the _____ of association and (if the variables are ordinal or higher) the _____ of association between two variables.

5. Lambda (λ) is appropriate for variables at which level of measurement?

6. Gamma (γ) is appropriate for variables at which level of measurement?

7. Pearson's r and the coefficient of determination (r^2) are appropriate for variables at which level of measurement?

8. List the range of values for each of the following measures:
   - Lambda
   - Gamma

9. The closer to _____ or _____ (value), the stronger the relationship between the variables.

10. The closer to _____ (value), the weaker the relationship between the variables.

11. A value of _____ or _____ indicates a perfect (or the strongest) relationship between the variables.

12. What does a positive association between two variables indicate? What does a negative association between two variables indicate?

13. Does a negative value for gamma necessarily indicate that the variables are negatively associated? Explain.

14. Name one measure of association you might use to examine the strength of association between the variable CHATT (ordinal) and SEX (nominal) without recoding these two variables.

NAME _____

CLASS _____

INSTRUCTOR _____

DATE _____

To complete these exercises, load your AdventuresPLUS.SAV data file.

Run Crosstabs with column percentages for the variables RACE (independent) and POLHITOK (dependent). Then request the appropriate measure of association and answer Questions 1 through 7. Be sure to define the values 0, 8, and 9 as "missing" for POLHITOK.

1. What is the level of measurement for the variable RACE?

2. What is the level of measurement for the variable POLHITOK?

3. What measure of association is appropriate to examine the relationship between these variables?

4. Record the percentage of respondents who said "Yes" to POLHITOK for each category of the variable RACE.

	Race		Other
	White	Black	
Yes	_____	_____	_____

5. Do the column percentages change or move, signifying a relationship between RACE and POLHITOK? Explain.

6. The strength/value of _____ (measure of association) is _____ (value/ strength of measure of association).

7. Would you characterize the relationship between these two variables as weak, moderate, or strong? Explain.

Choose two nominal variables from your data set. Run Crosstabs with column percentages for the variables. Then request the appropriate measure of association and answer Questions 8 through 12. Reminder: Be sure to define the appropriate values as "missing" for each of your variables.

8. List the two variables you chose, designating one as the independent variable and one as the dependent variable. Then state briefly why you chose them and what you expect to find in terms of the relationship between these variables (i.e., construct a hypothesis relating your variables).

9. What measure of association is appropriate to examine the relationship between these variables?

10. Do the column percentages change or move, signifying a relationship between these variables? Explain.

11. The strength/value of _____ (measure of association) is _____ (value/strength of measure of association).

12. Would you characterize the relationship between these two variables as weak, moderate, or strong? Explain.

Run Crosstabs with column percentages for the variables RACWORK (independent) and DISCAFF (dependent). Then request the appropriate measure of association and answer Questions 13 through 20. Before proceeding with your analysis, remember to define the values 0, 6, 8, and 9 as "missing" for RACWORK. In addition, be sure to define 0, 8, and 9 as "missing" for DISCAFF.

13. What is the level of measurement for the variable RACWORK?

14. What is the level of measurement for the variable DISCAFF?

15. What measure of association is appropriate to examine the relationship between these variables?

16. Record the percentage of respondents who said "Very likely" and "Somewhat likely" to DISCAFF for each category of the variable RACWORK.

	RACWORK				
	All White	Mostly White	Half White-Half Black	Mostly Black	All Black
Very likely					
Somewhat likely					

17. Do the column percentages change or move, signifying a relationship between RACWORK and DISCAFF? Explain.

18. The strength/value of _____ (measure of association) is _____ (value/strength of measure of association).

19. Would you characterize the relationship between these two variables as weak, moderate, or strong? Explain.

20. What is the direction of association between these variables?

Choose two ordinal variables from your data set. Run Crosstabs with column percentages for the variables. Then request the appropriate measure of association and answer Questions 21 through 26. Reminder: Be sure to label appropriate values for each variable as "missing."

21. List the two variables you chose, designating one as the independent variable and one as the dependent variable. Then state briefly why you chose them and what you expect to find in terms of the relationship between these variables (i.e., construct a hypothesis relating your variables).

22. What measure of association is appropriate to examine the relationship between these variables?

23. Do the column percentages change or move, signifying a relationship between these variables? Explain.

24. The strength/value of _____ (measure of association) is _____ (value/strength of measure of association).

25. Would you characterize the relationship between these two variables as weak, moderate, or strong? Explain.

26. What is the direction of association between these variables?

Chapter 14   Correlation and Regression Analysis

In the previous chapter we carefully examined the nature and logic of statistical association through *proportionate reduction of error (PRE)*. The discussion in Chapter 13 was limited to *measures of association* for nominal and ordinal variables. Lambda (λ) and gamma (γ) were both introduced and explored. Of course, we understand that other measures of association can be utilized for nominal and ordinal variables, as presented in Table 13.4.

We begin our discussion in this chapter with Pearson's r, often referred to as a correlation coefficient. After enumerating the details of this statistic, we explore the logic of correlation. We will then focus on linear regression analysis, which includes a discussion of r^2, the coefficient of determination. You'll see that r^2 is, in fact, a PRE statistic, just like lambda and gamma. Last, we will revisit measures appropriate for mixed types of variables.

Pearson's *r*: A Measure Appropriate for Interval/Ratio Variables

Finally, we are going to work with **Pearson's r**, also known as a *product-moment correlation coefficient*, and sometimes simply referred to as a *correlation coefficient*. Pearson's r is a measure of association that reflects the PRE logic and is appropriate to continuous, interval/ratio (I/R) variables such as age, education, and income.

An Indication of Strength and Direction of Association

Like gamma, the value of Pearson's r ranges from –1.00 to +1.00, indicating both the strength and direction of the relationship between two I/R variables. Once again, a –1.00 indicates a perfect *negative association*, a +1.00 indicates a perfect *positive association*, and a 0.00 indicates no association.

Measure of association	Pearson's r
Type of variables	I/R × I/R
Values (strength)	−1.00 to +1.00
	0 = no association
	−1.00 = perfect (negative) association
	+1.00 = perfect (positive) association
Direction	+ indicates positive association (variables change or move in same direction, i.e., both ↓↓ OR both ↑↑)
	− indicates negative association (variables change or move in opposite directions, i.e., as one goes ↑, other goes ↓, OR as one goes ↓, other goes ↑)

Example 1: The Logic of Pearson's *r*

Although this measure also reflects the PRE logic, its meaning in that regard is not quite as straightforward as for the discrete variables analyzed by lambda and gamma. Although it made sense to talk about "guessing" someone's gender or employment status and being either right or wrong, there is little chance that we would ever guess someone's annual income in exact dollars or his or her exact age in days. Our best strategy would be to guess the mean income, and we'd be wrong almost every time. Pearson's *r* lets us determine whether knowing one variable would help us come closer in our guesses of the other variable and calculates how much closer we would come.

To understand *r*, let's take a simple example of eight young people and see whether there is a correlation between their heights (in inches) and their weights (in pounds). To begin, then, let's meet the eight subjects.

	Height	Weight
Eddy	68	144
Mary	58	111
Marge	67	137
Terry	66	153
Albert	61	165
Larry	74	166
Heather	67	92
Ruth	61	128

Take a minute to study the heights and weights. Begin with Eddy and Mary, at the top of the list. Eddy is both taller and heavier than Mary. If we were forced to reach a conclusion about the association between height and weight based on only these two observations, we would conclude that there is a positive correlation: The taller you are, the heavier you are. We might even go a step further and note that every additional inch of height corresponds to about 3 pounds of weight.

On the other hand, if you needed to base a conclusion on observations of Eddy and Terry, see what that conclusion would be. Terry is 2 inches shorter but 9 pounds heavier than Eddy.

Our observations of Eddy and Terry would lead us to just the opposite conclusion: The taller you are, the lighter you are.

Sometimes, it's useful to look at a *scattergram,* which graphs the cases at hand in terms of the two variables.[1] The diagram below presents the eight cases in this fashion. Notice that a general pattern of increasing height seems to be associated with increasing weight, although a couple of cases don't fit that pattern.

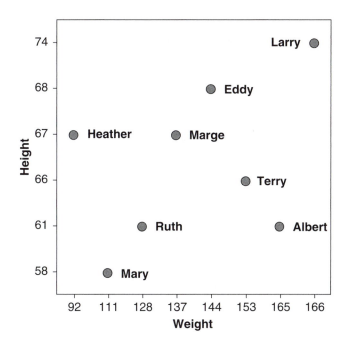

Pearson's *r* allows for the fact that the relationship between height and weight may not be completely consistent; nevertheless, it lets us discover any prevailing tendency in that regard. In the gamma logic presented above, we might consider a strategy of guessing who is heavier or lighter on the basis of who is taller or shorter, assuming either a positive (taller means heavier) or negative (taller means lighter) relationship between the two variables. With *r,* however, we'll take account of *how much* taller or heavier.

To calculate *r,* we need to know the mean value of each variable. As you recall, this is calculated by adding all the values on a variable and dividing by the number of cases. If you do these calculations in the case of height, you'll discover that the eight people, laid end to end, would stretch to 522 inches, for a mean height of 65.25 inches. Do the same calculation for their weights and you'll discover that the eight people weigh a total of 1,096 pounds, for a mean of 137 pounds.

From now on, we are going to focus less on the actual heights and weights of our eight people and deal more with the extent to which they differ from the means. The table below shows how much each person differs from the means for height and weight. Notice that plus and minus signs have been used to indicate whether a person is above or below the mean. (If you want to check your calculations in this situation, you should add all the deviations from height and notice that they total zero; the same is true for deviations from mean weight.)

---

[1]We are going to review how you can instruct SPSS Statistics to produce a scattergram (also called a scatterplot) toward the end of this chapter.

	Height	Weight	H-dev	W-dev
Eddy	68	144	+2.75	+7
Mary	58	111	−7.25	−26
Marge	67	137	+1.75	0
Terry	66	153	+0.75	+16
Albert	61	165	−4.25	+28
Larry	74	166	+8.75	+29
Heather	67	92	+1.75	−45
Ruth	61	128	−4.25	−9
Means	65.25	137		

As our next step, we want to determine the extent to which heights and weights vary from their means overall. Although we have shown the plus and minus signs above, it is important to note that both +2.00 and −2.00 represent deviations of 2 inches from the mean height. For reasons that will become apparent shortly, we are going to capture both positive and negative variations by squaring each of the deviations from the means. The squares of both +2.00 and −2.00 are the same: 4.00. The table below shows the squared deviations for each person on each variable. We've also totaled the squared deviations and calculated their means.

	Height	Weight	H-dev	W-dev	Sq H-dev	Sq W-dev
Eddy	68	144	+2.75	+7	7.5625	49
Mary	58	111	−7.25	−26	52.5625	676
Marge	67	137	+1.75	0	3.0625	0
Terry	66	153	+0.75	+16	0.5625	256
Albert	61	165	−4.25	+28	18.0625	784
Larry	74	166	+8.75	+29	76.5625	841
Heather	67	92	+1.75	−45	3.0625	2,025
Ruth	61	128	−4.25	−9	18.0625	81
Mean	65.25	137			Totals=	4,712
					179.5000	

Now we're going to present a couple of steps that would require more complicated explanations than we want to subject you to in this book; so if you can simply hear what we say without asking why, that's sufficient at this point. (If you are interested in learning the logic of the intervening steps, that's great. You should check discussions of variance and standard deviations in statistics textbooks.) Dividing the sum of the squared deviations by 1 less than the number of cases ($N − 1$) yields a quantity that statisticians call the *variance*. With a large number of cases, this quantity is close to the mean of the sum of squared deviations.

The variances in this case are 25.643 for height and 673.143 for weight. The square root of the variance is called the *standard deviation*. (Perhaps you are already familiar with these concepts, or perhaps you have heard the terms but didn't know what they mean.) Thus, the standard deviation for height is 5.063891; for weight, it is 25.94499.

Now we are ready to put all these calculations to work for us. We are going to express all the individual deviations from mean height and mean weight in units equal to the standard deviations. For example, Eddy was +2.75 inches taller than the average. Eddy's new deviation from the mean height becomes +0.54 (+2.75 ÷ 5.064). His deviation from the mean weight becomes +0.27 (+7 ÷ 25.945).

Our purpose in these somewhat complex calculations is to standardize deviations from the means of the two variables because the values on those variables are of very different scales. Whereas Eddy was 1.75 inches taller and 7 pounds heavier than the mean, we didn't have a way of knowing whether his height deviation was greater or less than his weight deviation. By dividing each deviation by the standard deviation for that variable, we can now see that Eddy's deviation on height is actually greater than his deviation in terms of weight. These new measures of deviation are called **z scores**. The table below presents each person's z score for both height and weight.

	Height	*Weight*	*z height*	*z weight*	*z cross*
Eddy	68	144	0.54	0.27	0.15
Mary	58	111	−1.43	−1.00	1.43
Marge	67	137	0.35	0.00	0.00
Terry	66	153	0.15	0.62	0.09
Albert	61	165	−0.84	1.08	−0.91
Larry	74	166	1.73	1.12	1.93
Heather	67	92	0.35	−1.73	−0.60
Ruth	61	128	−0.84	−0.35	0.29
Total					2.38

You'll notice that there is a final column of the table labeled "z cross." This is the result of multiplying each person's z score on height by the z score on weight. You'll notice that we've begun rounding the numbers to two decimal places. That level of precision is sufficient for our present purposes.

Thanks to your perseverance, we are finally ready to calculate Pearson's r product-moment correlation. By now, it's pretty simple.

r = sum of (z scores for height × z scores for weight) divided by $N − 1$

In our example, this amounts to

$r = 2.38 ÷ 8 − 1 = .34$

There is no easy, commonsense way to represent the meaning of r. Technically, it has to do with the extent to which variations in one variable can explain variations in the other. In fact, if you square r you get the coefficient of determination, .12 in this case. It can be interpreted as follows: 12% of the variation in one variable can be accounted for by the variation in the other. Recall that the variation in a variable reflects the extent to which individual cases deviate (or vary) from the mean value.

Reverting to the logic of PRE, this means that knowing a person's height reduces by 12% the extent of our errors in guessing how far he or she is from the mean weight.

In large part, r's value comes with use. When you calculate correlations among several pairs of variables, the resulting rs will tell which pairs are more highly associated with each other than is true of other pairs.

Interpreting Pearson's r and the Coefficient of Determination (r^2)

We can say for certain that for the measures of association we are considering in this chapter, a value of 1.00 (both positive and negative in the case of Pearson's r) indicates a perfect association or the strongest possible relationship between two variables. Conversely, a value of 0.00 indicates no relationship between the variables.

We can also say that the closer the value is to 1.00 (both positive and negative in the case of Pearson's r), the stronger the relationship. And the closer the value is to 0.00, the weaker the relationship.

That said, in order to give you a sense of one way you can approach these statistics, we include in Table 14.1 some general guidelines for interpreting the *strength of association* between variables. As the title of the table clearly indicates, these are merely loose guidelines with arbitrary cutoff points that must be understood within the context of our discussion above regarding the absence of any absolute "rules" of interpretation. Ultimately, determining whether or not a relationship is noteworthy depends on what association values other researchers have found.

Table 14.1   Some General Guidelines for Interpreting Strength of Association (Pearson's r, Coefficient of Determination)

Strength of Association	Value of Measures of Association (Pearson's r, Coefficient of Determination)
None	0.00
Weak/uninteresting association	±.01 to .09
Moderate/worth noting	±.10 to .29
Evidence of strong association/extremely interesting	±.30 to .99
Perfect/strongest possible association	±1.00

Note: This table has been adapted from a more in-depth discussion in Healey et al. (1999).

Instructing SPSS Statistics to Calculate Pearson's r

Here's the really good news. Your reward for pressing through all the calculations above in order to gain some understanding of what r represents is that you may never have to do it again. SPSS Statistics will do it for you.

Let's consider the possible relationship between two of the continuous variables in the data set: AGE and RINCOM06. If you think about it, you should expect that people tend to earn more money as they grow older, so let's check the correlation between age and respondents' incomes. (Note: RINCOM06 is the respondent's personal income as of 2010; INCOME06 is total family income as of 2010.) You might be reluctant to calculate the deviations, squared deviations, and so on for the 2,044 respondents in your data set (if not, you need a hobby), but computers thrive on such tasks.

Demonstration 14.1: Recoding RINCOM06 → RECINC

Before we tell SPSS Statistics to take on the task of computing correlations for us, we need to do a little housekeeping. First, request a frequency distribution for RINCOM06 in SPSS Statistics.

rincom06 RESPONDENTS INCOME

		Frequency	Percent	Valid Percent	Cumulative Percent
Valid	1 UNDER $1 000	29	1.4	2.4	2.4
	2 $1 000 TO 2 999	43	2.1	3.6	6.0
	3 $3 000 TO 3 999	33	1.6	2.7	8.7
	4 $4 000 TO 4 999	23	1.1	1.9	10.6
	5 $5 000 TO 5 999	26	1.3	2.2	12.8
	6 $6 000 TO 6 999	28	1.4	2.3	15.1
	7 $7 000 TO 7 999	19	.9	1.6	16.7
	8 $8 000 TO 9 999	34	1.7	2.8	19.6
	9 $10000 TO 12499	72	3.5	6.0	25.5
	10 $12500 TO 14999	38	1.9	3.2	28.7
	11 $15000 TO 17499	51	2.5	4.2	32.9
	12 $17500 TO 19999	37	1.8	3.1	36.0
	13 $20000 TO 22499	56	2.7	4.7	40.7
	14 $22500 TO 24999	51	2.5	4.2	44.9
	15 $25000 TO 29999	66	3.2	5.5	50.4
	16 $30000 TO 34999	78	3.8	6.5	56.9
	17 $35000 TO 39999	67	3.3	5.6	62.5
	18 $40000 TO 49999	102	5.0	8.5	71.0
	19 $50000 TO 59999	88	4.3	7.3	78.3
	20 $60000 TO 74999	98	4.8	8.2	86.4
	21 $75000 TO $89999	63	3.1	5.2	91.7
	22 $90000 TO $109999	36	1.8	3.0	94.7
	23 $110000 TO $129999	28	1.4	2.3	97.0
	24 $130000 TO $149999	14	.7	1.2	98.2
	25 $150000 OR OVER	22	1.1	1.8	100.0
	Total	1202	58.8	100.0	
Missing	0 IAP	684	33.5		
	26 REFUSED	145	7.1		
	98 DK	13	.6		
	Total	842	41.2		
Total		2044	100.0		

There are two problems with using RINCOM06 the way it is currently coded. First, we need to be sure codes 0 (IAP), 13 (REFUSED), 98 (DK), and 99 (NA) are not included in our analysis. They may be eliminated by recoding them "system-missing." This may have already been done for you, but it is certainly important to take a moment to check. Second, the code categories are not equal in width. Code Category 8 includes respondents whose incomes were between $8,000 and $9,999, whereas Code Category 9 includes incomes between $10,000 and $12,499. The categories form an ordinal scale that is not appropriate for use with Pearson's *r*.

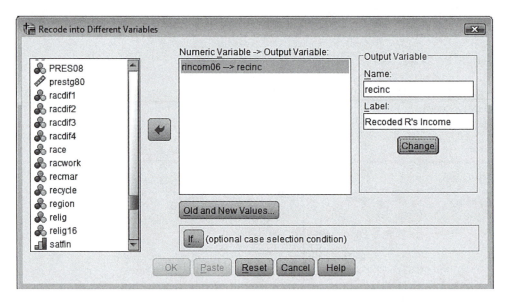

We can use Recode to improve on the coding scheme used for recording incomes.[2] If we assume that incomes are spread evenly across categories, then we can simply substitute the midpoints of the interval widths for the codes used in RINCOM06. That way, we can approximate an I/R scale and rid ourselves of the problems created by the interval widths not being equal. Code 25, $150,000 or more, has no upper limit. We just took an educated guess that the midpoint would be about $160,000. Even though we don't have each respondent's actual income, this approach will enable us to use Pearson's *r* to search for relationships between income and other variables. As SPSS Statistics commands, the recoding looks like this:

Transform → Recode → Into Different Variables		
Old Variable → New Variable		
RINCOM06 → RECINC		
Old Values	*and*	*New Values*
1	→	500
2	→	2,000
3	→	3,500
4	→	4,500
5	→	5,500
6	→	6,500
7	→	7,500
8	→	9,000
9	→	11,250
10	→	13,750

[2]The authors greatly appreciate the suggestion of this analysis by Professor Gilbert Klajman, Montclair State University.

Old Values	and	New Values
11	→	16,250
12	→	18,750
13	→	21,250
14	→	23,750
15	→	27,500
16	→	32,500
17	→	37,500
18	→	45,000
19	→	55,000
20	→	67,500
21	→	82,500
22	→	100,000
23	→	120,000
24	→	140,000
25	→	160,000
System or user missing	→	System-missing

Once you have completed the recode, access the **Variable View** tab and define your new variable RECINC (i.e., set the width, decimals, etc.). Also, don't forget to save your recoded variable on your AdventuresPLUS.SAV file so we can use it later on.

Now go ahead and run Frequencies for RECINC.

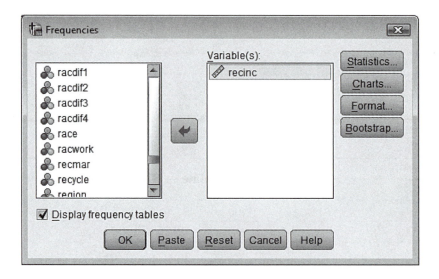

Here's what the recoded variable looks like:

recinc Recoded R's Income

		Frequency	Percent	Valid Percent	Cumulative Percent
Valid	500	29	1.4	2.4	2.4
	2000	43	2.1	3.6	6.0
	3500	33	1.6	2.7	8.7
	4500	23	1.1	1.9	10.6
	5500	26	1.3	2.2	12.8
	6500	28	1.4	2.3	15.1
	7500	19	.9	1.6	16.7
	9000	34	1.7	2.8	19.6
	11250	72	3.5	6.0	25.5
	13750	38	1.9	3.2	28.7
	16250	51	2.5	4.2	32.9
	18750	37	1.8	3.1	36.0
	21250	56	2.7	4.7	40.7
	23750	51	2.5	4.2	44.9
	27500	66	3.2	5.5	50.4
	32500	78	3.8	6.5	56.9
	37500	67	3.3	5.6	62.5
	45000	102	5.0	8.5	71.0
	55000	88	4.3	7.3	78.3
	67500	98	4.8	8.2	86.4
	82500	63	3.1	5.2	91.7
	100000	36	1.8	3.0	94.7
	120000	28	1.4	2.3	97.0
	140000	14	.7	1.2	98.2
	160000	22	1.1	1.8	100.0
	Total	1202	58.8	100.0	
Missing	System	842	41.2		
Total		2044	100.0		

Demonstration 14.2: Using SPSS Statistics to Compute Pearson's *r*

Now make sure codes 98 and 99 are defined as "missing" for AGE.

With the housekeeping out of the way, you need only move through this menu path to launch SPSS Statistics on the job of computing *r*. Unlike the measures of association for nominal and ordinal variables we have discussed, Pearson's *r* is *not* available in the Crosstabs window. This stands to reason, because you typically would not want to create a crosstabulation of I/R variables; the table would be too large to be of any real use.

Instead, Pearson's *r* is appropriately available in the Bivariate Correlations dialog box. To access this window, simply follow the steps below:

Analyze → Correlate → Bivariate

This will bring you to the following window:

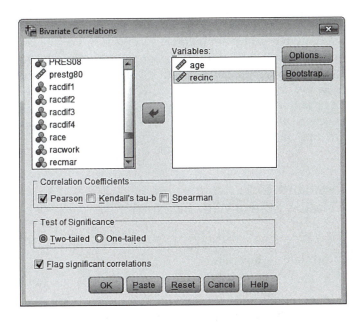

Transfer AGE and RECINC to the "Variables:" field. (Be sure to use AGE and not AGECAT, because we want the uncoded variable.) Below the "Variables:" list, you'll see that we can choose from among three forms of correlation coefficients. Our discussion above has described Pearson's *r*, so click **Pearson** if it is not selected already.

Did you ever think about what we would do if we had an AGE for someone but we did not know his or her RECINC? Because we would have only one score, we would have to throw that person out of our analysis. But suppose we had three variables and we were missing a score on a case. Would we throw out just the pair that had a missing value, or would we throw out the whole case? SPSS Statistics lets us do it either way. Click on **Options**, and you will see that we can choose to exclude cases either pairwise or listwise. If we exclude pairwise, we discard a pair only when either score is missing, but with listwise exclusion, we discard the entire case if only one pair is missing. We will use pairwise exclusion.

Given the time and money it takes to collect data, it's usually best to keep as much of it as we can. Unless there is a specific reason for using listwise exclusion, it is best to preserve as much of the data as possible by using pairwise exclusion.

Once you have selected **Exclude cases pairwise,** click on **Continue** and then **OK.** Your reward will be a *correlation matrix*, a table that shows the correlations among all variables (including, as we will see, the correlation of each item with itself).

Correlations

		age AGE OF RESPONDENT	recinc Recoded R's Income
age AGE OF RESPONDENT	Pearson Correlation	1	.189**
	Sig. (2-tailed)		.000
	N	2041	1201
recinc Recoded R's Income	Pearson Correlation	.189**	1
	Sig. (2-tailed)	.000	
	N	1201	1202

**. Correlation is significant at the 0.01 level (2-tailed).

The Pearson's *r* product-moment correlation between AGE and RECINC is .189.[3] Notice that the correlation between AGE and itself is perfect (1), which makes sense if you think about it. You now know that the .189 is a measure of the extent to which deviations from the mean income can be accounted for by deviations from the mean of age. By squaring *r* (.189 × .189 = .036), we learn that about 3.6% of the variation in income can be accounted for by the variation in how old people are. Squaring *r* yields r^2, also known as the coefficient of determination. The *coefficient of determination* (r^2) is a PRE measure that can be interpreted in the same way as gamma or lambda.

We are going to ignore the references to "Significance" until Chapter 15. This indicates the *statistical significance* of the association.

SPSS Statistics Command 14.1: Producing a Correlation Matrix With Pearson's *r*

Click **Analyze** → **Correlate** → **Bivariate**

→ Highlight variable name → Click right-pointing arrow to transfer variable to "Variables:" field

→ Repeat previous step until all variables have been transferred

→ Click **Pearson**

→ Click **Options**

→ Exclude cases either pairwise OR listwise

→ **Continue** → **OK**

Demonstration 14.3: Requesting Several Correlation Coefficients

What else do you suppose might account for differences in income? If you think about it, you might decide that education is a possibility. Presumably, the more education you get, the more money you'll make. Your understanding of *r* through SPSS Statistics will let you check it out.

---

[3]See Table 14.1 for some general guidelines you can use to interpret the value of Pearson's *r*.

The Correlations command allows you to request several correlation coefficients at once. Go back to the Bivariate Correlations window and add EDUC to the list of variables being analyzed.[4] Once again, exclude cases pairwise before executing the command.

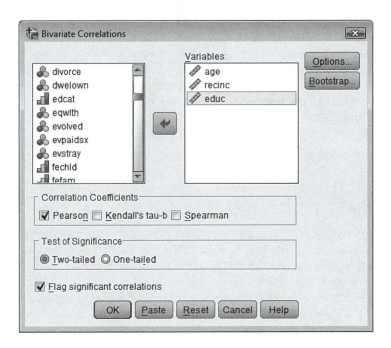

Here's what you should get in your Output window. (Hint: If you didn't get this, go back and make sure you defined the value 99 as "missing" for the variable EDUC, and then try it again.)

Correlations

		age AGE OF RESPONDENT	recinc Recoded R's Income	educ HIGHEST YEAR OF SCHOOL COMPLETED
age AGE OF RESPONDENT	Pearson Correlation	1	.189**	-.049*
	Sig. (2-tailed)		.000	.027
	N	2041	1201	2036
recinc Recoded R's Income	Pearson Correlation	.189**	1	.405**
	Sig. (2-tailed)	.000		.000
	N	1201	1202	1200
educ HIGHEST YEAR OF SCHOOL COMPLETED	Pearson Correlation	-.049*	.405**	1
	Sig. (2-tailed)	.027	.000	
	N	2036	1200	2039

**. Correlation is significant at the 0.01 level (2-tailed).
*. Correlation is significant at the 0.05 level (2-tailed).

Take a moment to examine your new correlation matrix. It is slightly more complex than the previous example. The fact that each variable correlates perfectly with itself should offer assurance that we are doing something right.

The new matrix also tells us that there is a stronger correlation between EDUC and RECINC: .405. Squaring the r ($.405^2 = .164$) yields the coefficient of determination (r^2) and tells us that 16% of the variance in income can be accounted for by how much education people have.

---

[4] Make sure to define the values 97, 98, and 99 for EDUC as "missing" before continuing with this demonstration.

Writing Box 14.1

Respondents' years of education are more strongly related to income than are their ages. While age is only moderately related to income ($r = .189$), highest year of school completed is more strongly related to income ($r = .405$). Interestingly, a weak inverse relation ($r = -.049$) exists between age and highest year of school, telling us that younger adults are more likely to obtain more years of schooling than are older people. (Note that this is despite the fact that the youngest adults in the sample are temporally limited by their age to a ceiling of years of education that they could have.) This, in part, accounts for the weaker relationship found between age and income compared with that between education and income.

A Note of Caution

We'll be using the Correlations command and related statistics as the book continues. In closing this discussion, we'd like you to recall that Pearson's r is appropriate only for I/R (interval and ratio) variables. The same is true for r^2, the coefficient of determination, which is also for use only with I/R variables. It would not be appropriate in the analysis of nominal variables such as RELIG and MARITAL, for example. But what do you suppose would happen if we asked SPSS Statistics to correlate r for those two variables? Again, exclude cases pairwise before you execute the command.

Correlations

		marital MARITAL STATUS	relig RS RELIGIOUS PREFERENCE
marital MARITAL STATUS	Pearson Correlation	1	.142**
	Sig. (2-tailed)		.000
	N	2043	1855
relig RS RELIGIOUS PREFERENCE	Pearson Correlation	.142**	1
	Sig. (2-tailed)	.000	
	N	1855	1855

**. Correlation is significant at the 0.01 level (2-tailed).

As you can see, SPSS Statistics does not recognize that we've asked it to do a stupid thing. It stupidly complies. It tells us that there is a somewhat significant (see next chapter) relationship between a person's marital status and the religion he or she belongs to, but the correlation calculated here has no real meaning.

SPSS Statistics has been able to do the requested calculation because it stores "Married" as 1 and "Widowed" as 2 and stores "Protestant" as 1, "Catholic" as 2, and so on, but these numbers have no numerical meaning in this instance. Catholics are not "twice" Protestants, and widowed people are not "twice" married people.

Here's a thought experiment we hope will guard against this mistake: (a) Write down the telephone numbers of your five best friends; (b) add them up and calculate the "mean" telephone number; (c) call that number and see if an "average" friend answers. Or go to a Chinese restaurant with a group of friends and have everyone in your party select one dish by its number on the menu. Add all those numbers and calculate the mean. When the waiter comes, get several orders of the "average" dish and see if you have any friends left.

Pearson's *r* is designed for the analysis of relationships among continuous, I/R variables. We have just entrusted you with a powerful weapon for understanding. Use it wisely. Remember: Statistics don't mislead—those who calculate statistics stupidly mislead.

Regression Analysis

The discussion of Pearson's *r* correlation coefficient opens the door for discussion of a related statistical technique that is also appropriate for I/R-level variables: *regression*. When we looked at the scattergram of weight and height in the hypothetical example that introduced the discussion of correlation, you will recall that we tried to "see" a general pattern in the distribution of cases. Regression makes that attempt more concrete.

Example 1: The Logic of Regression

To begin, let's imagine an extremely simple example that relates the number of hours spent studying for an examination and the grades that students got on the exam. Here are the data in a table:

Student	Hours	Grade
Fred	0	0
Mary	2	25
Sam	4	50
Edith	6	75
Earl	8	100

First question: Can you guess which of us prepared this example? Second question: Can you see a pattern in the data presented? The pattern, pretty clearly, is this: The more you study, the better the grade you will get. Let's look at these data in the form of a graph. (This is something you can do by hand, using graph paper, but we used SPSS's scattergram.)

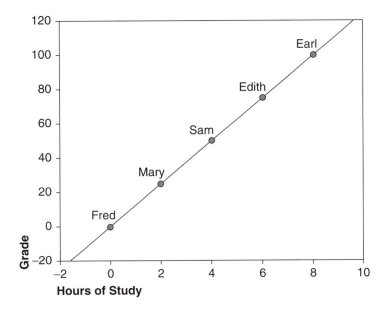

As you can see, the five people in this example fall along a straight line across the graph. This line is called the ***regression line***. As you may recall from plane geometry, it is possible to represent a straight line on a graph in the form of an equation. In this case, the equation would be as follows:

Grade = 12.5 × Hours

To determine a person's grade using this equation, you need only multiply the number of hours he or she studied by 12.5. Multiply Earl's 8 hours of study by 12.5 and you get his grade of 100. Multiply Edith's 6 hours by 12.5 and you get 75. Multiply Fred's 0 hours by 12.5 and, well, you know Fred.

Whereas correlation considers the symmetrical association between two variables, regression adds the notion of causal direction. One of the variables is the dependent variable—grades, in this example—and the other is the independent variable or cause—hours of study. Thus, the equation we just created is designed to predict a person's grade based on how many hours he or she studied. If we were to tell you that someone not included in these data studied 5 hours for the exam, you could predict that that person got a 62.5 on the exam (5 × 12.5).

If all social science analyses produced results as clear as these, you probably wouldn't need SPSS Statistics or a book such as this one. In the example we just discussed, there is a perfect relationship between the two variables, without deviation or error. We call this a ***deterministic relationship***. In practice, however, the facts are usually a bit more complex, and SPSS Statistics is up to the challenge.

Given a set of data with an imperfect relationship between two variables, SPSS Statistics can discover the line that comes closest to passing through all the points on the graph. To understand the meaning of the notion of coming close, we need to recall the squared deviations found in our calculation of Pearson's *r*.

Suppose Sam had gotten 70 on the exam, for example. Here's what he would look like on the graph we just drew.

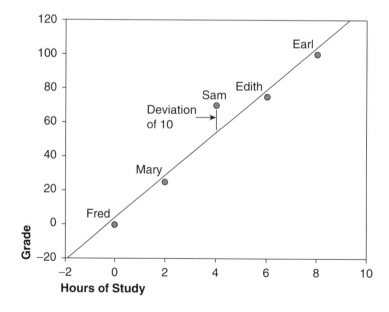

Notice that the improved Sam does not fall on the original regression line: His grade represents a deviation of 10 points. With a real set of data, most people fall to one side or the other of any line we might draw through the data. SPSS Statistics, however, is able to determine the line that would produce the smallest deviations overall—measured by squaring all the individual deviations and adding them. This calculated regression line is sometimes called the *least-squares regression line*.

Requesting such an analysis from SPSS Statistics is fairly simple. To use this technique to a real advantage, you need more instruction than is appropriate for this book. However, we wanted to introduce you to regression because it is a popular technique among many social scientists.

Demonstration 14.4: Regression

To experiment with the regression technique, let's make use of a new variable: SEI. This variable rates the socioeconomic prestige of respondents' occupations on a scale from a low of 0 to a high of 100 based on other studies that have asked a sample from the general population to rate different occupations.

Here's how we would ask SPSS Statistics to find the equation that best represents the influence of EDUC on SEI. You should realize that there are a number of ways to request this information, but we'd suggest you do it as follows. (Just do it this way and nobody gets hurt, okay?)

Under **Analyze**, select **Regression → Linear**. Here's what you get.

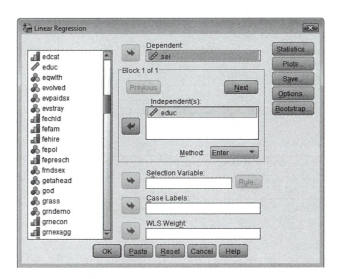

Select **SEI** as the "Dependent:" variable and **EDUC** as the "Independent(s):" variable.

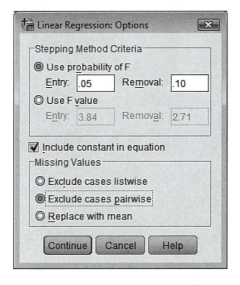

Click **Options** and select pairwise exclusion, and then hit **Continue**. Click **OK**, and SPSS Statistics is off and running. Here's the output you should get in response to this instruction.

Model Summary

Model	R	R Square	Adjusted R Square	Std. Error of the Estimate
1	.596[a]	.355	.354	15.3952

a. Predictors: (Constant), educ HIGHEST YEAR OF SCHOOL COMPLETED

ANOVA[a]

Model		Sum of Squares	df	Mean Square	F	Sig.
1	Regression	243683.191	1	243683.191	1028.152	.000[b]
	Residual	443210.199	1870	237.011		
	Total	686893.390	1871			

a. Dependent Variable: sei RESPONDENT SOCIOECONOMIC INDEX

b. Predictors: (Constant), educ HIGHEST YEAR OF SCHOOL COMPLETED

Coefficients[a]

Model		Unstandardized Coefficients		Standardized Coefficients	t	Sig.
		B	Std. Error	Beta		
1	(Constant)	.212	1.562		.136	.892
	educ HIGHEST YEAR OF SCHOOL COMPLETED	3.624	.113	.596	32.065	.000

a. Dependent Variable: sei RESPONDENT SOCIOECONOMIC INDEX

The key information we are looking for is contained in the final table, titled "Coefficients": the *intercept*, labeled "Constant" (.212), and the *slope*, labeled "educ" (3.624). These are the data we need to complete our regression equation. Use them to fill the equation as follows:

$$Y = a + bX$$

$$SEI = .212 + (EDUC \times 3.624)$$

This means that we would predict the occupation prestige ranking of a high school graduate (12 years of schooling) as follows:

$$SEI = .212 + (12 \times 3.624) = 43.7$$

On the other hand, we would predict the occupational prestige of a college graduate (16 years of schooling) as

$$SEI = .212 + (16 \times 3.624) = 58.196$$

Notice that along with the regression coefficient and constant we are also provided other information, including both Pearson's r (.596) and r^2 (.355), the coefficient of determination.

SPSS Statistics Command 14.2: Regression

Click **Analyze** → **Regression** → **Linear**

→ Highlight dependent variable and click right-pointing arrow to transfer it to "Dependent:" field

→ Highlight independent variable and click right-pointing arrow to transfer it to "Independent(s):" field

→ Click **OK**

Demonstration 14.5: Presenting Data Graphically— Producing a Scatterplot With a Regression Line

Another way to explore the relationship between SEI and EDUC is to instruct SPSS to produce a scatterplot with a regression line. *Scatterplot* is a term SPSS Statistics uses to identify what we referred to earlier in our discussion of Pearson's *r* as a scattergram. A scatterplot/ scattergram is simply a graph with a *horizontal (x) axis* and a *vertical (y) axis*. As you saw earlier, each dot on the graph represents the point of intersection for each individual case. It is the "scatter" of dots taken as a whole that, along with the regression line, indicates the strength and *direction of association* between the variables.

Instructing SPSS Statistics to produce a scatterplot with a regression line allows us to see the distribution and array of cases while saving us the time-consuming task of creating the graph on our own (which, as you can imagine with about 1,500 cases, would be quite a daunting task).

To produce a scatterplot, all you have to do is open the Scatterplot window by clicking **Graphs, Legacy Dialogs,** and then choosing **Scatter/Dot** in the drop-down menu.

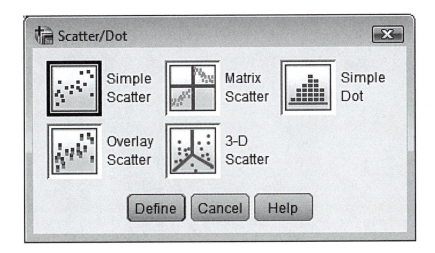

When you have opened the Scatter/Dot box, as shown above, select **Simple Scatter** if it is not already highlighted, and then click **Define**. You should now be looking at the Simple Scatterplot dialog box, as shown below.

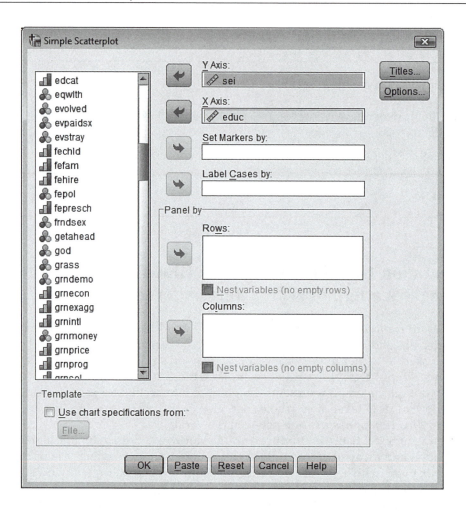

Following the example we used above, place **EDUC** (the independent variable) along the x-axis. Then place **SEI** (the dependent variable) along the y-axis. When you are finished with this process, you can hit **OK,** and the following graph should appear on your screen.

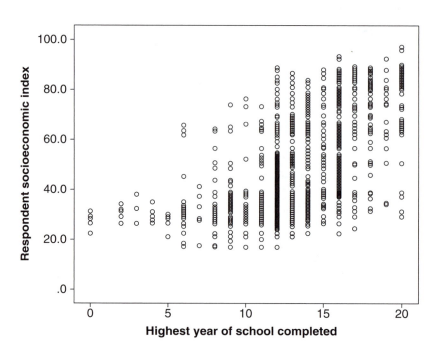

As requested, our independent variable EDUC is arrayed along the *x* (horizontal) axis and our dependent variable SEI is arrayed along the *y* (vertical) axis. You can also see the splatter of dots on the graph, each representing a particular case.

Now double-click on the graph, and the SPSS Chart Editor will appear.

We are going to use the Chart Editor to add our regression line, which, as we saw earlier, is simply a single straight line that will run through our scatterplot and come as close as possible to touching all the "dots," or data points.

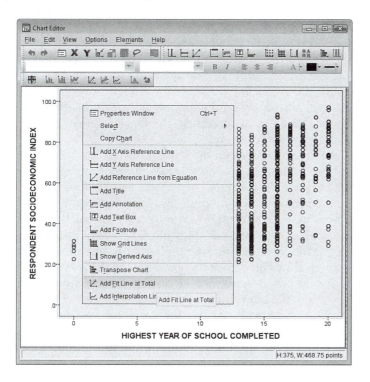

With the Chart Editor active (double-click the graph to activate the Chart Editor), right-click on the chart to show the Properties window.

Click on **Add Fit Line at Total** to insert the regression line. Then close the Chart Editor. You'll find the result of that action below.

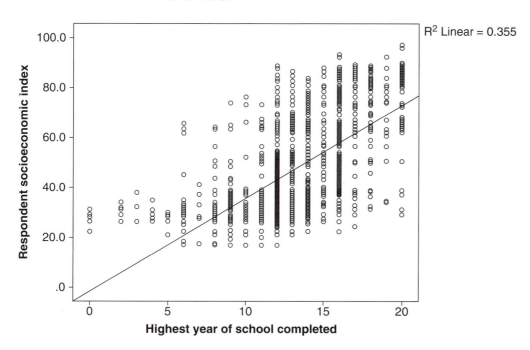

An Indication of Direction and Strength of Association

The scatterplot and regression line serve as a visual indication of the direction and strength of association between EDUC and SEI.

The direction of association can be determined based on the way the regression line sits on the scatterplot. Generally, when the line moves from the bottom left-hand side of the graph to the upper right-hand side of the graph (uphill), this indicates a positive association between the variables (they both either increase or decrease in the same direction), whereas a regression line that runs from the upper left to the lower right side of the scatterplot (downhill) indicates a negative association between the items (as one increases, the other decreases).

The strength of association can be estimated based on the extent to which the dots (cases) are scattered around the regression line. The closer the dots are to the regression line, the stronger the association between the variables (in the case of a "perfect" or deterministic relationship, the dots will sit exactly on the regression line, indicating the strongest possible association between the variables). Conversely, the farther spread out or scattered the dots are, the weaker the relationship between the items.

We want to emphasize that although this type of visual representation allows us to estimate the strength of association between variables, there is no agreement regarding what constitutes a "strong" versus a "moderate" or "weak" association.

Consequently, your best option may be to use the scatterplot with a regression line in conjunction with a measure such as Pearson's r, and/or r^2, which allows you to obtain a value summarizing the association between variables more precisely than just the scatterplot and regression line do. Keep in mind that because both options are appropriate for continuously distributed I/R variables, they can be used to examine the same types of items.

Now try to see if you can describe the strength and direction of association between EDUC and SEI as shown in the scatterplot with a regression line above.

What is the direction of association? Are the variables positively or negatively related (i.e., does the regression line run uphill or downhill)? What can you estimate in terms of the strength of association between EDUC and SEI? Are the dots or cases clustered around the regression line, or are they spread out?

Writing Box 14.2

The positive relationship between education and socioeconomic status can be seen by inspection of the scattergram. While the many cases are widely dispersed, there is a discernible pattern concentrating on the lower-left and upper-right quadrants. Adding the regression line confirms that assessment.

SPSS Statistics Command 14.3:
Producing a Scatterplot With a Regression Line

Click **Graphs** → **Legacy Dialogs** → **Scatter/Dot**

→ **Simple Scatter** → **Define**

→ Highlight independent variable → Click right-pointing arrow to transfer to x-axis

→ Highlight dependent variable → click right-pointing arrow to transfer to y-axis

→ Click **OK**

To Add Regression Line:

→ Double-click on scatterplot to open the SPSS Statistics Chart Editor

   → With the Chart Editor active, right-click on the chart to display the Properties window

      → Click **Add Fit Line at Total** to insert the regression line. Then close the Chart Editor.

That's enough fun for now. We'll return to regression later, when we discuss multivariate analysis.

Measures of Association for Interval and Ratio Variables

The measures of association we have focused on in this chapter are appropriate for variables measured at the interval and/or ratio level of measurement (I/R variables). Table 14.2 lists the measures we have reviewed along with their appropriate level of measurement.

Table 14.2 Measures of Association Reviewed in This Chapter

Level of Measurement	Measure of Association (Range of Values) of Each Variable
Interval/ratio	Pearson's r [−1 to +1]
	r^2 [0 to 1]
	Linear regression

Keep in mind, however, that there are situations where one might need to analyze variables that are not I/R along with those that are I/R. While there are other statistical techniques beyond the scope of this book (e.g., logistic regression for dichotomous dependent variables and I/R independent variables), there are some ways to incorporate different variables into linear regression analysis. We have to first make the variables "appropriate," as explained below.

Analyzing the Association Between Variables at Different Levels of Measurement

In some cases, you may discover that you want to analyze the association between variables at different levels of measurement: for instance, nominal by ordinal, nominal by I/R, and so on. While there are specialized bivariate measures of association for use when two variables have different levels of measurement, the safest course of action for the novice is to use a measure of association that meets the assumptions for the variable with the lowest level of measurement.

This usually means that one variable will have to be recoded to reduce its level of measurement. For instance, when we examined the relationship between church attendance (ordinal) and age (I/R) in Chapter 13, we recoded people's ages into ordinal categories.

While it would not be appropriate to use Pearson's r to correlate our CHATT and AGE, it is perfectly legitimate to use gamma to correlate CHATT with AGECAT.

With linear regression, there is a technique that allows us to incorporate into the equation an independent variable that is not originally an I/R variable. This technique is known as "creating a set of *dummy* variables." There is a brief discussion of this technique in Chapter 7 of *Using IBM SPSS Statistics for Research Methods and Social Science Statistics* (Wagner, 2012). Single dummy variables are addressed later on in *Adventures in Social Research* (the book you are reading right now!), in Chapter 17.

While it is beyond the scope of this chapter to review all the statistical measures appropriate for each particular situation, those of you who are interested in pursuing the relationship between mixed variables may want to consult a basic statistics or research methods text for a discussion of appropriate measures.[5] Keep in mind that a variety of basic statistics texts are now easily accessible on the World Wide Web.

In addition, several websites help students identify appropriate statistical tests. The following is an example of one of the numerous sites available: HyperStat Online (http://davidmlane.com/hyperstat/index.html).

Conclusion

In this chapter, we've seen a number of statistical techniques that can be used to summarize the degree of relationship or association between two variables. We've seen that the appropriate technique depends on the level of measurement represented by the variables involved. Lambda is appropriate for nominal, gamma for ordinal, and Pearson's r product-moment correlation and regression for I/R variables.

We realize that you may have had trouble knowing how to respond to the results of these calculations. How do you decide, for example, if a gamma of .25 is high or low? Should you get excited about it or yawn and continue looking? The following chapter on statistical significance offers one basis for making such decisions.

Main Points

- In this chapter, we looked at new ways to examine the relationships among interval/ratio (I/R) variables.
- Measures of association summarize the strength (and, in some cases, the direction) of association between two variables in contrast to the way percentage tables lay out the details.
- This chapter focuses on only some of the many measures available: Pearson's r, r^2 (the coefficient of determination), and regression.
- The coefficient of determination (r^2) is based on the logic of proportionate reduction of error (PRE).
- Pearson's r is appropriate for two I/R items; its values range from –1 to +1. The coefficient of determination, r^2, is also appropriate for two I/R items; it is a PRE statistic whose values range from 0 to 1.
- The closer the value is to positive or negative 1, the stronger the relationship between the items (with +1 or –1 indicating a perfect association). The closer the value is to 0,

---

[5]Among the many basic statistics texts you may want to consult are Freeman's *Elementary Applied Statistics* (1968), Siegel and Castellan's *Nonparametric Statistics for the Behavioral Sciences* (1998), and Frankfort-Nachmias et al.'s *Social Statistics for a Diverse Society* (2006). Alreck and Settle's *The Survey Research Handbook* (1985, pp. 287–362) contains a table titled "Statistical Measures of Association" (Table 10-5, p. 303), which, depending on the type of independent and dependent variables you are working with (i.e., categorical or continuous), specifies appropriate measures of association. The table is accompanied by a discussion of statistical analysis and interpretation.

the weaker the relationship between the items (with 0 indicating no association between the items).
- There are no set rules for interpreting strength of association.
- Like Pearson's r, linear regression is also appropriate for I/R variables.
- Another way to explore the strength and direction of association between two I/R variables is to produce a scatterplot with a linear regression line.
- A variety of statistics are appropriate for examining the relationship between mixed types of variables.

Key Terms

Measures of association

Variance

Proportionate reduction of error (PRE)

Standard deviation

z scores

Correlation matrix

Positive association

Statistical significance

Negative association

Regression

Coefficient of determination (r^2)

Strength of association

Least-squares regression line

Direction of association

Intercept

Slope

Regression line

Pearson's r

Horizontal (x) axis

Scattergram

Vertical (y) axis

Deterministic relationship

SPSS Statistics Commands Introduced in This Chapter

14.1   Producing a Correlation Matrix With Pearson's r

14.2   Regression

14.3   Producing a Scatterplot With a Regression Line

Review Questions

1. If two variables are strongly associated, does that necessarily mean they are causally related?

2. What is PRE?

3. List two measures of association for interval/ratio (I/R) variables.

4. Measures of association give us an indication of the _____ of association and (if the variables are ordinal or higher) the _____ of association between two variables.

5. Pearson's r and the coefficient of determination (r^2) are appropriate for variables at which level of measurement?

6. List the range of values for each of the following measures:

   - Pearson's r
   - Coefficient of determination (r^2)

7. The closer to _____ or _____ (value), the stronger the relationship between the variables.

8. The closer to _____ (value), the weaker the relationship between the variables.

9. A value of _____ or _____ indicates a perfect (or the strongest) relationship between the variables.

10. What does a positive association between two variables indicate? What does a negative association between two variables indicate?

11. Linear regression is appropriate for two variables at which level of measurement?

12. What type of graph did we review in this chapter that allows us to examine the strength and direction of association between two I/R variables?

NAME _____

CLASS _____

INSTRUCTOR _____

DATE _____

To complete these exercises, load your AdventuresPLUS.SAV data file.

Produce a correlation matrix, exclude cases pairwise, and request Pearson's *r* for the variables AGE (independent) and SEI (dependent). Then answer Questions 1 through 4. Remember to define the values 98 and 99 as "missing" for the variable AGE.

1. What is the level of measurement for the variable AGE?

2. What is the level of measurement for the variable SEI?

3. Pearson's *r* for AGE and SEI is _____.

4. Would you characterize the relationship between these two variables as weak, moderate, or strong? Explain.

Produce a scatterplot with a regression line for the variables EDUC (independent) and TVHOURS (dependent). Then answer Questions 5 through 10. Before proceeding with your analysis, be sure to define the values 97, 98, and 99 as "missing" for the variable EDUC. In addition, define the values –1, 98, and 99 as "missing" for the variable TVHOURS.

5. What is the level of measurement for the variable EDUC?

6. What is the level of measurement for the variable TVHOURS?

7. What is the regression equation for EDUC and TVHOURS?

   $Y =$ _____ + _____ X

8. What is the direction of the relationship between the variables? _____ As years of education increases, does the number of hours of television watched increase or decrease?

9. The relationship of the "dots" or cases to the regression line suggests what about the strength of the relationship between these two variables?

10. Produce a correlation matrix with Pearson's r, exclude cases pairwise, and then further describe the strength of the relationship between EDUC and TVHOURS by noting what the value of Pearson's r is and whether the relationship is weak, moderate, or strong.

11. Now choose two variables from your data set that you think may be associated. After defining appropriate values for each variable as "missing," analyze the relationship between the variables by instructing SPSS to calculate the appropriate measure of association. If you want to recode either of your variables, feel free to do so. Remember, if you choose two variables at different levels of measurement, you can follow our suggestion toward the end of the chapter that you use a measurement that meets the assumptions for the variable with the lowest level of measurement. If you choose two interval/ratio variables, also produce a scatterplot with a regression line.

Once you have completed your analysis, print your output and attach it to this sheet. Then, take a few moments to describe your findings in the space provided below. In particular, you may want to consider noting the following:

- What variables you chose and why
- The levels of measurement of each
- Whether you had to recode either item or designate any values as "missing"
- What measure of association you instructed SPSS to calculate
- The value of that measure
- What it indicates about the relationship between these variables (strength and direction)

If you are having any trouble producing an analysis of your findings in prose, you may want to reread the writing boxes throughout Chapters 13 and 14.

Chapter 15   Tests of Significance

Thus far, in Chapters 10 through 14, we've looked at the relationships between pairs of variables. In all that, you may have been frustrated over the ambiguity of what constitutes a "strong" or a "weak" relationship. As we noted in the last chapter, there is no absolute answer to this question. The strength and significance of a relationship between two variables depend on many things.

If you are trying to account for differences among people on some variable, such as prejudice, the explanatory power of one variable, such as education, needs to be contrasted with the explanatory power of other variables. Thus, you might be interested in knowing whether education, political affiliation, or region of upbringing has the greatest impact on a person's prejudice.

Sometimes the importance of a relationship is based on practical policy implications. Thus, the impact of some variable in explaining (and potentially reducing) auto theft rates, for example, might be converted to a matter of dollars. Other relationships might be expressed in terms of lives saved, students graduating from college, and so forth.

Statistical Significance

In this chapter, we're going to introduce you to another standard for judging the relationships among variables—one commonly used by social scientists. Whereas in Chapters 13 and 14 we discussed measures of association that allow us to examine the strength (and in the case of ordinal and interval/ratio [I/R] variables, the direction) of association, in this chapter we explore tests that will allow you to estimate *statistical significance*. Tests of statistical significance allow us to estimate the likelihood that a relationship between variables in a sample actually exists in the population as opposed to being an illusion due to chance or sampling error.

Whenever analyses are based on random samples selected from a population rather than on data collected from everyone in that population, there is always the possibility that what we learn from the samples may not truly reflect the whole population. Thus, we might discover in a sample that women are more religious than men, but that could be simply an artifact of our sample: We happened to pick too many religious women and/or too few religious men. Tests of significance allow us to estimate the likelihood that our finding, a relationship between gender and religiosity in this case, could have happened by chance. If the chances of our finding are very unlikely—say, only about 5 in 100—then we have the confidence needed to generalize our finding from the sample to the population from which it was drawn.

Significance Tests: Part of the Larger Body of Inferential Statistics

Significance tests are part of a larger body of statistics known as inferential statistics. *Inferential statistics* can probably best be understood in contrast to *descriptive statistics*, with which you are already familiar. Together, descriptive and inferential statistics constitute two of the main types of statistics that social researchers use. Up to this point (Chapters 5 through 14), our focus has been largely on descriptive statistics that allow us to describe or summarize the main features of our data or the relationships between variables in our data set. In contrast, inferential statistics allows us to go a step further by making it possible to draw conclusions or make inferences that extend beyond the items in our particular data set to the larger population. In short, we can use inferential statistics to help us learn what our sample tells us about the population from which it was drawn. In the case of tests of significance, for instance, we can estimate whether an observed association between variables in our sample is generalizable to the larger population.

Statistical Significance Versus Measures of Association

Social scientists often test the statistical significance of relationships discovered among variables. Although these tests do not constitute a direct measure of the strength of a relationship, they tell us the likelihood that the observed relationship could have resulted from the vagaries of probability sampling, which we call *sampling error*. These tests relate to the strength of relationships in that the stronger an observed relationship is, the less likely it is that it could be the result of sampling error. Correspondingly, it is more likely that the observed relationship represents something that exists in the population as a whole.

You will find that social scientists use measures of association and tests of significance in conjunction with each other because, together, they allow us to address three important questions about the relationships between variables:

1. *How strong is the relationship?* (Measures of association/descriptive statistics such as lambda [λ], gamma [γ], etc.; Chapter 13)

2. In the case of ordinal and I/R variables, *what is the direction of association?* (Measures of association/descriptive statistics such as gamma [γ], Pearson's *r*, the coefficient of determination [r^2], etc.; Chapters 13 and 14)

3. *Is the relationship statistically significant?* (Tests of significance/inferential statistics such as chi-square [χ^2], *t* test, etc.; Chapter 15)

In this chapter, we turn our attention to Question 3 and some of the measures that can be used to estimate statistical significance, primarily chi-square (χ^2), *t* tests, and ANOVA (analysis of variance).

Chi-Square (χ^2)

To learn the logic of statistical significance, let's begin with *chi-square (χ^2)*, a measure based on the kinds of crosstabulations we've examined in previous chapters. Chi-square is a test of significance that is most appropriate for nominal items, although it can be used with ordinal variables or a combination of nominal and ordinal variables. Chi-square, one of the most widely used tests of significance, estimates the probability that the association between variables is a result of random chance or sampling error by comparing the actual or observed distribution of responses with the distribution of responses we would expect if there were absolutely no association between two variables.

To help make this clearer, let's take some time now to look at the logic of statistical significance in general and chi-square in particular.

The Logic of Statistical Significance: Chi-Square (χ^2)

For a concrete example, let's return to one of the tables that examines the relationship between religion and abortion attitudes.

Let's reexamine the relationship between religious affiliation and unconditional support for abortion. Do a crosstabulation of ABANY (row variable) and RELIG (column variable), and be sure to request column percentages.[1]

ANALYZE → DESCRIPTIVE STATISTICS → CROSSTABS

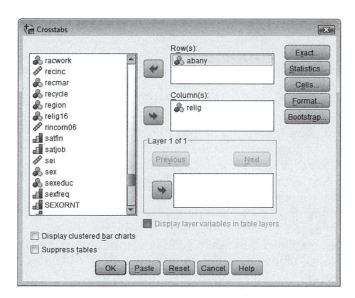

Click the **Cells** button.

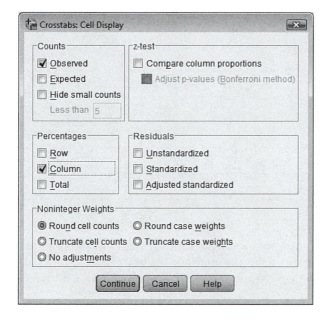

---

[1] Define the following values as "missing" for each variable: ABANY—0, 8, and 9; RELIG—0 and 5 to 99.

Here is the output that will be produced:

abany ABORTION IF WOMAN WANTS FOR ANY REASON * relig RS RELIGIOUS PREFERENCE Crosstabulation

			relig RS RELIGIOUS PREFERENCE				
			1 PROTESTANT	2 CATHOLIC	3 JEWISH	4 NONE	Total
abany ABORTION IF WOMAN WANTS FOR ANY REASON	1 YES	Count	216	103	21	137	477
		% within relig RS RELIGIOUS PREFERENCE	36.7%	36.1%	75.0%	62.8%	42.6%
	2 NO	Count	372	182	7	81	642
		% within relig RS RELIGIOUS PREFERENCE	63.3%	63.9%	25.0%	37.2%	57.4%
Total		Count	588	285	28	218	1119
		% within relig RS RELIGIOUS PREFERENCE	100.0%	100.0%	100.0%	100.0%	100.0%

The question this table (which contains the actual or observed frequencies) is designed to answer is whether a person's religious affiliation affects his or her attitude toward abortion. You'll recall that we concluded that it does: Catholics and Protestants are the most opposed to abortion, and Jews and those with no religion are the most supportive. The question we now confront is whether the observed differences point to some genuine pattern in the U.S. population at large or whether they result from a quirk of sampling.

To assess the *observed relationship* as shown in the table above, we are going to begin by asking what we should have expected to find if there were no relationship between religious affiliation and abortion attitudes. An important part of the answer lies in the rightmost column in the preceding table. It indicates that 43% of the whole sample supported a woman's unconditional right to an abortion (Yes) and 57% did not (No).

If there were no relationship between religious affiliation and abortion attitudes, we should expect to find 43% of the Protestants approving (Yes), 43% of the Catholics approving, 43% of the Jews approving, and so forth. But we recall that the earlier results did not match this perfect model of no relationship, so the question is whether the disparity between the model and our observations would fall within the normal degree of sampling error.

To measure the extent of the disparity between the model and what's been observed, we need to calculate the number of cases we'd expect in each cell of the table if there were no relationship. The table below shows how to calculate the expected cell frequencies.

ABANY	Protestant	Catholic	Jewish	None
Yes	588	285	28	218
	× 0.43	× 0.43	× 0.43	× 0.43
No	588	285	28	218
	× 0.57	× 0.57	× 0.57	× 0.57

Make sure you know how we constructed this table before moving ahead.

Consequently, if there were no relationship between religious affiliation and abortion attitudes, we would expect 43% of the 588 Protestants ($588 \times 0.43 = 253$) to approve and 57% of the 588 Protestants ($588 \times 0.57 = 335$) to disapprove.

If you continue this series of calculations, you should arrive at the following set of *expected cell frequencies.*

ABANY	Protestant	Catholic	Jewish	None
Yes	253	123	12	94
No	335	162	16	124

The next step in calculating chi-square is to calculate the difference between expected and observed values in each cell of the table. For example, if religion had no effect on abortion, we would have expected to find 253 Protestants approving; in fact, we observed only 216. Thus, the discrepancy in that cell is –37. The discrepancy for Catholics approving is –20 (observed – expected = 103 –123). The table below shows the discrepancies for each cell.

ABANY	Protestant	Catholic	Jewish	None
Yes	–37	–20	9	43
No	37	20	–9	–43

Finally, for each cell we square the discrepancy and divide it by the expected cell frequency. For the Protestants approving of abortion, then, the squared discrepancy is 1,369 (–37 × –37). Dividing that by the expected frequency of 253 yields 5.41. When we repeat this for each cell, we get the following results.

ABANY	Protestant	Catholic	Jewish	None
Approve	5.41	3.25	6.75	19.67
Disapprove	4.09	2.47	5.06	14.91

Chi-square is the sum of all these latest cell figures: 61.61. We have calculated a summary measure of the discrepancy between what we would have expected to observe if religion did not affect abortion and what we actually observed. Now the only remaining question is whether that resulting number should be regarded as large or small. Statisticians often speak of the **goodness of fit** in this context: How well do the observed data fit a model of two variables being unrelated to each other? The answer to this latest question takes the form of a probability: the probability that a chi-square this large could occur as a result of sampling error. A probability of .05 in this context would mean that it should happen 5 times in 100 samples. A probability of .001 would mean that it should happen only 1 time in 1,000 samples.

To evaluate our chi-square of 61.61, we need to look it up in a table of chi-square values, which you'll find in the back of most statistics textbooks or by searching the Internet. Such tables have several columns marked by different probabilities (e.g., .30, .20, .10, .05, .01, .001, etc.). The tables also have several rows representing different **degrees of freedom (df).**

If you think about it, you'll probably see that the larger and more complex a crosstabulation is, the greater the likelihood that there will be discrepancies from the perfect model of expected frequencies. We take account of this with one final calculation.

Degrees of freedom are calculated from the data table as (rows – 1) × (columns – 1). In our table, there are four columns and two rows, giving us (3 × 1) degrees of freedom. Thus, we would look across the third row in the table of chi-square values, which would look, in part, like this:

df	.05	.01	.001
3	7.815	11.341	16.268

These numbers tell us that a chi-square as high as 7.815 from a table such as ours would occur only 5 times in 100 samples if there were no relationship between religious affiliation and abortion attitudes among the whole U.S. population. A chi-square as high as 11.341 would happen only once in 100 samples, and a chi-square as high as 16.268 would happen only once in 1,000 samples.

Thus, we conclude that our chi-square of 61.61 could result from sampling error less than once in 1,000 samples. This is often abbreviated as $p < .001$: The probability is less than 1 in 1,000.

They have no magical meaning, but the .05 and .001 levels of significance are often used by social scientists as a convention for concluding that an observed relationship reflects a similar relationship in the population rather than arising from sampling error. Most social scientists agree that relationships with significance values of .05 or less are so unlikely to have occurred by chance that they can be called significant. The lower the probability, the more statistically significant the relationship is. Accordingly, if a relationship is significant at the .001 level, we can be more confident of our conclusion than if it were significant at the higher .05 level.

Going back to our example, then, if the value of chi-square is greater than the value printed in the reference table for the appropriate degree of freedom and at the probability level of .05 or less, then the relationship between the variables can be considered statistically significant.

As we noted, in our example the value of chi-square is 61.61, which is greater than the value printed in the reference table for the appropriate *df* at the probability of .05, .01, and .001. This, then, tells us that the relationship between our variables (ABANY and RELIG) can be considered statistically significant.

There you have it: far more than you ever thought you'd want to know about chi-square. By sticking it out and coming to grasp the logical meaning of this statistical calculation, you've earned a reward.

Demonstration 15.1: Instructing SPSS Statistics to Calculate Chi-Square

Rather than going through all the preceding calculations, we could have simply modified our Crosstabs request slightly (and after seeing how easy it is to instruct SPSS Statistics to run chi-square, you will probably wish we had just done this much earlier). In the Crosstabs window, click **Statistics** and select **Chi-Square** in the upper-left corner of the Statistics window.

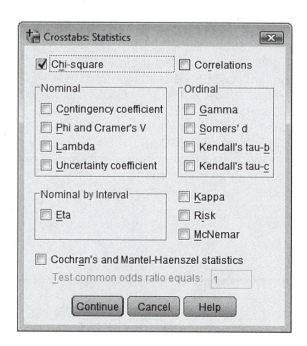

Then click **Continue** and **OK** to run the Crosstabs request.

Chi-Square Tests

	Value	df	Asymp. Sig. (2-sided)
Pearson Chi-Square	61.683[a]	3	.000
Likelihood Ratio	61.530	3	.000
Linear-by-Linear Association	48.179	1	.000
N of Valid Cases	1119		

a. 0 cells (0.0%) have expected count less than 5. The minimum expected count is 11.94.

Reading Your Output

We are interested primarily in the first row of figures in this report. Notice that the 61.683 value of chi-square is slightly different from our hand calculation. This is because of our rounding in our cell calculations, and it shouldn't worry you.

Notice that we're told that there are three degrees of freedom. Finally, SPSS Statistics has calculated the probability of getting a chi-square this high with three degrees of freedom and has run out of space after three zeros to the right of the decimal point. Thus, the probability is far less than .001, as we determined by checking a table of chi-square values.

The reference to a minimum expected frequency of 11.94 is worth noting. Because the calculation of chi-square involves divisions by expected cell frequencies, it can be greatly inflated if any of them are very small. By convention, adjustments to chi-square should be made if more than 20% of the expected cell frequencies are below 5. You should check a statistics text if you want to know more about this.

SPSS Statistics Command 15.1: Producing Crosstabs With Chi-Square

Click **Analyze** → **Descriptive Statistics** → **Crosstabs**
→ Highlight dependent variable → Click arrow to transfer to "Row(s):" field
→ Highlight independent variable → Click arrow to transfer to "Column(s):" field
→ Click **Cells**

→ Click **Column** in Percentages box → **Continue**

→ **Statistics**
→ **Chi-Square** → **Continue** → **OK**

Practice Running Chi-Square

While it is fresh in your mind, why don't you have SPSS Statistics calculate some more chi-squares for you? You may recall that sex had little impact on abortion attitudes as indicated by ABANY. Why don't you see what the chi-square is? Once you have done so, compare your interpretation of your findings to that in Writing Box 15.1.

Writing Box 15.1

With a chi-square of just 1.438, we find once again that gender has little impact on attitudes toward abortion. With a probability of .23, the observed relationship is not statistically significant.

To experiment more with chi-square, you might rerun some of the other tables relating various demographic variables to abortion attitudes. Notice how chi-square offers a basis for comparing the relative importance of different variables in determining attitudes on this controversial topic.

Significance and Association

It bears repeating here that tests of significance are different from measures of association, although they are related to each other. The stronger an association between two variables, the more likely it is that the association will be judged statistically significant—that is, not a simple product of sampling error. Other factors also affect statistical significance, however. As we've already mentioned, the number of degrees of freedom in a table is relevant. So is the size of the sample: The larger the sample, the more likely it is that an association will be judged significant.

Researchers often distinguish between statistical significance (examined in this section) and *substantive significance.* The latter refers to the importance of an association, and it can't be determined by empirical analysis alone. As we suggested at the outset of this chapter, substantive significance depends on practical and theoretical factors. All this notwithstanding, social researchers often find statistical significance a useful device in gauging associations.

That said, you may still be wondering how you should interpret findings that show, for instance, that there is a strong but not statistically significant relationship between two variables. Or, conversely, what you should say if you find that there is a fairly weak but nonetheless statistically significant relationship between two items. Table 15.1 is intended to help as you approach the sometimes daunting task of interpreting tests of significance and association. While there are no "rules" of interpretation, this table can be looked at as a general guide of sorts to help you as you begin to bring together the concepts of association and significance.

Table 15.1 A Guide to Interpreting Tests of Association and Significance

Strength of Association	Statistical Significance	Interpretation	Comment
Strong	Significant	Knowledge of the independent variable improves prediction of dependent variable, and relationship can be generalized from the sample to the population.	Look what I've discovered.
Strong	Nonsignificant	Although knowledge of the independent variable improves prediction of the dependent variable in the sample, the relationship cannot be generalized to the population.	Close, but no banana.
Weak	Significant	Knowledge of the independent variable is of little help in predicting the dependent variable but may be generalized to the population.	That's no big thing.
Weak	Nonsignificant	Knowing about the independent variable provides little help in the prediction of the dependent variable in the sample and cannot be generalized to the population.	Back to the drawing board.

However, we do want to add one note of caution. This table is meant to serve only as a general guide to interpreting tests of association and significance. Consequently, you may (and probably should) find that you do not agree with the interpretation offered in every case. For instance, if your findings show that the association between two variables is strong and statistically significant, you can look across the first row of the table to get some guidance in interpreting these results.

Whereas chi-square operates on the logic of the contingency table, to which you've grown accustomed by using the Crosstabs procedure, we're going to turn next to a test of significance based on means.

t Tests

Who do you suppose lives longer, men or women? Whichever group lives longer should, as a result, have a higher average age at any given time. Regardless of whether you know the answer to this question for the U.S. population as a whole, let's see if our General Social Survey (GSS) data can shed some light on the issue.

We could find the average ages of men and women in our GSS sample with the simple command path:

Analyze → Compare Means → Means

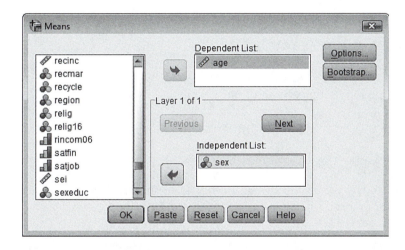

Because age is the characteristic on which we want to compare men and women, **AGE** is the dependent variable and **SEX** the independent variable.[2] Transfer those variables to the appropriate fields in the window. Then click **OK**.

Report

age AGE OF RESPONDENT

sex RESPONDENTS SEX	Mean	N	Std. Deviation
1 MALE	47.78	890	17.044
2 FEMALE	48.11	1151	18.159
Total	47.97	2041	17.678

[2]Define the values 98 and 99 as "missing" for the variable AGE.

As you can see, our sample reflects the general population in that women have a mean age of 48.11, compared with the mean age of 47.78 for men. The task facing us now parallels the one pursued in the discussion of chi-square. Does the observed difference, quite small as it is, reflect a pattern that exists in the whole population, or is it simply a result of a sampling procedure? Would another sample indicate that men are older than women or that there is no difference at all?

In the previous section, we ran Crosstabs with chi-square to estimate the statistical significance of the observed association between gender and religiosity, two nominal variables. We will not rely on crosstabs and chi-square here because our dependent variable, AGE, is an I/R item that contains many categories. Consequently, the Crosstabs procedure would produce a table too large and unwieldy to analyze easily. Instead, we will rely on one of the most commonly used inferential statistics—*t* tests. While there are three types of *t* tests available on SPSS Statistics, we are going to utilize the ***independent-samples* t *test* (*two-sample* t *test*).**

Given that we've moved very deliberately through the logic and calculations of chi-square, we are going to avoid such details in the present discussion. The **t** *test*, which is best suited for dependent variables at the I/R level of measurement, examines the distribution of values on one variable (AGE) among different groups (men and women—two categories of one variable, SEX) and calculates the probability that the observed difference in means results from sampling error alone. As with chi-square, it is customary to use the value of .05 or less to identify a statistically significant association.

Demonstration 15.2: Instructing SPSS Statistics to Run Independent-Samples *t* Test

To request a *t* test from SPSS Statistics in order to examine the relationship between AGE and SEX, you enter the following command path:

Analyze → Compare Means → Independent-Samples T Test

You will notice that the drop-down menu lists the three types of *t* tests available on SPSS Statistics: "One-Sample T Test," "Independent-Samples T Test," and "Paired-Samples T Test." As noted above, we are interested in the "Independent-Samples T Test" option. After following the command path above, you should see the following screen.

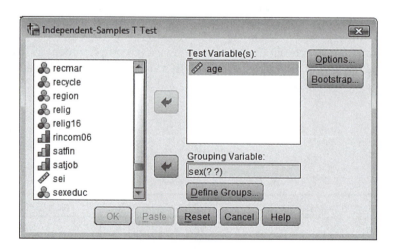

In this window, we want to enter **AGE** as the test variable and **SEX** as the grouping variable. This means that SPSS Statistics will group respondents by sex and then examine and compare the mean ages of the two gender groups.

Notice that when you enter the grouping variable, SPSS Statistics puts "SEX(??)" in that field. Although the comparison groups are obvious in the case of SEX, it might not be so obvious with other variables, so SPSS Statistics wants some guidance. Click **Define Groups**.

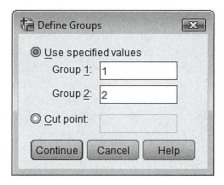

Type **1** (male) into the "Group 1:" field and **2** (female) into the "Group 2:" field. Click **Continue**. Notice that the two question marks next to the grouping variable, SEX, are gone. In their place, you should see the new values to be used for comparison: 1 and 2.

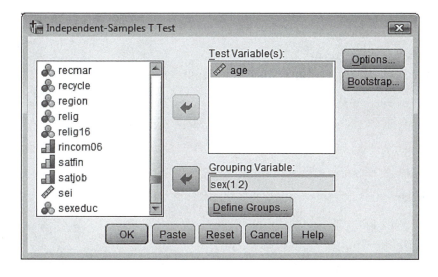

Now, click **OK**.

Group Statistics

	sex RESPONDENTS SEX	N	Mean	Std. Deviation	Std. Error Mean
age AGE OF RESPONDENT	1 MALE	890	47.78	17.044	.571
	2 FEMALE	1151	48.11	18.159	.535

Independent Samples Test

		Levene's Test for Equality of Variances		t-test for Equality of Means						95% Confidence Interval of the Difference	
		F	Sig.	t	df	Sig. (2-tailed)	Mean Difference	Std. Error Difference	Lower	Upper
age AGE OF RESPONDENT	Equal variances assumed	7.695	.006	-.421	2039	.674	-.332	.789	-1.880	1.216
	Equal variances not assumed			-.424	1964.48	.671	-.332	.783	-1.868	1.203

Reading Your Output

The program gives you much more information than you need for present purposes, so let's identify the key elements. Some of the information is a repeat of what we got earlier from the Means command: means, standard deviations, and standard errors for men and women. In the "Group Statistics" box under "Mean," for example, we see once again that the average age is 47.78 for men and 48.11 for women.

The results regarding significance that we are most interested in now are given in the box below, labeled "Independent Samples Test," under the heading "t-test for Equality of Means." If you look there under the subheading "Sig. (2-tailed)," in the top row labeled "Equal variances assumed," you will see the probability we are looking for: .674.

As you will anticipate, .674 in this context indicates a probability of 674 in 1,000. The "2-tailed" notation requires just a little more explanation.

In our sample, the average age for women is 0.33 years higher than for men (48.11 − 47.78). SPSS Statistics has calculated that about 674 times in 1,000 samples, sampling error might produce a difference this great in either direction. That is, if the average age of men and the average age of women in the population were exactly the same and we were to select 1,000 samples like this one, we could expect 674 of those samples to show women at least 0.33 years older than men or men as much as 0.33 years older than women.

When you don't have theoretical reasons to anticipate a particular relationship, it is appropriate for you to use the "2-tailed" probability in evaluating differences in means such as these. For our purposes, we'll stick with the 2-tailed test.

SPSS Statistics Command 15.2:
Running *t* Test (Independent-Samples *t* Test)

Click **Analyze** → **Compare Means** → **Independent-Samples T Test**

→ Highlight test variable in variable list

→ Click arrow pointing to the "Test Variable(s):" box

→ Highlight name of grouping variable in variable list

→ Click arrow pointing to the "Grouping Variable:" box

→ Click **Define Groups**

　　→ Define "Group 1" and "Group 2"

→ **Continue** → **OK**

Demonstration 15.3: *t* Test—EDUC by SEX

Some of the variables in your GSS data set allow you to explore this issue further. For example, it would be reasonable for better-educated workers to earn more than poorly educated workers; so if the men in our sample have more education than the women, that might explain the difference in pay. Let's see.

Return to the T Test window and substitute EDUC for AGE as the test variable.[3] Leave "SEX(1 2)" as the grouping variable.

---

[3]Define the values 97, 98, and 99 as "missing" for the variable EDUC.

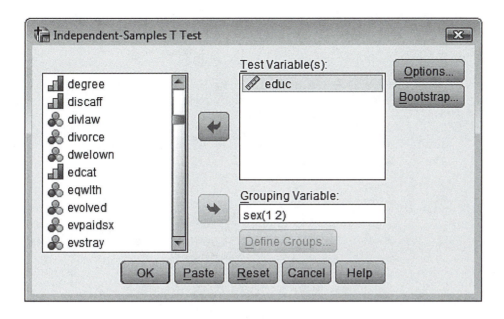

Run the new *t* test.

Group Statistics

	sex RESPONDENTS SEX	N	Mean	Std. Deviation	Std. Error Mean
educ HIGHEST YEAR OF SCHOOL COMPLETED	1 MALE	889	13.45	3.258	.109
	2 FEMALE	1150	13.47	3.064	.090

Independent Samples Test

		Levene's Test for Equality of Variances		t-test for Equality of Means						95% Confidence Interval of the Difference	
		F	Sig.	t	df	Sig. (2-tailed)	Mean Difference	Std. Error Difference		Lower	Upper
educ HIGHEST YEAR OF SCHOOL COMPLETED	Equal variances assumed	3.012	.083	-.196	2037	.844	-.028	.141		-.303	.248
	Equal variances not assumed			-.195	1849.49	.846	-.028	.142		-.306	.250

What conclusion do you draw from these latest results? Notice first that the men and women in our sample have very similar mean numbers of years of education (men = 13.45; women = 13.47). The difference is small and not statistically significant at a probability of .844.

With such a small difference between men's and women's educational backgrounds, it is unlikely that education can be used as a "legitimate reason" for women earning less than men. That's not to say that there aren't other legitimate reasons that may account for the difference in pay. For instance, it is often argued that women tend to concentrate in less-prestigious jobs than do men: nurses rather than doctors, secretaries rather than executives, teachers rather than principals.

Leaving aside the reasons for such occupational differences, that might account for the differences in pay. As you may recall, your Adventures.SAV GSS data contain a measure of socioeconomic status (SEI). We used that variable in our experimentation with correlations. Let's see if the women in our sample have lower-status jobs, on average, than do the men.

Demonstration 15.4: *t* Test—SEI by SEX

Go back to the T Test window and replace EDUC with SEI.

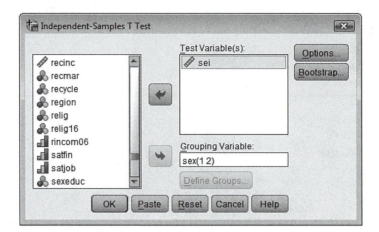

Run the procedure, and you should get the following result.

Group Statistics

	sex RESPONDENTS SEX	N	Mean	Std. Deviation	Std. Error Mean
sei RESPONDENT SOCIOECONOMIC INDEX	1 MALE	824	49.704	19.4623	.6780
	2 FEMALE	1051	48.434	18.9112	.5833

Independent Samples Test

		Levene's Test for Equality of Variances		t-test for Equality of Means						
									95% Confidence Interval of the Difference	
		F	Sig.	t	df	Sig. (2-tailed)	Mean Difference	Std. Error Difference	Lower	Upper
sei RESPONDENT SOCIOECONOMIC INDEX	Equal variances assumed	.005	.946	1.425	1873	.154	1.2699	.8913	-.4782	3.0179
	Equal variances not assumed			1.420	1743.56	.156	1.2699	.8944	-.4843	3.0241

The mean difference in occupational prestige ratings of men and women is 1.27 on a scale from 0 to 100. SPSS Statistics tells us that such a difference could be expected just as a consequence of sampling error in 154 samples in 1,000 ($p = .154$). How would you interpret this finding? Do you think this difference is statistically significant enough to have any social significance?

Writing Box 15.2

Our comparison of mean scores on occupational prestige ($p = .154$) suggests that the observed difference could easily be due to sampling error and is unlikely a difference that exists in the population. The small difference and the lack of statistical significance lead us to conclude that there is no difference between men's and women's occupational prestige.

To pursue this line of inquiry further, you will need additional analytic skills that will be covered shortly in the discussion of multivariate analysis.

Analysis of Variance

The *t* test is limited to the comparison of two groups at a time (for example, male and female). If we wanted to compare the levels of education of different religious groups, we'd have to compare Protestants and Catholics, Protestants and Jews, Catholics and Jews, and so forth. And if some of the comparisons found significant differences and other comparisons did not, we'd be hard-pressed to reach an overall conclusion about the nature of the relationship between the two variables.

The *analysis of variance (ANOVA)* is a technique that resolves the shortcoming of the *t* test. It examines the means of subgroups in the sample and analyzes the variances as well. That is, it examines more than whether the actual values are clustered around the mean or spread out from it.

If we were to ask ANOVA to examine the relationship between RELIG and EDUC, it would determine the mean years of education for each of the different religious groups, noting how they differed from one another. Those "between-group" differences would be compared with the "within-group" differences (variance): how much Protestants differed among themselves, for example. Both sets of comparisons are reconciled by ANOVA to calculate the likelihood that the observed differences are merely the result of sampling error.

Demonstration 15.5: Instructing SPSS Statistics to Run ANOVA

To get a clearer picture of ANOVA, ask SPSS Statistics to perform the analysis we've been discussing. You can probably figure out how to do that, but here's a hint.

Analyze → General Linear Model → Univariate[4]

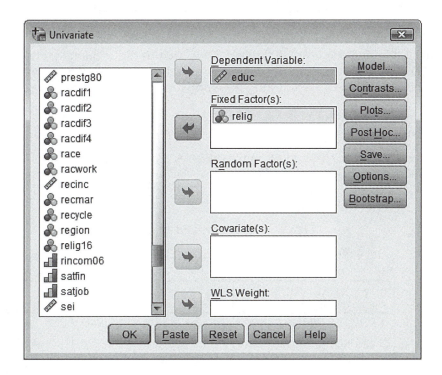

---

[4]If you have used earlier versions of SPSS Statistics, you may notice a difference here. The simple factorial ANOVA procedure has been replaced with the GLM univariate procedure (General Factorial). This allows for ANOVA tables but does not require a defined range of factor variables.

Put **EDUC** into the "Dependent Variable:" field and **RELIG** in the "Fixed Factor(s):" field. Next, open the Univariate Options box by clicking on the **Options** button.

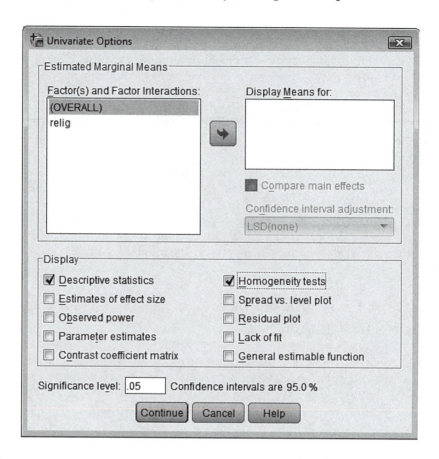

To enable viewing the years of education for each of the religious groups, put a check mark by "Descriptive statistics." To request an equality of variance test, put a check mark next to "Homogeneity tests" by clicking the empty box. Then click **Continue** and **OK**.

Reading Your Output

Here's the SPSS Statistics report on the analysis. Again, we've gotten more information than we need for our present purposes.

Descriptive Statistics

Dependent Variable: educ HIGHEST YEAR OF SCHOOL COMPLETED

relig RS RELIGIOUS PREFERENCE	Mean	Std. Deviation	N
1 PROTESTANT	13.40	2.946	973
2 CATHOLIC	12.98	3.572	480
3 JEWISH	15.11	2.989	37
4 NONE	13.99	3.093	362
Total	13.44	3.173	1852

Levene's Test of Equality of Error Variances[a]

Dependent Variable: educ HIGHEST YEAR OF SCHOOL COMPLETED

F	df1	df2	Sig.
2.641	3	1848	.048

Tests the null hypothesis that the error variance of the dependent variable is equal across groups.

a. Design: Intercept + relig

Tests of Between-Subjects Effects

Dependent Variable: educ HIGHEST YEAR OF SCHOOL COMPLETED

Source	Type III Sum of Squares	df	Mean Square	F	Sig.
Corrected Model	315.798[a]	3	105.266	10.617	.000
Intercept	93531.796	1	93531.796	9433.742	.000
relig	315.798	3	105.266	10.617	.000
Error	18322.185	1848	9.915		
Total	353094.000	1852			
Corrected Total	18637.983	1851			

a. R Squared = .017 (Adjusted R Squared = .015)

The first table of interest is labeled "Descriptive Statistics." It displays the means, standard deviations, and numbers of respondents in each religious category. Note that the Jewish respondents (mean = 15.11) have greater than 1 year of education higher than any of the other categories. The means for those in the "Protestant" (13.40) and "None" (13.99) categories are close, while Catholics, still close, have the lowest mean at 12.98.

The second table, "Levene's Test of Equality of Error Variances," tests whether or not the variances between the groups are the same. In our table, we are checking to see if the variability in education is the same for each of the categories of religion. If they differed significantly in variability, then our test of differences in means could be distorted by variability between the groups rather than differences between the means. The Levene statistic tests for equal amounts of variation within the groups.

If the amounts of variation are significantly different, the ANOVA should not be used. We know the significance level of the Levene test by looking at the probability labeled "Sig." on the table. If the significance level is .05 or less, the variances are considered significantly different and the ANOVA test should not be used. If the significance level is greater than .05, the variances are considered equal and the ANOVA can be used. In our example, the significance level is .048.

Once the variances are determined to be equal, we can proceed to the table titled "Tests of Between-Subjects Effects" and simply look at the row titled "Corrected Model." This refers to the amount of variance in EDUC that can be explained by variations in RELIG. Because our present purpose is to learn about tests of statistical significance, let's move across the row to the statistical significance of the explained variance. You can see that the significance level for the corrected model F value (10.617) is .000.[5] This means that if religion and education were unrelated to each other in the population, we might expect samples that would generate this amount of explained variance less than once in 1,000 samples.

---

[5]Make sure the following values for each variable are defined as "missing": AGE—0, 98, 99; EDUC—97, 98, 99; SEI—.0, 99.8, 99.9.

Perhaps you will find it useful to think of ANOVA as something like a statistical broom. We began by noting a lot of variance in educational levels of our respondents; imagine people's educations spread all over the place. In an attempt to find explanatory patterns in that variance, we use ANOVA to sweep the respondents into subgroups based on religious affiliation (stay with us on this). The questions are whether variations in education are substantially less within each of the piles than we originally observed in the whole sample and whether the mean years of education in each of the subgroups are quite different from one another.

Imagine a set of tidy piles that are quite distant from one another. ANOVA provides a statistical test of this imagery.

SPSS Statistics Command 15.3: ANOVA (GLM Univariate)

Click **Analyze → General Linear Model → Univariate**

→ Highlight dependent variable → Click arrow pointing to "Dependent Variable:" field

→ Highlight factor variable → Click arrow pointing to "Fixed Factor(s):" field

→ Click **Options**

    → Check box next to "Homogeneity tests" → **Continue**

→ **OK**

It is also possible for ANOVA to consider two or more independent variables, but that goes beyond the scope of this book. We have introduced you to ANOVA because we feel it is useful and we wanted to open up the possibility of your using this popular technique; however, you will need more specialized training in the performance of analysis of variance to use it effectively.

Writing Box 15.3

Protestants, Catholics, and those who indicate no religious preference are relatively similar in years of education—13.4, 13, and 14 years, respectively—while Jews average 15.11 years, significantly more than the other groups.

A Statistical Toolbox: A Summary

As we near the end of our discussion of bivariate analysis, we think it may be useful as a review, and a summary of sorts, to provide you with a table listing descriptive and inferential statistics by their appropriate levels of measurement. You will note that those statistics we reviewed in depth are listed in boldface. In addition, we included a few references to some basic statistics we did not cover but you may find useful as you pursue your own research.

Please keep in mind that Table 15.2 is not exhaustive. It references only a few of the many statistics that social researchers find useful when working with items at various levels of measurement.

Table 15.2 A Statistical Toolbox

	Level of Measure		
Statistics for Measuring	Nominal	Ordinal	I/R
Descriptive statistics			
Central tendency	**Mode**	**Median**	**Mean**
Dispersion		**Range**	**Variance**
		Interquartile range	**Standard deviation**
Association (PRE and non-PRE)	**Lambda** (λ) Cramer's *V*	**Gamma** (γ) Somers' *d*	r^2 **Pearson's *r***
	Phi	Kendall's tau-*b*	beta
	Contingency		
	Coefficient		
Inferential statistics			
Tests of significance	**Chi-Square** (χ^2)	**Chi-Square** (χ^2)	**Independent samples**
			t **test**
			ANOVA

Conclusion

This chapter has taken on the difficult question of whether the observed relationship between two variables is important or not. It is natural that you would want to know whether and when you have discovered something worth writing home about.

No one wants to shout, "Look what I've discovered!" and have others say, "That's no big thing." Ultimately, there is no simple test of the substantive significance of a relationship between variables. If we found that women earn less than men, who could say if that amounts to a lot less or just a little bit less? In a more precise study, we could calculate exactly how much less in dollars, but we would still not be in a position to say absolutely whether that amount was a lot or a little. If women made a dollar a year less than men on the average, we'd all probably agree that this was not an important difference. If men earned 100 times as much as women, on the other hand, we'd probably all agree that this was a big difference. However, few of the differences we discover in social science research are that dramatic.

In this chapter, we've examined a very specific approach social scientists often take in addressing the issue of significance. As distinct from notions of *substantive* significance, we have examined *statistical* significance. In each of the measures we've examined—chi-square (χ^2), *t* test, and analysis of variance—we've asked how likely it would be that sampling error could produce the observed relationship if there were actually no relationship in the population from which the sample was drawn.

This assumption of "no relationship" is sometimes referred to as the *null hypothesis*. The tests of significance we've examined all deal with the probability that the null hypothesis is correct. If the probability is relatively high, then we conclude that there is no relationship between the two variables under study in the whole population. If the probability is small that the null hypothesis could be true, then we conclude that the observed relationship reflects a genuine pattern in the population.

Main Points

- In this chapter, we introduced tests of statistical significance.
- As opposed to measures of association that focus on the strength and direction of the relationship, tests of significance allow us to estimate whether the relationship can be considered statistically significant.
- Statistical significance generally refers to the likelihood that an observed relationship between variables in a sample could have occurred as a result of chance or sampling error.
- Tests of significance also allow you to determine whether or not an observed association between items in a sample is likely to exist in the population.
- In this chapter, we moved from focusing primarily on descriptive statistics to inferential statistics.
- The three tests of significance that we reviewed are part of the larger body of inferential statistics: chi-square, *t* tests, and ANOVA.
- Chi-square is most appropriate for nominal variables, although it can be used with ordinal variables.
- Traditionally, a chi-square with a probability of .05 or less is considered significant.
- The *t* test is most appropriate if the dependent variable is at the interval/ratio level of measurement.
- ANOVA builds on the shortcomings of the *t* test.
- There is an important difference between substantive and statistical significance.

Key Terms

Statistical significance	*t* test
Substantive significance	Chi-square (χ^2)
Inferential statistics	Analysis of variance (ANOVA)
Independent-samples *t* test (two-sample *t* test)	Goodness of fit
Descriptive statistics	Null hypothesis
Sampling error	Degrees of freedom (*df*)
Expected cell frequencies	Observed relationship

SPSS Statistics Commands Introduced in This Chapter

15.1   Producing Crosstabs With Chi-Square

15.2   Running *t* Test (Independent-Samples *t* Test)

15.3   ANOVA (GLM Univariate)

Review Questions

1. What does the term *statistical significance* mean?

2. If you are interested in estimating the strength of association between variables, would you rely on tests of significance or measures of association?

3. If you are interested in determining whether or not an observed relationship between variables in your sample is likely to exist in the population, would you rely on tests of significance or measures of association?

4. What are inferential statistics?

5. How do they differ from descriptive statistics?

6. Chi-square (χ^2) is considered appropriate for variables at what level of measurement?

7. Chi-square is based on a comparison of _____ frequencies and _____ frequencies.

8. If a chi-square has a probability level greater than .05, is it generally considered significant by traditional social science standards?

9. Would a chi-square with a probability of .01 be considered significant by traditional social science standards?

10. Would we have more confidence if an association was significant at the .05 or the .001 level?

11. The t test is considered appropriate if your dependent variable is at what level of measurement?

12. Does the t test allow us to determine whether the means, the variances, or both the means and variances of two groups are statistically different from each other?

13. How does ANOVA resolve the shortcomings of the t test? Explain.

14. Does ANOVA examine the means, variances, or both the means and variances of subgroups in a sample?

15. In this chapter, we have focused primarily on statistical significance. How does this differ from substantive significance?

SPSS STATISTICS LAB EXERCISE 15.1

NAME _____

CLASS _____

INSTRUCTOR _____

DATE _____

To complete the following exercises, you need to load the AdventuresPLUS.SAV data file.

1. Define the values 0, 8, and 9 as "missing" for AFFRMACT. Then recode AFFRMACT as follows to create a new variable AFFREC. When you have done that, be sure to set the decimal places for AFFREC to 0 and add value labels.

Old Values		New Values	Labels
1 to 2	→	1	Support
3 to 4	→	2	Oppose

System- or user-missing → System-missing

Run Crosstabs listing RACE and SEX as the independent variables and AFFREC as the dependent variable. Request column percentages and chi-square. Then answer Questions 2 through 8.

2. Record the percentage of respondents who said they "Support" preferences (AFFREC) for each independent variable (RACE and SEX).

	White	Black	Other	Chi-Square Asymp. Sig.
Percentage				
Support				
	Male	Female		Chi-Square Asymp. Sig.
Percentage				
Support				

3. (RACE) Do the percentages change or move, signifying a relationship between RACE and AFFREC? Explain how you know.

4. (RACE) What is the significance of chi-square? Is it less than, equal to, or more than .05?

5. (RACE) Is the relationship between RACE and AFFREC statistically significant?

6. (SEX) Do the percentages change or move, signifying a relationship between SEX and AFFREC? Explain how you know.

7. (SEX) What is the significance of chi-square? Is it less than, equal to, or more than .05?

8. (SEX) Is the relationship between SEX and AFFREC statistically significant?

9. Choose two variables from the data set (nominal and/or ordinal) that you think may be related. List the variables and explain briefly how you think they are related.

10. Explore the relationship between the variables using Crosstabs and chi-square. Print your output and describe your findings in a few sentences, focusing in particular on whether the relationship between the variables you chose is statistically significant.

Run the *t* test, specifying AGE, SEI, and EDUC as the test variables and RACE as the grouping variable. Keep in mind that you will have to ask SPSS Statistics to limit the comparison to Whites and Blacks, omitting the "Other" category. Once you have run the procedure, complete Questions 11 and 12.

11. Fill in the blanks with the appropriate information.

		RACE	
		White	*Black*
AGE	Mean	_____	_____
	Sig. (2-tailed)		
EDUC	Mean	_____	_____
	Sig. (2-tailed)		
SEI	Mean	_____	_____
	Sig. (2-tailed)		

	What Is the Mean Difference?	*Is the Mean Difference Significant?*
AGE		
EDUC		
SEI		

12. On the basis of these results, what conclusions, if any, can you draw about racial inequality in the United States? How much inequality is there? Is there inequality in education, the workforce, or both? Do your data indicate that Whites have a higher average age, suggesting that they live longer? What, if anything, does this say about racial inequality today?

We want to find out if there is a relationship between the amount of TV people watch and their social class, using the variables CLASS and TVHOURS from your AdventuresPLUS.SAV file. To start the process, make sure any codes that indicate missing information, such as DK, NA, and so on, have been defined as "missing" values.

Use the Univariate procedure (hint: **Analyze → General Linear Model → Univariate**) to produce an ANOVA with TVHOURS as the dependent variable and CLASS as the fixed factor. On the Options screen, mark both "Descriptive statistics" and "Homogeneity tests."

13. From your results, what is the average number of hours per day spent watching TV by each of the social classes? How many respondents were in each class?

Class	Mean Hours Watching TV	Number of Respondents
Lower class		
Working class		
Middle class		
Upper class		

14. Find the F values and significance levels for both the Levene test and the independent variable.

	F Value	Significance Level
Levene test		
Independent variable		

15. Describe the relationship between hours spent watching TV and social class.

16. Using the information you gathered above, explain why the relationship you described above can or cannot be generalized from the Adventures.SAV sample to the United States population.

Chapter 16   Suggestions for Further Bivariate Analyses

By now, you've amassed a powerful set of analytic tools. In a world where people make casual assertions about sociological topics, you're now in a position to determine the facts. You can determine how the U.S. population feels about a variety of topics, and with your new bivariate skills, you can begin to explain why people feel as they do.

Remember, in Chapters 10 through 15, we introduced you to a number of techniques that make this type of analysis possible, including crosstabs (Chapters 10–12); measures of association—lambda (λ), gamma (γ), Pearson's r, regression—(Chapters 13–14); and tests of significance—chi-square (χ^2), t test, ANOVA—(Chapter 15). Our goal in this chapter is to encourage you to use these techniques to explore some of the other variables contained on your data files.

To get you started, we are going to suggest some analyses you might undertake. In Chapter 9, we suggested some topics—drawn from the items on your AdventuresPLUS.SAV file—that you might pursue with the techniques of univariate analysis. Let's start by returning to those topics.

Demonstration 16.1: Desired Family Size

Your AdventuresPLUS.SAV file contains the variable CHLDIDEL, which asked respondents what they considered the ideal number of children for a family. A little more than half the respondents said that two was best. Because that is also the number of children that would represent population stabilization, you might want to begin by recoding this variable to create two response categories. If you are having difficulty with this, here is a "hint" to point you in the right direction: Select **Transform** → **Recode** → **Into Different Variables**.

Then recode CHLDIDEL to create a new variable, CHLDNUM, with the following values and labels:[1]

CHLDIDEL	New Variable	CHLDNUM
Old Values	New Values	Label
0 to 2	1	0 to 2
3 to 8	2	3 or more

---

[1]Before proceeding, define the following values as "missing" for CHLDIDEL: –1 and 9.

Once you've recoded CHLDIDEL into the more manageable variable CHLDNUM, you can use some of the bivariate techniques you learned to examine the causes of differences.

As a start, you might want to see if the amount of education a person has is related to opinions about ideal family size. The better-educated members of the population are, for instance, generally more concerned about environmental issues. Now you can explore if they are also more committed to small families. How about age? Support for small families is a fairly recent development in the United States, against a historical backdrop of large farm families. Does this mean that young people would be more supportive of small families than would older people? You find out.

Religion and race are good candidates for shaping opinions about ideal family size, because the nature of family life is often central to subcultural patterns. We saw that Catholics and Protestants were resistant to abortion. How do they feel about limiting family size in general? You can know the answer in a matter of minutes.

Several family variables may very well relate to attitudes toward ideal family size. Marital status and whether respondents have ever been divorced might be relevant. Can you see why that would be worth exploring? What would you hypothesize? The data set contains SIBS (the number of brothers and sisters the respondent has), which may be relevant. You might want to see if the experience of having brothers and sisters has any impact on opinions about what's best in family size.

We found that gender was basically unrelated to abortion attitudes. How about ideal family size? Do you think men and women differ in their images of the perfect family? If you think so, in which direction do you think that difference goes? Finally, what about the variables we examined in relation to abortion attitudes? Take a moment to see if they are at all related to opinions about ideal family size.

Once you have explored one or more of these possibilities, take a moment to describe your findings in a few sentences or paragraphs.

In Writing Box 16.1, we provide you with an example of how a social scientist examining the relationship between ideal family size (CHLDNUM) and attitudes toward abortion (ABINDEX) might describe his or her findings in prose. We created ABINDEX in Chapter 8, so it should be saved on your AdventuresPLUS.SAV file. If not, simply go back to the chapter and re-create it before proceeding.

Take a moment to consider which measure of association and test of statistical significance are appropriate in this case. If you have trouble making this determination, refer back to the discussion in Chapters 13, 14, and 15.

Now that you have done the hard work, you are ready to determine if there is a relationship between education and attitudes toward abortion. After you have completed your analysis and written a short description of your findings, compare it to the discussion in Writing Box 16.1.

Writing Box 16.1

The relationship between the number of children people have and support for abortion is indeed statistically significant by the chi-square measurement; the level of significance is .048. Gamma (−.193) provides evidence of a moderate association worth noting. Respondents who strongly approve of abortion are less likely to want more than two children. Likewise, those who strongly disapprove of abortion are more likely to say that their ideal number of children is three or more.

Child Training

In Chapter 9, we took an initial look at different opinions about what was important in the development of children. The key variables were as follows:

OBEY	To obey
POPULAR	To be well liked or popular
THNKSELF	To think for himself or herself
WORKHARD	To work hard
HELPOTH	To help others when they need help

If you examined these variables, then you discovered some real differences in how people want their children to turn out. Now let's see what causes those differences, because opinions on this topic can reflect some more general attitudes and worldviews.

Once again, such demographic variables as sex, age, race, and religion might make a difference. OBEY, for example, reflects a certain authoritarian leaning. Perhaps it is related to political variables, such as PARTYID and POLVIEWS; perhaps it is not. There's only one way to find out.

HELPOTH measures an altruistic dimension. That's something that religions often encourage. Maybe there's a relation between this variable and some of the religion variables.

Also consider the variable THNKSELF, which values children's learning to think for themselves. What do you expect will influence this? Education perhaps? How about age and sex? Do you think older respondents would be relatively cool with children's thinking for themselves? Would men or women be more supportive? Don't rule out religious and political variables. Some of these results are likely to confirm your expectations; some are not.

When it comes to the value of children's thinking for themselves, you may find that some of the other attitudinal variables in the General Social Survey (GSS) data set are worth looking at. Consider those who have told us that they are permissive on premarital sex and homosexuality. Do you think they would be more or less likely to value children's learning to think for themselves?

You might want to pursue any number of directions in looking for the causes of different attitudes toward the qualities most valued in children.

Attitudes About Sexual Behavior

You may want to focus on the three sexual variables. What do you suppose would cause differences of opinion regarding premarital sexual relations and homosexuality? What do you suppose determines who goes to X-rated movies? You have the ability and the tools to find out for yourself.

Near the end of the movie *Casablanca*, the police chief instructs his officers to "round up the usual suspects." You might do well to round up the usual demographic variables as a way of beginning your examination of sexual attitudes: age, gender, race, religion, education, social class, and marital status, for example.

Before examining each of these relationships, take some time to think about any links you might logically expect. Will men or women be more permissive about homosexuality? Will married, single, or divorced people be more supportive of premarital sex? How do you expect that young and old people will differ? As you investigate these attitudes, be careful about assuming that the three items are just different dimensions of the same orientation. The kinds of people who are permissive about premarital sex are not necessarily the same ones who are permissive about homosexuality.

Demonstration 16.2: Investigating Sexual Permissiveness Further

Another possibility is to use the index of sexual permissiveness (SEXPERM) we created in Chapter 9 to explore why some people are more permissive than others when it comes to attitudes toward sex.

Writing Box 16.2 contains an example of how a social scientist examining this issue might convey his or her findings in prose. In this case, we examined the possibility that two basic demographic variables may be related to sexual permissiveness: AGE (as represented by the recoded variable AGECAT) and SEX. After making sure to define appropriate values for each variable as "missing," determine which measure of association and test of statistical significance are appropriate to examine the relationship between SEXPERM, AGECAT, and SEX. After exploring the relationship between these variables, take some time to describe your findings in prose. Then compare your analysis to the short description in Writing Box 16.2.

Writing Box 16.2

The relationship between gender (SEX) and sexual permissiveness (SEXPERM) is neither strong ($\lambda = .02$) nor significant ($p = .299$) when measured by lambda. A significant chi-square ($p = .035$), however, suggests that there is a statistically significant gender difference along lines of sexual permissiveness. Respondents' ages are also related to permissiveness. Not only does chi-square ($p = .000$) indicate a statistically significant relationship, but so does gamma ($\gamma = -.247$; $p = .000$), showing a moderate relationship worth noting between age and permissiveness.

Prejudice

Several items in your AdventuresPLUS.SAV file address different aspects of racial prejudice about African Americans. RACMAR measures respondents' attitudes toward the legality of interracial marriage, whereas RACDIF1, RACDIF2, RACDIF3, and RACDIF4 measure attitudes about Black–White differences.

Certainly, RACE is the most obvious variable to examine, and you probably won't be surprised at what you find. Don't stop there, however. There are other variables that provide even more dramatic relationships. Education, politics, and social class offer fruitful avenues for understanding the roots of attitudes on these variables. You may be surprised by the impact of religious variables.

As a different approach, you might look at the opinion that homosexuality is morally wrong as prejudice against gays and lesbians. It's worth checking whether responses to that item are related to prejudice against African Americans.

Additional Resources

The National Opinion Research Center maintains a website that has not only a codebook for the GSS but also a fairly extensive bibliography. This is a particularly useful resource if you want to identify studies, reports, books, and articles that use the GSS variables with which you are working.

You can access the site at http://www.norc.org/GSS+Website/. From that site, you have several options. You may, for instance, search the reference section directly by clicking

"Bibliography" under the "Publications" menu at the top of the main page. This, however, may not be particularly useful unless you know exactly what you are looking for.

A better alternative may be to search by either subject or mnemonic (an index of abbreviated variable names) under the "Browse GSS Variables" menu on the main page. For example, if you are interested in pursuing the subject of child training further, simply click **Subject Index** → **C** (for *Children*) → **Children**. Here you will see a list of subjects and mnemonics relevant to "Children." If, for instance, you are interested in desirable qualities of children, you can click on that option and choose one of the abbreviated variable names. If you click on OBEY, for instance, you will get not only a box with the question and responses by year but also the option of exploring trends and links.

If, for instance, you scroll to the bottom of the screen on the OBEY variable webpage (as is the case with most GSS variable pages), you will see a link (under the "Notes" heading) to a fairly extensive bibliography referencing studies, articles, reports, and books pertaining to child training in general (and the variable OBEY in particular).

If you wish to search an abbreviated variable name directly, once you access the site, click on **Browse GSS Variables** → **Mnemonic Index** → **O** (for OBEY) → **OBEY** → **Links**. Once again, if you scroll to the bottom of your screen, you will see a bibliography of studies that pertain to child training and utilize the variable OBEY.

Conclusion

The preceding suggestions should be enough to keep you busy, but you shouldn't feel limited by them. The most fruitful guides to your analyses should be your own personal interests. Keep in mind that the Adventures.SAV (or AdventuresPLUS.SAV) file contains a number of additional variables, covering issues such as the environment, mass media use, national government spending priorities, gender roles, law enforcement, teen sex, the Internet, as well as sexual attitudes and behavior. Think about which topics from the Adventures.SAV file interest or concern you. Now you have a chance to learn something about them on your own. You don't have to settle for polemical statements about "the way things are." You now have the tools you need to find out for yourself.

In examining these bivariate relationships, you may want to begin with Crosstabs, because that technique gives you the most detailed view of the data. At the same time, you should use this exercise as an opportunity to experiment with the other bivariate techniques we've examined. Try chi-squares where appropriate, for example. As you find interesting relationships between variables, you may want to test their statistical significance to get another view of what they mean.

What you've learned so far may be sufficient for most of your day-to-day curiosities. Now you can learn what public opinion really is on a given topic, and you can determine what kinds of people hold differing views on that topic. In the remaining chapters of this book, however, we are going to show you an approach to understanding that goes much deeper. As we introduce you to multivariate analysis, you're going to have an opportunity to sample a more complex mode of understanding than most people are even aware of.

Main Points

- Now that you are capable of describing both what Americans think about a variety of issues and why, we suggest some additional bivariate analyses for you to pursue on your own.
- We focus on four topics drawn from the items in your Adventures.SAV file as examples of the types of investigations you may want to pursue: desired family size, child training, attitudes about sexual behavior, and prejudice.
- Don't be limited by these suggestions, however. Pursue topics and issues that interest or concern you. After all, this is your adventure.

■ Keep in mind that your Adventures.SAV file contains 150 General Social Survey items covering a number of important and controversial issues and topics in American life.

■ When pursuing relations between two variables, you should begin with Crosstabs and then experiment with the other techniques we reviewed in this section, including measures of association and tests of statistical significance.

Key Terms

No new terms were introduced in this chapter.

SPSS Statistics Commands Introduced in This Chapter

No new commands were introduced in this chapter.

Review Questions

1. Use the National Opinion Research Center website discussed at the end of this chapter ("Additional Resources") to identify an article or study that pertains to each of the following topics. Then list the names, authors, and publication dates of the articles.

   a. Desired family size
   b. Child training
   c. Attitudes toward sexual behavior
   d. Prejudice

2. Write a summary of one of the articles you found in response to Question 1.

3. Discuss how you might apply the techniques and procedures we covered in Part III (Chapters 10–15) on bivariate analysis (e.g., crosstabs, measures of association, and tests of statistical significance) to examine each of the following topics.

   a. Desired family size
   b. Child training
   c. Attitudes toward sexual behavior
   d. Prejudice

NAME _____

CLASS _____

INSTRUCTOR _____

DATE _____

In Lab Exercise 9.1, we asked you to use univariate techniques to examine one of the topics from your AdventuresPLUS.SAV file in more depth: sex roles, law enforcement, environment, mass media (use and confidence), national government spending priorities, teen sex, affirmative action, the environment, sexual attitudes and behaviors, and so on.

In this lab exercise, you will be given an opportunity to expand your analysis of this issue by applying some of the bivariate techniques discussed in Chapters 10 through 15.

   1. List the general topic/issue you examined in Lab Exercise 9.1.

   2. List the variables in your AdventuresPLUS.SAV file that pertain to this topic/issue.

   3. In Lab Exercise 9.1, did you either RECODE or create an INDEX based on any of the variables you were working with? If so, list the abbreviated variable name(s) below.

   4. Choose one of the variables listed in response to Question 2 or 3 above. Make sure it is an item that you are interested in examining further, and list it in the space below.

Dependent variable _____

   5. List the names of two other items (independent variables) from the AdventuresPLUS.SAV file that you think may be causally related to or associated with your dependent variable.

Independent Variable 1 _____

Independent Variable 2 _____

   6. Write two hypotheses linking your dependent variable (Question 4) and independent variables (Question 5).

Hypothesis 1:

List the dependent variable in Hypothesis 1 _____

List the independent variable in Hypothesis 1 _____

Hypothesis 2:

List the dependent variable in Hypothesis 2 _____

List the independent variable in Hypothesis 2 _____

7. From the list of bivariate techniques below, circle the procedures appropriate for examining the relationship in Hypothesis 1 (Question 6) further. (Hint: You should circle at least one technique in each "group": (a) Chapters 10–12; (b) Chapters 13–14; (c) Chapter 15.)

   a. Chapters 10 through 12: Crosstabs

      Crosstabs

   b. Chapters 13 and 14: Measures of association

      Lambda

      Gamma

      Correlation matrix with Pearson's r

      Coefficient of determination (r^2)

      Regression

      Scatterplot with a regression line

   c. Chapter 15: Tests of significance

      Chi-square

      t test

      ANOVA

8. From the list of bivariate techniques reviewed in Chapters 10 through 15, circle the procedures appropriate for examining the relationship in Hypothesis 2 (Question 6) further. (Hint: You should circle at least one technique in each "group": (a) Chapters 10–12; (b) Chapter 13–14; (c) Chapter 15.)

   a. Chapters 10 through 12: Crosstabs

      Crosstabs

   b. Chapter 13 and 14: Measures of association

      Lambda

      Gamma

      Correlation matrix with Pearson's r

      Coefficient of determination (r^2)

      Regression

      Scatterplot with a regression line

   c. Chapter 15: Tests of significance

      Chi-square

      t test

      ANOVA

9. Now use the techniques you identified as appropriate in response to Question 7 to examine the relationship between the variables in Hypothesis 1. After you have completed your analysis, do the following:

   ■ Print and attach a copy of your output to this sheet.
   ■ Summarize your findings in prose below.

10. Now use the techniques you identified as appropriate in response to Question 8 to examine the relationship between the variables in Hypothesis 2. After you have completed your analysis, do the following:

   ■ Print and attach a copy of your output to this sheet.
   ■ Summarize your findings in prose below.

11. Choose a variable from your AdventuresPLUS.SAV file and list it below.

12. Then access the National Opinion Research Center website (http://www.norc.org/ GSS+Website/) discussed in the "Additional Resources" section at the end of this chapter. Locate an article, study, or report that references this variable. Remember to first look up the variable, then find the link to research at the bottom of that page, under "Notes." Read the article, and then write a critique that includes the following:

   ■ Article title
   ■ Author
   ■ Main point(s) of article
   ■ How the author uses/refers to the variable
   ■ What you learned from the article (as it pertains to the variable)
   ■ How you might explore this variable further in future research

Part IV Multivariate Analysis

Now that you've mastered the logic and techniques of bivariate analysis, we are going to take you one step further: to the examination of three or more variables at a time, known as multivariate analysis.

In Chapter 17, we'll delve more deeply into religious orientations to gain a more comprehensive understanding of this variable. In Chapter 18, we will pick up some loose threads of our bivariate analysis and pursue them further with our new analytic capability.

In Chapter 19, we will set as our purpose the prediction of attitudes toward abortion. We'll progress, step-by-step, through a number of variables previously found to have an impact on abortion attitudes, and we'll accumulate them in a composite measure that will offer a powerful predictor of opinions.

Finally, Chapter 20 launches us into uncharted areas of social research, which you should now be empowered to chart for yourself.

Chapter 17 **Multiple Causation**

Examining Religiosity in Greater Depth

In Part III, we focused primarily on the relationship between two variables: an independent variable and a dependent variable. If we continued to limit ourselves solely to the examination of two variables at a time, our understanding of the social world would remain incomplete, not to mention dissatisfying.

Bivariate analysis alone cannot help us understand the social world, because in the "real world," two or more factors often have an impact on, influence, or cause variation in a single dependent variable. Consequently, to understand the complexities of the social world, we need to introduce a more sophisticated form of statistical analysis that allows us to examine the impact of more than one independent variable on a single dependent variable.

Social scientists refer to this type of analysis as **multivariate analysis**, the *simultaneous* analysis of three or more variables. Multivariate analysis is the next step after bivariate analysis. By helping us move beyond the limitations of univariate and bivariate analysis, it allows us to develop a more complete understanding of the complexities of the social world.

We are going to begin our introduction to multivariate analysis by looking at the simplest of outcomes, multiple causation.

Multiple Causation

In Chapter 10, we discussed several variables, including gender and age, that might affect the levels of respondents' religiosity. It is often the case with social phenomena that people's attitudes and behaviors are affected by more than one factor. The task of the social scientist, then, is to discover all those factors that influence the dependent variable under question and discover how those factors work together to produce a result. If both age and gender affect religiosity independently, perhaps a combination of the two would predict it even better.

Demonstration 17.1: The Impact of Age and Sex on Religiosity

To begin our multivariate analysis, let's see how well we can predict religiosity if we consider AGE and SEX simultaneously. Does religiosity increase with age among both men and women separately? Moreover, do the two variables have a cumulative effect on religiosity? That is, are older women the most religious and younger men the least?

To begin our exploration of this topic, let's open our AdventuresPLUS.SAV file and access the Crosstabs dialog box. In this case, we want to use CHATT as the dependent/row variable and AGECAT (recoded AGE) as the independent/column variable.

Now select **SEX** in the list of variables. Notice that the arrows activated would let you transfer SEX to the row or column fields—but don't do that! Instead, transfer it to the third field, near the bottom of the window.

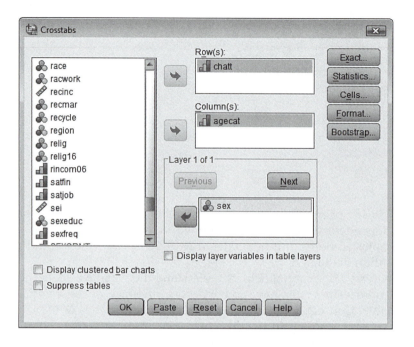

We have now told SPSS Statistics to examine the relationship between AGECAT and CHATT while controlling for SEX. The result of this simple act is that now we will get two crosstabs, one showing the relationship between AGECAT and CHATT for men and another showing the association between AGECAT and CHATT for women.

Now all you need to do to run Crosstabs is make sure that cells are percentaged by column and then execute the command (i.e., click **Cells → Percentages: Column → Continue → OK**).

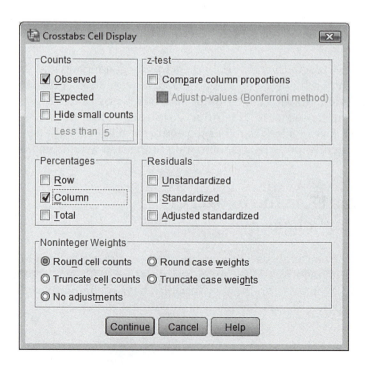

As we noted, this command produces more than one table. We have asked SPSS Statistics to examine the impact of AGECAT on CHATT separately for men and women. Thus, we are rewarded with the following three-variable crosstabulation.

chatt Recoded Church Attendance * agecat Recoded Age * sex RESPONDENTS SEX Crosstabulation

sex RESPONDENTS SEX				1 Under 21	2 21-39 years old	3 40-64 years old	4 65 and older	Total
1 MALE	chatt Recoded Church Attendance	1 About weekly	Count	10	46	105	57	218
			% within agecat Recoded Age	34.5%	16.7%	24.6%	36.5%	24.6%
		2 About monthly	Count	5	44	63	21	133
			% within agecat Recoded Age	17.2%	16.0%	14.8%	13.5%	15.0%
		3 Seldom	Count	9	106	146	41	302
			% within agecat Recoded Age	31.0%	38.5%	34.2%	26.3%	34.0%
		4 Never	Count	5	79	113	37	234
			% within agecat Recoded Age	17.2%	28.7%	26.5%	23.7%	26.4%
	Total		Count	29	275	427	156	887
			% within agecat Recoded Age	100.0%	100.0%	100.0%	100.0%	100.0%
2 FEMALE	chatt Recoded Church Attendance	1 About weekly	Count	8	91	169	120	388
			% within agecat Recoded Age	27.6%	22.6%	35.3%	51.1%	33.9%
		2 About monthly	Count	9	85	67	23	184
			% within agecat Recoded Age	31.0%	21.1%	14.0%	9.8%	16.1%
		3 Seldom	Count	9	126	154	49	338
			% within agecat Recoded Age	31.0%	31.3%	32.2%	20.9%	29.5%
		4 Never	Count	3	101	89	43	236
			% within agecat Recoded Age	10.3%	25.1%	18.6%	18.3%	20.6%
	Total		Count	29	403	479	235	1146
			% within agecat Recoded Age	100.0%	100.0%	100.0%	100.0%	100.0%
Total	chatt Recoded Church Attendance	1 About weekly	Count	18	137	274	177	606
			% within agecat Recoded Age	31.0%	20.2%	30.2%	45.3%	29.8%
		2 About monthly	Count	14	129	130	44	317
			% within agecat Recoded Age	24.1%	19.0%	14.3%	11.3%	15.6%
		3 Seldom	Count	18	232	300	90	640
			% within agecat Recoded Age	31.0%	34.2%	33.1%	23.0%	31.5%
		4 Never	Count	8	180	202	80	470
			% within agecat Recoded Age	13.8%	26.5%	22.3%	20.5%	23.1%
	Total		Count	58	678	906	391	2033
			% within agecat Recoded Age	100.0%	100.0%	100.0%	100.0%	100.0%

Notice that the table is divided into three parts, male, female, and total. The "Total" part of the table contains the same information that would have been provided had we not selected SEX as a "layer." For our purposes, we can create a summary table, as follows, that is easier to read to consider more easily these differences across age, with a specific focus on gender.

Percentage Who Attend Worship Services About Weekly

	Under 21	21 to 39	40 to 64	65 and Older
Men	—	17	25	37
Women	—	23	35	51

A dash (—) in this table indicates that there are too few cases for meaningful percentages. We required at least 10 cases, a common standard.

There are two primary observations to be made regarding this table. First, women are more likely to attend worship services than are men within each age group. Second, the previously observed relationship between AGECAT and CHATT remains present for both men and women. In other words, the introduction of a third or control variable (SEX) did not alter or diminish the strength of the relationship between AGECAT and CHATT. So age and church attendance do appear to be related both among women and among men.

The SPSS Statistics procedure we just ran is sometimes referred to as controlling for a third variable, and it is often used to elaborate on bivariate relationships. *Elaboration* is a technique that helps us examine the relationship between two variables (an independent and dependent) while controlling for a third variable.

In this example, for instance, we specified CHATT as our dependent variable, AGECAT as our independent variable, and SEX as our *control variable* (or third variable). This allowed us to examine the relationship between church attendance and age while controlling for sex. Consequently, our output was in two tables: one showing the relationship between CHATT and AGECAT for men and the other showing the relationship between CHATT and AGECAT for women.[1]

SPSS Statistics Command 17.1: Running Crosstabs With a Control or Third Variable

Analyze → Descriptive Statistics → Crosstabs

→ Highlight dependent variable → Click arrow pointing to "Row(s):" field

→ Highlight independent variable → Click arrow pointing to "Column(s):" field

→ Highlight control/third variable → Click arrow pointing toward bottom field ("Layer 1 of 1")

→ **Cells** → "Column" box under "Percentages" → **Continue** → **OK**

Demonstration 17.2: Family Status and Religiosity

If you read the excerpt by Glock et al. on the *Adventures in Social Research* student study website (http://www.sagepub.com/babbie8estudy/), you will recall that, according to social deprivation theory, "family status" is also related to religiosity. Those who had "complete families" (spouse and children) were the least religious among the 1,952 Episcopal Church members, suggesting that those lacking families were turning to the church for gratification.

Using Crosstabs, set CHATT as the row variable and MARITAL as the column variable.[2] Here's what you should get:

chatt Recoded Church Attendance * marital MARITAL STATUS Crosstabulation

| | | | marital MARITAL STATUS | | | | | |
			1 MARRIED	2 WIDOWED	3 DIVORCED	4 SEPARATED	5 NEVER MARRIED	Total
chatt Recoded Church Attendance	1 About weekly	Count	312	92	75	15	113	607
		% within marital MARITAL STATUS	35.1%	51.4%	22.0%	23.1%	20.1%	29.8%
	2 About monthly	Count	141	22	48	9	98	318
		% within marital MARITAL STATUS	15.9%	12.3%	14.1%	13.8%	17.4%	15.6%
	3 Seldom	Count	266	38	112	26	199	641
		% within marital MARITAL STATUS	30.0%	21.2%	32.8%	40.0%	35.3%	31.5%
	4 Never	Count	169	27	106	15	153	470
		% within marital MARITAL STATUS	19.0%	15.1%	31.1%	23.1%	27.2%	23.1%
Total		Count	888	179	341	65	563	2036
		% within marital MARITAL STATUS	100.0%	100.0%	100.0%	100.0%	100.0%	100.0%

---

[1]While a complete discussion of the elaboration model is beyond the scope of this text, this technique is explored in detail in most texts on social statistics and research methods. See, for instance, Chapter 16 of Earl Babbie's *The Practice of Social Research* (9th ed., 2001).

[2]Define the value 9 as "missing" for the variable MARITAL.

These data do not confirm the earlier finding. Those who are currently married, for instance, are more likely than everyone else, except those who are widowed, to attend church about weekly. It does not appear that those deprived of conventional family status (i.e., those divorced or never married) are turning to the church for an alternative source of gratification. Perhaps the explanation for this lies in historical changes.

In the years separating these two studies, there have been many changes with regard to family life in the United States. Divorce, single-parent families, unmarried couples living together—these and other variations on the traditional family have become more acceptable and certainly more common. It would make sense, therefore, that people who lacked regular family status in 2010 did not feel as deprived as such people may have in the early 1950s.

Demonstration 17.3: Family Status and Religiosity, Controlling for Age

Before setting this issue aside, however, we should take a minute to consider whether the table we've just seen is concealing anything. In particular, can you think of any other variable that is related to both attendance at worship services and marital status? If so, that variable might be clouding the relations between marital status and religiosity.

The variable we are thinking of is age. We've already seen that age is related to church attendance. It is also probably related to marital status in that young people (lower in church attendance) are the most likely to be "never married." And older people (higher in church attendance) are the most likely to be widowed. It is possible, therefore, that the widowed are higher in church attendance because they're mostly older, and those never married are lower in church attendance only because they're young. This kind of reasoning lies near the heart of multivariate analysis, and the techniques you've mastered allow you to test this possibility.

Return to the Crosstabs window and add AGECAT as the control or third variable. Once you've reviewed the resulting tables, see if you can construct the following summary table.

Percentage Who Attend Church About Weekly

	Married	Widowed	Divorced	Separated	Never Married
Under 21	—	—	—	—	26
21 to 39	28	—	23	—	15
40 to 64	35	46	19	29	29
65 and older	45	54	32	—	—

Again, remember that dashes in this table indicate that there are too few cases for meaningful percentages.

Once again, these findings do not seem to confirm the theory that those lacking families turn to the church for gratification whereas those with families are the least religious. The widowed and married appear to be among the most religious in each category. The lack of data for so many cells in our table is unfortunate in that those additional figures might have helped complete the puzzle.

You can also observe in this table that the effect of age on church attendance is maintained regardless of marital status. Older respondents are more likely to attend religious services than are younger ones, except among the widowed, where there is a small decline. Social scientists often use the term *replication* for the analytic outcome we've just observed.

Having discovered that church attendance increases with age overall, we've now found that this relationship holds true regardless of marital status. That's an important discovery in terms of the generalizability of what we have learned about the causes of religiosity.

Demonstration 17.4: Social Class and Religiosity

In the earlier study, Glock and his colleagues also found that religiosity increased as social class decreased; that is, those in the lower class were more religious than those in the upper class. This fit nicely into the deprivation thesis, that those deprived of status in the secular society would turn to the church as an alternative source of gratification. The researchers indicated, however, that this finding might be limited to the Episcopalian Church members under study. They suggested that the relationship might not be replicated in the general public. You have the opportunity to check it out.

Let's begin with our measure of subjective social class. Run Crosstabs with column percentages, requesting CHATT as the row variable and CLASS as the column variable.[3] Here's what you should get:

chatt Recoded Church Attendance * class SUBJECTIVE CLASS IDENTIFICATION Crosstabulation

			class SUBJECTIVE CLASS IDENTIFICATION				
			1 LOWER CLASS	2 WORKING CLASS	3 MIDDLE CLASS	4 UPPER CLASS	Total
chatt Recoded Church Attendance	1 About weekly	Count	45	259	278	20	602
		% within class SUBJECTIVE CLASS IDENTIFICATION	25.7%	27.5%	32.8%	34.5%	29.8%
	2 About monthly	Count	28	155	124	10	317
		% within class SUBJECTIVE CLASS IDENTIFICATION	16.0%	16.5%	14.6%	17.2%	15.7%
	3 Seldom	Count	47	304	273	14	638
		% within class SUBJECTIVE CLASS IDENTIFICATION	26.9%	32.3%	32.2%	24.1%	31.6%
	4 Never	Count	55	224	172	14	465
		% within class SUBJECTIVE CLASS IDENTIFICATION	31.4%	23.8%	20.3%	24.1%	23.0%
Total		Count	175	942	847	58	2022
		% within class SUBJECTIVE CLASS IDENTIFICATION	100.0%	100.0%	100.0%	100.0%	100.0%

These data certainly do not confirm the earlier findings. In fact, the findings seem to run contrary to our expectations. Examine the first row, for instance, and you see that those in the upper class are more likely to attend church on about a weekly basis than are those in the lower and working classes. To be sure of this conclusion, you might want to rerun the table, controlling for sex and for age.

At the same time, you can test the generalizability of the previously observed effects of sex and age on church attendance. Do they hold up among members of different social classes? Once you have examined this possibility, compare your findings to those in Writing Box 17.1 on the next page.

---

[3]Define the values 0 and 5 to 9 as "missing" for the variable CLASS.

Writing Box 17.1

The finding that women attend church more regularly than men is replicated in each of the social class groups. Once again, both men (26%) and women (42%) in the upper classes appear somewhat more likely to attend religious services about weekly than those in the lower classes. Though, middle class men are slightly more likely to attend religious services about weekly (27%) than are upper class men.

By the same token, the relationship between age and church attendance is also replicated among the several social class groups, with some minor variations. That is, older respondents attend church more often than younger ones, regardless of their social class. One possible exception to this seems to be those under 21 from the working class: They attend with greater frequency than all other age groups among those from the working class. The sample size for that age/class group is small, but this certainly warrants further investigation and may echo a recent trend toward youth involvement in church organizations in the United States.

Other Variables to Explore

Notice that our analyses so far in this chapter have used CHATT as the dependent variable: the measure of religiosity. Recall our earlier comments on the shortcomings of single-item measures of variables. Perhaps our analyses have been misleading by seeking to explain church attendance. Perhaps different conclusions might be drawn if we studied beliefs in an afterlife or frequency of prayer. Why don't you test some of the earlier conclusions by using other measures of religiosity. If you are really ambitious, you can create a composite index of religiosity and look for causes.

Similarly, we have limited our preceding investigations in this chapter to the variables examined by Glock and his colleagues. Now that you have gotten the idea about how to create and interpret multivariate tables, you should broaden your exploration of variables that might be related to religiosity. What are some other demographic variables that might affect religiosity? Or you might want to explore the multivariate relationships between religiosity and some of the attitudinal variables we've been exploring: political philosophies, sexual attitudes, and so forth. In each instance, you should examine the bivariate relationships first and then move on to the multivariate analyses.

Chi-Square and Measures of Association

Thus far, we've introduced the logic of multivariate analysis through the use of Crosstabs and controlling for a third variable. You've already learned some other techniques that can be used in your examination of several variables simultaneously.

Chi-Square (χ^2)

First, we should remind you that you may want to use a chi-square test of statistical significance when you use Crosstabs. It's not required, but you may find it useful as an independent assessment of the relationships you discover.

Measures of Association

Second, you may also want to experiment using an appropriate measure of association, such as lambda or gamma, to test the strength and, in certain cases, the direction of association.

Multiple Regression

You may recall our fairly brief discussion of regression at the end of Chapter 14. At that point, we discussed a form of regression known as simple linear regression or just *linear regression*, which involves one independent variable and one dependent variable. Regression can also be a powerful technique for exploring multivariate relationships.

When you are conducting multivariate analysis involving one dependent variable and more than one independent variable, the technique is referred to as *multiple regression*. In both cases, regression is appropriate for two or more interval/ratio (I/R) or continuous variables. The linear regression methods addressed in this book are based on *ordinary least-squares (OLS) regression*, meaning that this regression method minimizes the squared deviations between the observed data and the regression line.

To use either linear or multiple regression effectively, you need much more instruction than we propose to offer in this book. Still, we want to give you a brief overview of multiple regression, in much the same way as we did when we introduced linear regression earlier.

In our previous use of regression (linear regression, Chapter 14), we examined the impact of EDUC on SEI, respondents' socioeconomic status scores. Now we'll open the possibility that other variables in the data set might also affect occupational prestige.

Dummy Variables

In addition to EDUC, we are also going to consider two additional independent variables: SEX and RACE. These variables were chosen because many argue that in today's workforce, men are still treated differently than women and Whites are still treated differently than African Americans.

You will notice, however, that both SEX and RACE are nominal variables, not I/R variables. Since we told you that regression is appropriate for I/R continuous variables, you may begin to wonder how we can propose to use two nominal variables in a regression equation. That's a very good question.

The answer lies in the fact that researchers sometimes treat such items as *dummy variables* appropriate to a regression analysis. In regard to the variable SEX, for example, the logic used here transforms gender into a measure of "maleness," with male respondents being 100% male and female respondents being 0% male.

Recoding SEX to Create a Dummy Variable: MALE

Let's recode SEX as described above into the new variable MALE. So take the following steps:

Transform → Recode → Into Different Variables

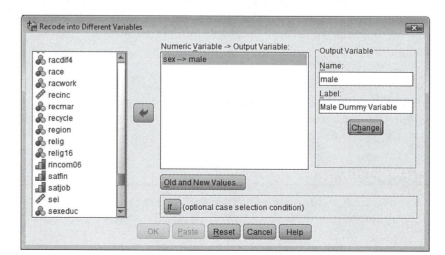

Select **SEX** as the numeric variable. Let's call the new variable **MALE**. Using the Old and New Values window, make these assignments.

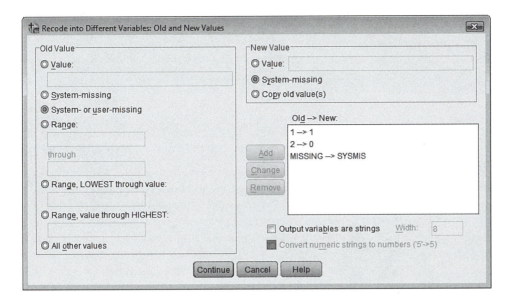

Execute the Recode command by clicking **Continue** and then **OK**.

Before moving on to the variable RACE, make sure you set the decimal places for MALE to 0. You may also want to give a brief description of MALE and define your values and labels (e.g., "male" and "not male"). It might also be a good idea to check the Data View window and verify that your coding is accurate: Cases with a 1 for SEX should also have a 1 for MALE, while cases with a 2 for SEX should have a 0 for MALE.

Recoding RACE to Create a Dummy Variable: WHITE

We will use the same basic procedure to recode RACE as we used to recode SEX. Open the Recode dialog box by selecting **Transform → Recode → Into Different Variables.**

Designate **RACE** as the numeric variable and name the new variable **WHITE**. Then use the Old and New Values window to accomplish your recode.

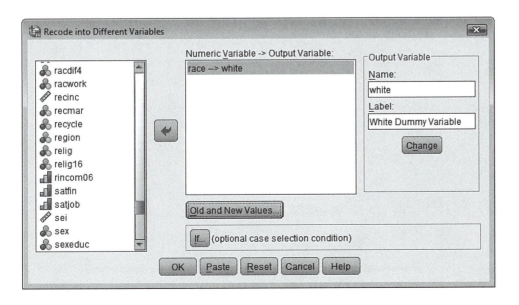

You will notice that unlike the variable SEX, RACE contains three values: 1 ("White"), 2 ("Black"), and 3 ("other"). Consequently, in this case we are going to recode RACE as follows:

Race		White
Old Values		New Values
1	→	1
2 to 3	→	0

With this coding scheme, the dummy code 1 designates 100% majority group status; 0 designates 0% majority group status. Once you have set the new values, click **Continue** and **OK** to execute the command.

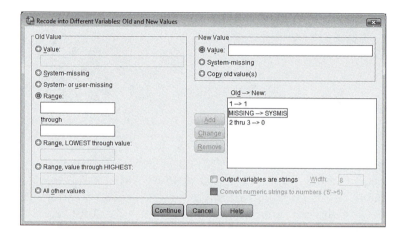

Before we ask SPSS Statistics to run our regression analysis, make sure you set the decimal places for your new variable (WHITE) to 0. In addition, you may want to provide a brief description of the variable and define the values and labels (e.g., "white" and "not white"). Again, it wouldn't hurt to take a look in the Data View window to verify that your coding was accurate: Cases with a 1 for RACE should also have a 1 for WHITE, and cases with a 2 or 3 for RACE should have a 0 for WHITE.

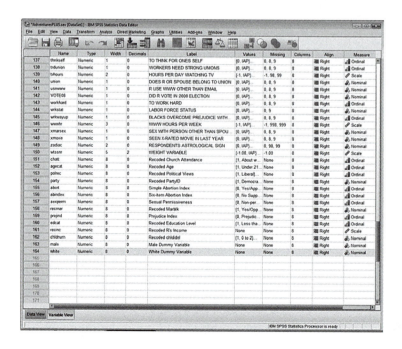

SPSS Statistics Command 17.2: Recoding to Create a Dummy Variable

Transform → Recode → Into Different Variables

→ Highlight name of variable recoding → Click arrow pointing toward "Numeric Variable" field

→ Type name of new variable in field under "Output Variable" labeled "Name:"

→ **Change**

→ **Old and New Values**

> → Recode **Old/Add New Values**

> → **Continue**

→ **OK**

Multiple Regression

Now that we have created our dummy variables, we are ready to request the multiple regression analysis.

Analyze → Regression → Linear takes us to the window we want. Select **SEI** and make it the dependent variable. Then place **EDUC, WHITE,** and **MALE** in the "Independent(s):" field. In the "Method:" menu, the "Enter" method is probably displayed by default. If not, select it by clicking on the down arrow.

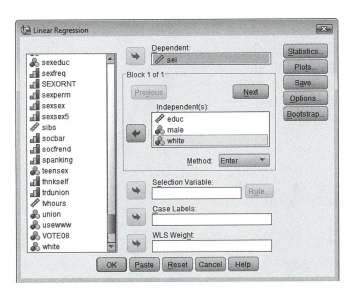

Run this command by clicking **OK**, and you will receive a mass of output. Without going into all the details, we are simply going to show you how it establishes the equation we asked for. We'll take the output a piece at a time. For our purposes, we'd like you to skip through the output on your screen until you find the following tables:

Model Summary

Model	R	R Square	Adjusted R Square	Std. Error of the Estimate
1	.601[a]	.361	.360	15.3305

a. Predictors: (Constant), white White Dummy Variable, male Male Dummy Variable, educ HIGHEST YEAR OF SCHOOL COMPLETED

ANOVA^a

Model		Sum of Squares	df	Mean Square	F	Sig.
1	Regression	247859.558	3	82619.853	351.536	.000^b
	Residual	439026.686	1868	235.025		
	Total	686886.244	1871			

a. Dependent Variable: sei RESPONDENT SOCIOECONOMIC INDEX

b. Predictors: (Constant), white White Dummy Variable, male Male Dummy Variable, educ HIGHEST YEAR OF SCHOOL COMPLETED

Coefficients^a

Model		Unstandardized Coefficients		Standardized Coefficients	t	Sig.
		B	Std. Error	Beta		
1	(Constant)	-3.169	1.683		-1.883	.060
	educ HIGHEST YEAR OF SCHOOL COMPLETED	3.619	.115	.587	31.501	.000
	male Male Dummy Variable	1.209	.714	.031	1.692	.091
	white White Dummy Variable	3.242	.852	.071	3.806	.000

a. Dependent Variable: sei RESPONDENT SOCIOECONOMIC INDEX

We have given SPSS Statistics three variables that we felt might help predict occupational prestige.

To create our equation, we take two numbers from the "Unstandardized Coefficients" column: the constant (–3.169) and the β (beta) value (called the slope) for EDUC (3.619), WHITE (3.242), and MALE (1.209). Locate those in your output. We use these numbers to create the following equation for SEI, based on the standard regression model for the predicted value of Y:

$$\hat{Y} = a + bX_1 + bX_2 + bX_3$$

$$SEI = -3.169 + (EDUC \times 3.619) + (WHITE \times 3.242) + (MALE \times 1.209)$$

How would you predict the occupational prestige of a White male with 10 years of education? Here is a hint to get you started:

$$SEI = -3.169 + (10 \times 3.619) + (\underline{\hspace{1cm}} \times \underline{\hspace{1cm}}) + (\underline{\hspace{1cm}} \times \underline{\hspace{1cm}}) = \underline{\hspace{1cm}}$$

The column titled "Standardized Coefficients" gives you a guide to the *relative impact* of the different variables. Take a minute to consider some independent variable that has no impact on the dependent variable. What slope would it be given? If you think about it, the only proper weight would be zero. That would mean that a person's value on that variable would never make any difference in predicting the dependent variable. By the same token, the larger the slope for any given variable, the larger its part in determining the resulting prediction.

It is possible that a variable that is supposed to be a better predictor, such as EDUC, could have a smaller slope than an item such as MALE, which is not supposed to be as good a predictor as SEI. How can this happen? The solution to this puzzle lies in the different scales used in the different variables. MALE goes only as high as 1 ("male"), whereas EDUC obviously goes much higher to accommodate differences in the levels of education of respondents. Slopes must be standardized before they can be compared. Standardized slopes are what the slopes would be if each of the variables used the same scale.

SPSS Statistics prints standardized slopes under the column "Standardized Coefficients." The data presented above indicate that EDUC (.587) has the greatest impact on SEI, followed distantly by WHITE (.071) and MALE (.031). Interpreted, this means that education has the greatest impact on socioeconomic status, followed by "maleness" and then "majorityness."

SPSS Statistics Command 17.3: Multiple Regression

> **Analyze → Regression → Linear**
>
> → Highlight name of dependent variable → Click arrow pointing toward "Dependent:" field
>
> → Highlight name of first independent variable → Click arrow pointing to "Independent(s):" field
>
> → Repeat last step as many times as necessary until all independent variables are listed in the "Independent(s):" field
>
> → Click down arrow next to box labeled "Method"
>
> → Select **Enter**
>
> → **OK**

Conclusion

In this chapter, we have given you an initial peek into the logic and techniques of multivariate analysis. As you've seen, the difference between bivariate and multivariate analysis is much more than a matter of degree. Multivariate analysis does more than bring in additional variables: It represents a new logic for understanding social scientific relationships.

For this contact, we've looked at how multivariate analysis lets us explore the nature of multiple causation, seeing how two or more independent variables affect a dependent variable. In addition, we've used multivariate techniques for the purpose of testing the generalizability of relationships.

In the latter regard, we have begun using multivariate techniques for the purpose of considering hidden relationships among variables, as when we asked whether the widowed attended church frequently just because they were mostly older people. We'll pursue this kind of detective work further in the chapters to come.

Main Points

- This chapter introduced a new, more sophisticated form of statistical analysis: multivariate analysis.
- Multivariate analysis is the simultaneous analysis of three or more variables.
- The Crosstabs procedure can be used to analyze the relationship between an independent variable and a dependent variable while controlling for a third variable.
- We examined multiple causes of religiosity in more depth by focusing on items such as AGECAT, SEX, MARITAL, and SEI.
- Other analytic techniques may aid in the examination of several items at once, including chi-square, measures of association, and regression.
- We introduced simple linear regression (ordinary least-squares method) in Chapter 14.
- In this chapter, we introduced another regression procedure: multiple regression.
- Nominal and ordinal items can be recoded to create dummy variables that are suitable for regression analysis or other analyses where interval/ratio variables are a prerequisite.

Key Terms

Multivariate analysis	Multiple regression
Control variable	Ordinary least-squares (OLS) regression
Elaboration	Dummy variables
Linear regression	Replication

SPSS Statistics Commands Introduced in This Chapter

17.1   Running Crosstabs With a Control or Third Variable

17.2   Recoding to Create a Dummy Variable

17.3   Multiple Regression

Review Questions

1. Describe the major differences between univariate, bivariate, and multivariate analysis.

2. If a researcher uncovers a relationship between class (independent) and prejudice (dependent variable), what technique introduced in this chapter might he or she use to examine the relationship further?

3. Describe the relationship between social class and church attendance. Do the findings support the "social deprivation" theory? Why or why not?

4. To what does "replication" refer?

5. Name at least two other techniques you learned (before reading Chapter 16) that you could use in your multivariate examinations.

6. Simple linear regression involves the analysis of _____ [number] independent and _____ [number] dependent variables.

7. Multiple regression involves the analysis of _____ [number] independent and _____ [number] dependent variables.

8. Regression is appropriate for variables at what level(s) of measurement?

9. What is a dummy variable?

10. Why did we recode SEX to create MALE for our multiple regression example? Why didn't we just use SEX as our measure of gender?

11. What is a better predictor of SEI—EDUC or MALE?

12. When reading your multiple regression output, which of the following gives a sense of the relative impact of the independent variable: unstandardized coefficients, standardized coefficients, t, or sig.?

SPSS STATISTICS LAB EXERCISE 17.1

NAME _____

CLASS _____

INSTRUCTOR _____

DATE _____

To complete the following exercises, you need to load the data file AdventuresPLUS.SAV.

1. Examine the simultaneous impact of class and race on support for national spending on crime. Before proceeding, designate the values indicated below as "missing" for each item. Then run Crosstabs with column percentages and request chi-square. Use your output to complete the summary table and answer the questions below:

Dependent variable—NATCRIME (define 0, 8, 9 as "missing")

Independent variable 1—CLASS (define 5–9, 0 as "missing")

Independent variable 2—RACE (define 0, 3 as "missing")

NATCRIME: Percentage who feel that the national government is spending too little fighting crime

	LOWER	*WORKING*	*MIDDLE*	*UPPER*	*SIG. [chi-square]*
WHITE					
BLACK					

2. For WHITES: The percentage who feel that the national government is spending too little fighting crime_____ [increases, decreases, neither increases nor decreases] as class increases. The value of chi-square indicates that there _____ [is/is not] a statistically significant relationship.

3. For BLACKS: The percentage who feel that the national government is spending too little fighting crime_____ [increases, decreases, neither increases nor decreases] as class increases. The value of chi-square indicates that there _____ [is/is not] a statistically significant relationship.

4. Summarize the major findings from the table. Explain the primary observations that can be made regarding your findings and the table above. You may want to note, for instance, the relationship between race and support for government spending on crime, as well as the relationship between class and support for government spending on crime. Also note whether the two variables have a cumulative effect on support for national spending on crime. That is, are lower-class Blacks more likely to feel that the government is spending too little fighting crime as opposed to upper-class Whites?

5. Examine the simultaneous impact of class and gender on support for national spending on welfare. Before proceeding, designate the values indicated below as "missing" for each item. Then run Crosstabs and request chi-square. Use your output to complete the summary table and answer the questions below:

Dependent variable—NATFARE (define 0, 8, 9 as "missing")

Independent variable 1—CLASS

Independent variable 2—SEX

NATFARE: Percentage who feel that the national government is spending too little on welfare.

	LOWER	WORKING	MIDDLE	UPPER	SIG. [chi-square]
MALE					
FEMALE					

6. Summarize the major findings from the table. Explain the primary observations that can be made regarding the table. You may want to note, for instance, the relationship between gender and support for government spending on welfare, as well as the relationship between class and support for government spending on welfare. Also note whether the two variables have a cumulative effect on support for national spending on welfare. That is, are lower-class women more likely to feel that the government is spending too little on welfare as opposed to upper-class men? In addition, be sure to discuss whether the findings are statistically significant.

Examine the simultaneous impact of two independent variables of your choice on support for the national government's spending on education (NATEDUC). Run Crosstabs, request chi-square, and then answer the following questions and complete the summary table (use as many spaces as necessary). Don't forget to define DK as "missing" for NATEDUC and the other variables you choose. In addition, if you need to recode one or both of your independent variables, indicate how you did that in Question 7.

7. List the two independent variables you chose.

8. Justify your choice of the two independent variables and explain how you expect them to be related to NATEDUC.

9. NATEDUC: Percentage who feel that the government is spending too little on education

_____	_____	_____	_____	Chi-square
_____				
_____				
_____				
_____				

10. Summarize your findings in detail below. Be sure to explain the primary observations that can be made regarding the table and discuss whether the findings are statistically significant.

11. Choose one dependent and two independent variables and write the names of the variables you chose below.

Dependent variable _____

Independent variable 1 _____

Independent variable 2 _____

12. Write two hypotheses explaining the relationship between each independent variable and the dependent variable.

Hypothesis 1:

Hypothesis 2:

Examine the simultaneous impact of the two independent variables on the dependent variable. Run Crosstabs with chi-square and an appropriate measure of association. Then create a summary table below detailing the relationship (Question 14). Remember to define DKs as "missing" and, if necessary, recode your variables and indicate how you did that in the space below Question 15.

13. Measure of association _____

14. Summary table:

15. Summarize your findings in detail below. Be sure to explain the primary observations that can be made regarding the table. Discuss the strength and, if possible, the direction of association. Also indicate whether the findings are statistically significant and whether they support your hypotheses.

If necessary, recode RACE to create the dummy variable WHITE, similar to the way we did it in the chapter. Then designate the following values as "missing":

EDUC—97 to 99; AGE—98 to 99; TVHOURS—1, 98 to 99

Now open the Linear Regression window and designate TVHOURS as the dependent variable and EDUC, WHITE, and AGE as the independent variables.

16. What is the constant for the regression equation?

Constant _____

17. What are β (beta) values for the slopes of each variable (EDUC, WHITE, and AGE)?

18. Fill in the blanks to create a prediction equation for TVHOURS. TVHOURS = _____ (EDUC × _____) + (WHITE × _____) + (AGE × _____)

19. How many hours per day would you predict that a 25-year-old White high school graduate would watch TV?

20. How many hours per day would you predict that a 48-year-old non-White college graduate would watch TV?

21. Which of the three independent variables is most strongly related to TVHOURS? Explain the logic of your choice.

22. Of the three independent variables in the regression equation, which are (is) statistically significant? How do you know?

23. What PRE measure is given in the linear regression output? _____ What is its value in this case? _____ Interpret the meaning of that value for the relationship between these variables.

Chapter 18   Dissecting the Political Factor

In Chapter 11, we began exploring some of the causes of political philosophies and party identification. Now you are equipped to dig more deeply. Let's start with the relationship between political philosophy and party identification. As you'll recall, our earlier analysis showed a definite relationship, although it was not altogether consistent. Perhaps we can clarify it.

Political Philosophy and Party Identification

On the whole, Democrats in our sample were more liberal than Independents or Republicans. Also, Republicans were the most conservative, although there wasn't as large a distinction between Democrats and Independents as you might have expected. Here's the basic table from Chapter 11 examining the relationship between political philosophy (POLREC) and party identification (PARTY).[1]

party Recoded PartyID * polrec Recoded Political Views Crosstabulation

			polrec Recoded Political Views			
			1 Liberal	2 Moderate	3 Conservative	Total
party Recoded PartyID	1 Democrat	Count	321	248	104	673
		% within polrec Recoded Political Views	56.8%	33.5%	15.9%	34.3%
	2 Independent	Count	194	366	225	785
		% within polrec Recoded Political Views	34.3%	49.5%	34.3%	40.0%
	3 Republican	Count	38	110	308	456
		% within polrec Recoded Political Views	6.7%	14.9%	47.0%	23.3%
	4 Other	Count	12	16	19	47
		% within polrec Recoded Political Views	2.1%	2.2%	2.9%	2.4%
Total		Count	565	740	656	1961
		% within polrec Recoded Political Views	100.0%	100.0%	100.0%	100.0%

For purposes of this analysis, let's focus on the percentages of those who identified themselves as conservative. In the table above, the percentage difference separating the Democrats and Republicans in calling themselves conservative amounts to 31 points. You'll recall, perhaps, that percentage differences are sometimes designated by the Greek letter *epsilon (ε).*

---

[1]In Chapter 7, we recoded the variables POLVIEWS and PARTYID to create POLREC and PARTY.

Demonstration 18.1: Controlling for Education

If you were to undertake a study of the political party platforms and/or the speeches of political leaders from the two major parties, you would conclude that Democrats are, in fact, somewhat more liberal than Republicans and that Republicans are, in fact, somewhat more conservative than Democrats. If the relationship between political philosophy and party identification is not as clear as we might like, then perhaps some of the respondents simply don't know how the two parties are generally regarded.

Who do you suppose would be the least likely to know the philosophical leanings of the two parties? Perhaps those with the least education would be unaware of them. If that were the case, then we should expect a clearer relationship between political philosophy and party identification among the more educated respondents than among the less educated.

Why don't you open your AdventuresPLUS.SAV file and run the SPSS Statistics command that lets you create the following three-variable summary table.[2]

Percentage Saying They Are Conservative

	Less Than High School	High School Graduate	Some College	College Graduate	Graduate Studies
Democrat	27	21	13	9	—
Independent	31	30	24	31	30
Republican	46	55	71	80	82
Epsilon (ε)	19	34	58	71	52 (*78)

As in other tables, the dashes here indicate that there were too few cases for meaningful percentages.

The clearest relationship between party and political philosophy appears among those with the highest levels of education (college graduates and post-college education). The epsilon difference between Democrats and Republicans among college graduates is 71. While the epsilon is only 52 for those who have attained the graduate studies level, there were too few Democrats at that education level who claimed to be conservative. If we used the percentage that was in the cell (4), the epsilon would have been the greatest in our table: 78 (noted in the table with an asterisk).

Notice, however, that Democrats and Republicans are separated by an epsilon of 34 percentage points among those who graduated high school and an epsilon of 58 percentage points among those with some college education. Independents of every education level say they are more conservative than Democrats but much less so than Republicans.

When we look at the relationship between two variables, such as political philosophy and party identification, among subgroups determined by some other variable, such as education, we often say that we are *controlling for a third variable.*

Social scientists use the expression "controlling for" in the sense of creating controlled conditions: only college graduates, only those with some college, and so on. We also speak of "holding education constant" in the sense that education is no longer a variable (it is a *constant*) when we look at only one educational group at a time.

Why don't you experiment with this logic, testing the generalizability of the relationship between political philosophy and party identification among other subgroups formed by holding other variables constant.

---

[2]To construct this summary table, run Crosstabs with column percentages, specifying EDCAT as the column (independent) variable, POLREC as the row (dependent) variable, and PARTY as the control variable.

Demonstration 18.2: The Mystery of Politics and Marital Status

In Chapter 11, we encouraged you to explore the relationship between marital status and politics. If you took us up on the invitation, you should have found an interesting relationship between marital status and political philosophy.

Recoding MARITAL

Because relatively few respondents were "Separated," we should combine them with some other group. It would seem to make sense to combine the separated with the divorced, reasoning that separation is often experienced as an interim step toward divorce.

Let's recode MARITAL into a new variable, MARITAL2, with **Transform** → **Recode** → **Into Different Variables**.

Once in the Recode window, you should enter **MARITAL** as the "Input Variable" and **MARITAL2** as the "Output Variable." Click on **Change** to move MARITAL2 to the recode list. Then, in the Old and New Values window, recode as follows:

MARITAL		MARITAL2	
Old Values		New Values	Labels
1	→	1	Married
2	→	2	Widowed
3 to 4	→	3	Divorced/Separated
5	→	4	Never Married

Don't forget to use **Add** to record the instructions. Then, you can **Continue** and **OK** your way to the recoded variable.

Before we move on, make sure you set the decimal places for your new variable to 0 and define the values and labels for MARITAL2 to prevent confusion among the categories in the following analyses.

POLREC by MARITAL2

Now create a Crosstab with MARITAL2 as the column variable and POLREC as the row variable. Ask SPSS Statistics for column percentages and chi-square. Here's what you should get:

polrec Recoded Political Views * marital2 Recoded Marital Crosstabulation

			marital2 Recoded Marital				
			1 Married	2 Widowed	3 Divorced/Sep arated	4 Never Married	Total
polrec Recoded Political Views	1 Liberal	Count	210	49	112	196	567
		% within marital2 Recoded Marital	24.4%	28.3%	28.4%	36.0%	28.7%
	2 Moderate	Count	316	67	151	212	746
		% within marital2 Recoded Marital	36.7%	38.7%	38.3%	38.9%	37.8%
	3 Conservative	Count	335	57	131	137	660
		% within marital2 Recoded Marital	38.9%	32.9%	33.2%	25.1%	33.5%
Total		Count	861	173	394	545	1973
		% within marital2 Recoded Marital	100.0%	100.0%	100.0%	100.0%	100.0%

Chi-Square Tests

	Value	df	Asymp. Sig. (2-sided)
Pearson Chi-Square	35.048[a]	6	.000
Likelihood Ratio	35.303	6	.000
Linear-by-Linear Association	31.974	1	.000
N of Valid Cases	1973		

a. 0 cells (.0%) have expected count less than 5. The minimum expected count is 49.72.

The chi-square test of statistical significance shows that this relationship is statistically significant. So why are married respondents more conservative than others? Similarly, why is it that those who have never been married are most likely to be liberal? Your multivariate skills will allow you to explore this matter in more depth than was possible before.

POLREC by MARITAL2 by AGECAT

Perhaps age is the key. Perhaps those who have never been married tend to be more liberal because they also tend to be younger. Here is a summary table created from the results of the Crosstab of POLREC by MARITAL2 by AGECAT. See if you can duplicate this yourself.

Percentage Who Say They Are Conservative

	Married	Widowed	Divorced/ Separated	Never Married
Under 21	—	—	—	—
21 to 39	33	—	22	27
40 to 64	40	26	33	25
65 and older	47	35	38	—

While there are a number of cells with too few cases to include in the analysis, this table does help clarify matters somewhat. The married are somewhat more conservative than the widowed, divorced/separated, and never married. The difference seems to be greater at older ages.

POLREC by MARITAL2 by SEX

How about sex? Perhaps it can shed some light on this relationship. Why don't you run the tables that would result in this summary?

Percentage Who Say They Are Conservative

	Married	Widowed	Divorced/Separated	Never Married
Men	41	—	44	26
Women	37	35	26	25

As before, the married remain relatively conservative, and the divorced and never married remain slightly less conservative. However, notice that divorced/separated men are more conservative than those who are married (44% vs. 41%). Although the epsilon is small, this is the type of interaction for which we are searching.

POLREC by MARITAL2 by EDCAT

To pursue this further, you might want to consider education. Here's the summary table you should generate if you follow this avenue.

Percentage Who Say They Are Conservative

	Married	Widowed	Divorced/Separated	Never Married
Less than high school	35	41	32	23
High school graduate	39	22	39	26
Some college	35	29	34	27
College graduate	46	50	29	28
Graduate studies +	40	—	27	20

Here, we seem to have dug a dry well. Education does not seem to clarify the relationship we first observed between marital status and political philosophy. This is the point in an analysis where you sometimes wonder if you should ever have considered this line of inquiry.

POLREC by MARITAL2 by RACE

See what happens when we introduce race as a control. It provides no clarification for our analysis.

	Married	Widowed	Divorced/Separated	Never Married
White	41	35	34	26
Black	29	—	28	27
Other	23	—	41	16

POLREC as Independent Variable

When we consistently fail to find a clear answer to a question—Why do people of different marital statuses differ in their political philosophies?—it is sometimes useful to reconsider the question itself. Thus far, we have been asking why marital status would affect political philosophy. Perhaps we have the question reversed.

What if political philosophy affects marital status? Is that a possibility? Perhaps those who are politically conservative are also socially conservative. Maybe it would be especially important for them to form and keep traditional families. During the past few presidential elections, the political conservatives made "traditional family values" a centerpiece of their campaigns.

Let's see what the table would look like if we percentaged it in the opposite direction. (Hint: Place POLREC as the column variable and MARITAL2 as the row variable this time.)

marital2 Recoded Marital * polrec Recoded Political Views Crosstabulation

| | | | \multicolumn{3}{c}{polrec Recoded Political Views} | |
			1 Liberal	2 Moderate	3 Conservative	Total
marital2 Recoded Marital	1 Married	Count	210	316	335	861
		% within polrec Recoded Political Views	37.0%	42.4%	50.8%	43.6%
	2 Widowed	Count	49	67	57	173
		% within polrec Recoded Political Views	8.6%	9.0%	8.6%	8.8%
	3 Divorced/Separated	Count	112	151	131	394
		% within polrec Recoded Political Views	19.8%	20.2%	19.8%	20.0%
	4 Never Married	Count	196	212	137	545
		% within polrec Recoded Political Views	34.6%	28.4%	20.8%	27.6%
Total		Count	567	746	660	1973
		% within polrec Recoded Political Views	100.0%	100.0%	100.0%	100.0%

Chi-Square Tests

	Value	df	Asymp. Sig. (2-sided)
Pearson Chi-Square	35.048[a]	6	.000
Likelihood Ratio	35.303	6	.000
Linear-by-Linear Association	31.974	1	.000
N of Valid Cases	1973		

a. 0 cells (.0%) have expected count less than 5. The minimum expected count is 49.72.

Notice that, again, the chi-square value is statistically significant and, in fact, identical to when we ran this analysis with the variables in the reverse order.

Look at the first row in the crosstab table. The percentage married increases steadily with increasing conservatism across the table. Singlehood, on the other hand, decreases just as steadily. Perhaps marital status is more profitably seen as a dependent variable in this context—affected to some extent by worldviews such as those reflected in political philosophy.

Sometimes, the direction of a relationship—which is the dependent and which is the independent variable—is clear. If we discover that voting behavior is related to gender, for instance, we can be sure that gender can affect voting, but how you vote can't change your gender. In other situations, such as the present one, the direction of a relationship is somewhat ambiguous. Ultimately, this decision must be based on theoretical reasoning. There is no way that the analysis of data can determine which variable is dependent and which is independent.

If you wanted to pursue the present relationship, you might treat marital status as a dependent variable, subjecting its relationship with political philosophies to a multivariate analysis. Try, for instance, rerunning the crosstab, using AGECAT as a control variable. After you examine your output, compare your findings to those in Writing Box 18.1 below.

Writing Box 18.1

Age has confirmed the original relationship between political orientations and marital status. Among those in each of the age categories, conservatives are consistently more likely to be married than liberals. It is also worth noting that the youngest respondents (those 21 and under) are most likely never to have been married.

Now, why don't you continue your analysis using SEX, EDCAT, and RACE as control variables. When you are done, create your own summary as we did in the writing box above.

Political Issues

In Chapter 11, we began looking for the causes of opinions on two political issues: GUNLAW (registration of firearms) and CAPPUN (capital punishment). Now that you have the ability to undertake multivariate analysis, you can delve more deeply into the causes of public opinion. Let's think a little about capital punishment for a moment. Here are some variables that might logically affect how people feel about the death penalty.

POLREC and PARTY are obvious candidates. Liberals are generally more opposed to capital punishment than are conservatives. Similarly, Republicans have tended to support it more than have Democrats. You might check to see how these two variables work together on death penalty attitudes.

Given that capital punishment involves the taking of a human life, you might expect some religious effects. How do the various religious affiliations relate to support for or opposition to capital punishment? What about belief in an afterlife? Do those who believe in life after death find it easier to support the taking of a life? How do religious and political factors interact in this arena? Those opposed to capital punishment base their opposition on the view that it is wrong to take a human life. The same argument is made by those who oppose abortion. Logically, you would expect those opposed to abortion also to oppose capital punishment. Why don't you check it out. You may be surprised by what you find.

Another approach to understanding opinions about capital punishment might focus on which groups in society are most likely to be victims of it. Men are more likely to be executed than are women. Blacks are executed disproportionately often in comparison with their numbers in the population.

Conclusion

These few suggestions should launch you on an extended exploration of the nature of political orientations. Whereas people often speak casually about political matters, you are now in a position to check out the facts and begin to understand why people feel as they do about political issues. Multivariate analysis techniques let you uncover some of the complexities that can make human behavior difficult to understand.

Main Points

- You can use your multivariate analysis skills to delve into the nature of political orientation.
- We began by exploring the relationship between political philosophy and party identification while controlling for education.
- You can test the generalizability of the relationship between two items by controlling for a third variable or holding a third variable constant.

- The direction of the relationship between variables (which is the dependent and which is the independent) is not always clear.
- In exploring the relationship between political philosophy and marital status, we took turns examining both as the independent variable.
- We discovered that marital status is perhaps better seen as the dependent variable in this context.
- You can use your multivariate analysis skills to discover why people hold the opinions they do on volatile political issues such as gun control and capital punishment.

Key Terms

Epsilon (ε)   Controlling for a third variable
Constant

SPSS Statistics Commands Introduced in This Chapter

No new commands were introduced in this chapter.

Review Questions

1. What does it mean to "control" for a third variable?

2. Is this the same as or different from "holding a variable constant"?

3. If we wanted to instruct SPSS Statistics to create crosstabs examining the relationship between POLREC and PARTY while controlling for SEX, where in the Crosstabs window would we transfer SEX (i.e., to which field: "Row(s)," "Column(s)," or the third box near the bottom of the window)?

4. In examining the relationship between political philosophy and party identification, what other variables (besides sex and education) might you want to control for? Name at least two.

5. When conducting multivariate analysis, how do we ultimately know which variable is the dependent one and which is the independent one?

6. If we were performing multivariate analysis and found a relationship between RACE and voting behavior, could RACE be the dependent variable? Why or why not?

7. Name three variables from your data set that you think may logically affect or have an impact on how people feel about gun control (GUNLAW), and explain why.

SPSS STATISTICS LAB EXERCISE 18.1

NAME _____

CLASS _____

INSTRUCTOR _____

DATE _____

To complete the following exercises, you need to load the data file AdventuresPLUS.SAV.

Designate 0 and 6 to 9 as "missing" for RACWORK. Then recode the variable to create RACWORK2:

RACWORK		*RACWORK2*	
Old Values		*New Values*	*Labels*
1 to 2	→	1	Mostly White
3	→	2	Half White, Half Black
4 to 5	→	3	Mostly Black

Now use crosstabs to examine the relationship between RACWORK2 (recoded RACWORK) and AFFREC (recoded AFFRMACT).[1] Designate RACWORK2 as the column variable and AFFREC as the row variable, and request column percentages and chi-square. Then use your output to answer the questions below.

1. The table shows that those who work mostly with Whites are _____ [more likely/less likely/not any more or less likely] than those who work mostly with Blacks to support affirmative action.

2. The significance of chi-square is _____ [less than/more than] .05, so the relationship between AFFREC and RACWORK2 _____ [is/is not] statistically significant.

3. List three variables from the data file AdventuresPLUS.SAV that you want to use to examine why those who work with mostly Whites are less supportive of affirmative action than are those who work with mostly Blacks.

   Variable 1 _____

   Variable 2 _____

   Variable 3 _____

4. Justify your choice of Variable 1 above (i.e., give theoretical reasons for choosing this variable).

5. Justify your choice of Variable 2 above (i.e., give theoretical reasons for choosing this variable).

6. Justify your choice of Variable 3 above (i.e., give theoretical reasons for choosing this variable).

7. Examine the relationship between AFFREC and RACWORK2 while controlling for Variable 1 above. Run Crosstabs with chi-square. If you had to recode Variable 1, make sure you explain how you did that. Then create a summary table based on the results of your crosstab and show it below (if you need help creating a summary table, see the tables in the chapter for guidance).

8. Summarize below the major findings from your table. Explain what, in your view, are the primary observations that can be made regarding the table. Be sure to note, for instance, whether the findings supported your expectations. Use chi-square as a criterion for judging whether the differences in your table are significant (if your findings are statistically significant, you may want to go back and measure the strength of the relationship and then discuss how you did that and what you found).

9. Examine the relationship between AFFREC and RACWORK2 while controlling for Variable 2 above. Run Crosstabs with chi-square. If you had to recode Variable 2, make sure you explain how you did that. Then create a summary table based on the results of your crosstab and show it below.

10. Summarize the major findings from your table. Explain what, in your view, are the primary observations that can be made regarding the table. Be sure to note, for instance, whether the findings supported your expectations. Use chi-square as a criterion for judging whether the differences in your table are significant (if your findings are statistically significant, you may want to go back and measure the strength of the relationship and then discuss how you did that and what you found).

11. Examine the relationship between AFFREC and RACWORK2 while controlling for Variable 3 above. Run Crosstabs with chi-square. If you had to recode Variable 3, make sure you explain how you did that. Then create a summary table based on the results of your crosstab and show it below.

12. Summarize the major findings from your table below. Explain what, in your view, are the primary observations that can be made regarding the table. Be sure to note, for instance, whether the findings supported your expectations. Use chi-square as a criterion for judging whether the differences in your table are significant (if your findings are statistically significant, you may want to go back and measure the strength of the relationship and then discuss how you did that and what you found).

Notes

1. We recoded AFFRMACT to create AFFREC in SPSS Statistics Lab Exercise 15.1. If you did not save AFFREC to your AdventuresPLUS.SAV file, refer back to the recode instructions before proceeding.

Chapter 19 A Powerful Prediction of Attitudes Toward Abortion

In previous analyses, we've seen how complex attitudes about abortion are. As we return to our analysis of this controversial topic, you have additional tools for digging deeper. Let's begin with the religious factor. Then we'll turn to politics and other variables.

Religion and Abortion

In Chapter 12, we found that both religious affiliation and measures of religiosity were related to abortion attitudes. The clearest relationships were observed in terms of the unconditional right to abortion, because only a small minority are opposed to abortion in all circumstances.

Protestants and Catholics are generally less supportive of abortion than are Jews and "Nones." And on measures of religiosity, opposition to abortion increases with increasing religiosity. The most religious are the most opposed to a woman's right to choose an abortion.

With your multivariate skills, you can examine this issue more deeply. Consider the possibility, for example, that one of these relationships is an artifact of the other.

Demonstration 19.1: Religious Affiliation and Church Attendance

To explore the possibility that one of the relationships we found in Chapter 12 is merely an artifact of another association, you need to open your AdventuresPLUS.SAV file and examine the relationship between religious affiliation (RELIG) and church attendance (CHATT).[1]

---

[1]For RELIG, the values 0 and 5 to 99 should be defined as "missing."

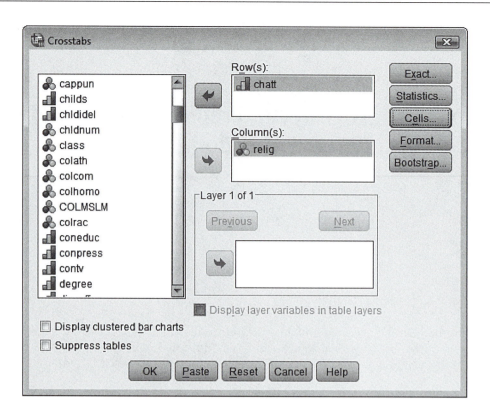

Remember to request cells be percentaged by column.

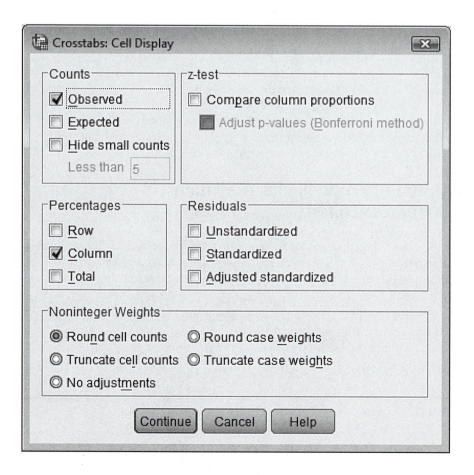

Here is what SPSS Statistics will produce for you.

chatt Recoded Church Attendance * relig RS RELIGIOUS PREFERENCE Crosstabulation

			relig RS RELIGIOUS PREFERENCE				Total
			1 PROTESTANT	2 CATHOLIC	3 JEWISH	4 NONE	
chatt Recoded Church Attendance	1 About weekly	Count	402	142	3	7	554
		% within relig RS RELIGIOUS PREFERENCE	41.4%	29.5%	8.1%	1.9%	29.9%
	2 About monthly	Count	176	97	10	11	294
		% within relig RS RELIGIOUS PREFERENCE	18.1%	20.1%	27.0%	3.0%	15.9%
	3 Seldom	Count	268	178	22	111	579
		% within relig RS RELIGIOUS PREFERENCE	27.6%	36.9%	59.5%	30.6%	31.2%
	4 Never	Count	126	65	2	234	427
		% within relig RS RELIGIOUS PREFERENCE	13.0%	13.5%	5.4%	64.5%	23.0%
Total		Count	972	482	37	363	1854
		% within relig RS RELIGIOUS PREFERENCE	100.0%	100.0%	100.0%	100.0%	100.0%

As you can see, there is a relationship between these two variables. Protestants and Catholics are the most likely to attend worship services weekly. Nearly 95% of those with no religious affiliation attend religious services seldom or never.

If we combine the two most frequent categories, we see that more than half of the whole sample attend church at least monthly. There are big differences, however, among the four religious groups.

Percentage Who Attend at Least Monthly	
Protestants	60
Catholics	50
Jews	35
None	5

Demonstration 19.2: Religious Affiliation, Church Attendance, and Abortion

Because religious affiliation and church attendance are related to each other and each is related to abortion attitudes, there are two possibilities for us to explore. For example, perhaps church attendance seems to affect abortion attitudes only because Protestants and Catholics (relatively opposed to abortion) attend more often. Or, conversely, perhaps Protestants and Catholics seem more opposed to abortion simply because they attend church more often.

We can test for these possibilities by running a multivariate table, taking account of all three variables.

Recoding RELIG and ATTEND Into Different Variables

To simplify our analysis, let's recode RELIG into two categories—"Christians" and "None," those respondents expressing no religious preference. While it would be interesting to compare Christians to those with other religious preferences, Christians (Protestants plus Catholics) and those responding "None" have sufficient numbers of cases for meaningful analysis. We will then recode ATTEND into two categories as well. We will recode the variables into the creatively named new variables RELIG2 and ATTEND2. So let's use **Transform → Recode → Into Different Variables**.

Make the recodes listed below, beginning with **RELIG** and then moving to **ATTEND**. Once you have recoded the items, access the **Variable View** tab and label your recoded variables. If you need to verify the process for recoding into different variables, consult SPSS Statistics Command 19.1.

Recode RELIG Old Values		New Values	Labels
1 to 2[2]	→	1	Christian
3	→	System-missing	—
4	→	2	None
Recode ATTEND Old Values		New Values	Labels
4 to 8	→	1	Often
0 to 3	→	2	Seldom

SPSS Statistics Command 19.1:
Recoding Into Different Variables

Transform → Recode → Into Different Variables

→ Highlight variable name → Click right-pointing arrow to move variable to "Numeric Variables:" field

→ Click **Old and New Values**

→ Define Old and New Values in the Recode Into Different Variables: Old and New box

→ Click **Add** to change Old → New Values

→ Click **Continue**

→ **OK**

Click **Variable View** tab → Click right side of rectangle that corresponds with recoded variable name and labels

→ Add new labels

---

[2]Previously, we declared all cases except those coded Protestant, Catholic, Jew, and None as missing. Now we are going to collapse Protestants and Catholics into a category named "Christian." Jewish respondents are being declared missing because they are too few for meaningful analysis. The code for "None" (4) is being changed to 2 simply to keep the code categories contiguous.

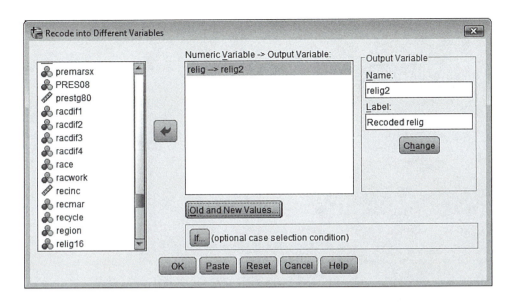

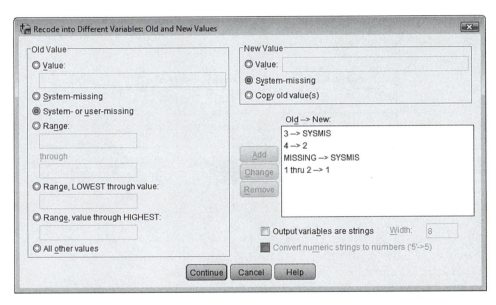

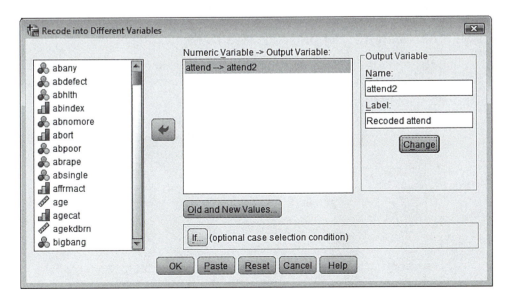

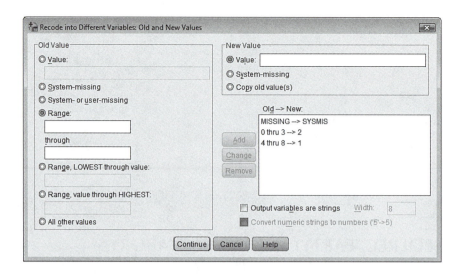

Crosstab Recoded Variables

Now run a Crosstab with column percentages. Designate ATTEND as the row variable and RELIG as the column variable.

attend2 Recoded attend * relig2 Recoded relig Crosstabulation

| | | | relig2 Recoded relig | | |
			1 Christian	2 None	Total
attend2 Recoded attend	1 Often	Count	817	18	835
		% within relig2 Recoded relig	56.2%	5.0%	46.0%
	2 Seldom	Count	637	345	982
		% within relig2 Recoded relig	43.8%	95.0%	54.0%
Total		Count	1454	363	1817
		% within relig2 Recoded relig	100.0%	100.0%	100.0%

As you can see, the relationship between religious affiliation and church attendance is still obvious after categories are collapsed on both variables.

Now let's review the relationships between each variable and abortion, again using the recoded variables.

Relationship Between ABORT and Recoded Items

Notice that the relationship between affiliation and abortion is now represented by an epsilon of 29 percentage points. The relationship between church attendance and abortion has an epsilon of 31 percentage points.

abort Simple Abortion Index * relig2 Recoded relig Crosstabulation

| | | | relig2 Recoded relig | | |
			1 Christian	2 None	Total
abort Simple Abortion Index	0 Yes/Approve	Count	296	136	432
		% within relig2 Recoded relig	35.3%	63.6%	41.0%
	1 Conditional Support	Count	297	54	351
		% within relig2 Recoded relig	35.4%	25.2%	33.3%
	2 No/Disapprove	Count	246	24	270
		% within relig2 Recoded relig	29.3%	11.2%	25.6%
Total		Count	839	214	1053
		% within relig2 Recoded relig	100.0%	100.0%	100.0%

abort Simple Abortion Index * attend2 Recoded attend Crosstabulation

| | | | attend2 Recoded attend | | Total |
			1 Often	2 Seldom	
abort Simple Abortion Index	0 Yes/Approve	Count	137	372	509
		% within attend2 Recoded attend	25.9%	56.5%	42.9%
	1 Conditional Support	Count	179	205	384
		% within attend2 Recoded attend	33.8%	31.2%	32.4%
	2 No/Disapprove	Count	213	81	294
		% within attend2 Recoded attend	40.3%	12.3%	24.8%
Total		Count	529	658	1187
		% within attend2 Recoded attend	100.0%	100.0%	100.0%

Politics (POLREC, PARTY) and Abortion (ABORT)

As we saw in Chapter 12, political philosophies and party identification have a strong impact on attitudes toward abortion. You might want to refresh your memory by rerunning those tables. (You now know how. Wow!)

abort Simple Abortion Index * polrec Recoded Political Views Crosstabulation

| | | | polrec Recoded Political Views | | | Total |
			1 Liberal	2 Moderate	3 Conservative	
abort Simple Abortion Index	0 Yes/Approve	Count	204	183	114	501
		% within polrec Recoded Political Views	60.9%	41.8%	29.9%	43.4%
	1 Conditional Support	Count	83	165	122	370
		% within polrec Recoded Political Views	24.8%	37.7%	32.0%	32.1%
	2 No/Disapprove	Count	48	90	145	283
		% within polrec Recoded Political Views	14.3%	20.5%	38.1%	24.5%
Total		Count	335	438	381	1154
		% within polrec Recoded Political Views	100.0%	100.0%	100.0%	100.0%

The impact of political philosophy on unconditional support for a woman's right to choose abortion results in an epsilon of 31 percentage points. As we saw earlier, however, political party identification—despite official party differences on the issue of abortion—does not have much of an effect.

Demonstration 19.3: The Interaction of Religion and Politics on Abortion Attitudes

In a multivariate analysis, we might next want to explore the possible interaction of religion and politics on abortion attitudes. For example, in Chapter 12, we found that Protestants and Catholics were somewhat more conservative than "Nones." Perhaps their political orientations account for the differences that the religious groups have on the issue of abortion.

With your multivariate skills, testing this new possibility is a simple matter. Take a minute to figure out the SPSS Statistics command that would provide for such a test. Then enter it and review the results. (Hint: Run Crosstabs with column percentages, POLREC as the column variable, ABORT as the row variable, and RELIG as the control/third variable.) Here's a summary table of the results you should find if you are working with the latest recode for RELIG. Be sure you can replicate this on your own.

Percentage Who Support Right to Abortion (Simple Abortion Index)

	Liberal	*Moderate*	*Conservative*
Christian	53	36	25
None	73	59	51
Epsilon (ε)	20	23	26

This table demonstrates the independent impact of religion and politics on abortion attitudes. In addition, what would you say it tells us about the religious effect observed earlier? Does it occur among all the political groups? What, if anything, do the epsilons for each political group tell us about the religious effect?

Overall, you can see that the joint impact of politics and religion is represented by an epsilon of 48 percentage points (73 – 25), a fairly powerful degree of prediction for these controversial opinions.

Demonstration 19.4: Constructing an Index of Ideological Traditionalism

To support our continued analysis, let's create an index to combine the religious and political factors. For the time being, we'll call it our "index of ideological traditionalism." You may recall from our earlier discussion that there are several different ways to construct indices. In Chapter 8, we introduced you to a method using the Count command. In this chapter, we are going to introduce you to another method using the *Compute Variable command*.

As you are now aware, a number of steps are involved in creating an index. However, all the work is worth it because in the end, we will have a composite measure that captures religious and political predispositions to support abortion. In other words, we will have built an index based on our two religious items (ATTEND and RELIG) and one political item (POLREC) that allows us to predict attitudes toward abortion.

In this case, our index will range from 0 (traditional) to 4 (nontraditional). Before we begin constructing our index, it may be useful to give you a brief overview of the seven steps involved in creating our new index of ideological traditionalism, which we will call IND.

Step 1:	Create new index, IND.
Steps 2 to 5:	Assign points on the index based on responses to the three component variables: POLREC, RELIG, and ATTEND. As noted, our index scores will run from 0 ("Opposed to abortion") to 4 ("Supportive of abortion"). We will assign points as follows:
Step 2:	If Liberal (1) on POLREC, get two points on IND.
Step 3:	If Moderate (2) on POLREC, get one point on IND.
Step 4:	If None (2) on RELIG, get one point on IND.
Step 5:	If attend religious services seldom (2) on ATTEND, get one point on IND.
Step 6:	Use COMPUTE to handle missing data.
Step 7:	Access **Variable View** tab and define IND (e.g., decimals, labels, etc.).

Now that you have reviewed the steps involved in creating our new index of ideological traditionalism, IND, let's go through each of the steps one at a time.

Step 1: Create IND

To begin, select **Transform → Compute Variable**.

In the Compute Variable box, create a new "Target Variable:" called **IND** (for index). We'll start by giving everyone a 0, so type 0 or use the calculator pad to move 0 into the "Numeric Expression:" box. Your expression should now read "IND = 0," as shown below.

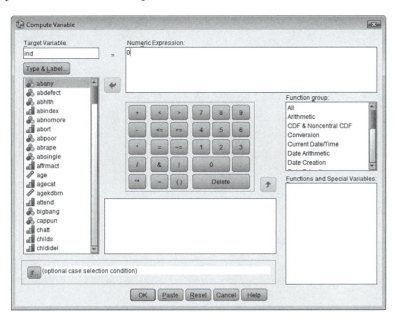

Run this instruction by selecting **OK**.

Step 2: Assign Points—If Liberal (1) on POLREC, Get Two Points on IND

Now we are ready to begin assigning points on our index. Open the Compute Variable box once again by selecting **Transform → Compute Variable**.

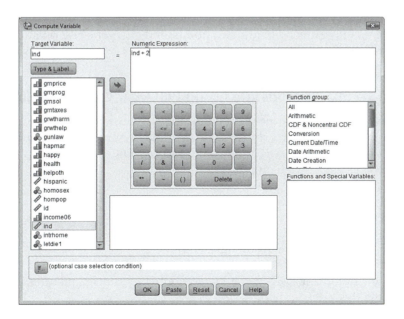

Let's begin by giving people two points if they have a 1 on POLREC (Liberal). To do this, change the "Numeric Expression" to **IND + 2**.

Now click **If** and, in the Compute Variable: If Cases box, select the "Include if case satisfies condition:" option. Then, type or click **POLREC = 1** into the rectangle box, as shown below.

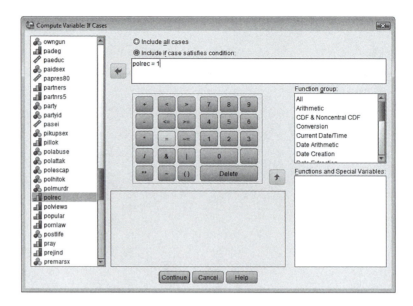

Run this instruction by selecting **Continue**, **OK**, and then **OK** again to change the existing variable IND. SPSS Statistics may verify if you wish to change the existing IND variable, as shown below. Click **OK**.

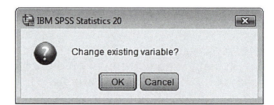

At this point, we have given liberals two points, and everyone else (including those with missing data) has zero points.

Step 3: Assign Points—If Moderate (2) on POLREC, Get One Point on IND

Now let's give the moderates one point on the index. Return to the Compute Variable window and make the following two changes: Change the "Numeric Expression" to **IND + 1**. Change the "If" statement to **POLREC = 2**. Select **Continue**, **OK**, and **OK** to run this instruction.

Thus far, we have established an index as follows:

- Liberals have two points.
- Moderates have one point.
- Conservatives have zero points.
- Those who are none of the above have zero points on the index.

You may want to look at IND in the Data window or run Frequencies anytime in this process to see how the index is shaping up.

Step 4: Assign Points—If "None" (2) on RELIG, Get One Point on IND

Now we are ready to add to the index. Go back to the Compute Variable window and make the following specifications:

■ Leave the "Numeric Expression" as **IND + 1**.
■ Change the "If" statement to **RELIG = 2**.

Run this instruction by clicking **Continue** → **OK** → **OK**. Now we've added one point for each respondent who indicated no religious preference.

Step 5: Assign Points—If Seldom (2) on ATTEND, Get One Point on IND

Finally, return to the Compute Variable window and make these specifications:

■ Leave the "Numeric Expression" as **IND + 1**.
■ Change the "If" statement to **ATTEND = 2**.

Run this expression and review the logic of what we have done.

We've now created an index that presumably captures religious and political predispositions to support a woman's right to an abortion. Scores on the index run from 0 ("Opposed to abortion") to 4 ("Supportive of abortion").

Step 6: Missing Data

There is one glitch in this index that we need to correct before moving on: those respondents for whom we have missing data. We don't know about the religious behavior or political orientations of those respondents for whom data were missing for ATTEND, POLREC, or RELIG. Since we can't provide an index score for respondents we don't know about, we must eliminate them from our analysis.

We can do this easily by using the NMISS function in the Compute procedure to set any cases to $SYSMIS (system-missing) if they have missing information.

To do this, select **Transform** → **Compute Variable**, and then follow the instructions below:

■ With IND in the "Target Variable" box, type **$SYSMIS** in the "Numeric Expression:" field, as shown below.

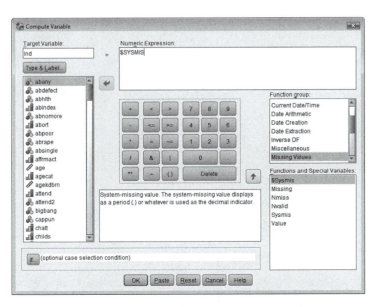

- Now click **If**.
- Click the button next to "Include if case satisfies condition."
- Type the following expression in the field directly below NMISS: (**relig2**, **attend2**, **polrec**) > 0.
- Click **Continue** → **OK** → **OK**.

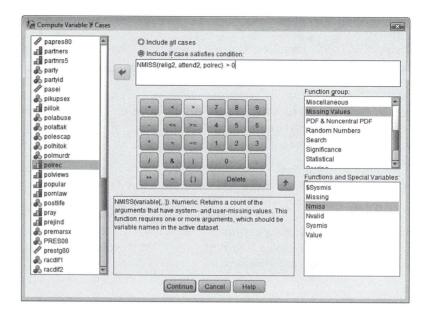

Step 7: Define IND

All that remains (thankfully!) is to define our new index. You can do that by accessing the **Variable View** tab and setting the decimals, width, and values (remember, for this index, 0 = traditional and 4 = nontraditional).

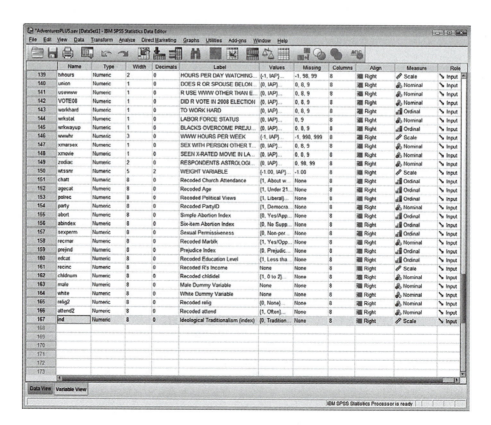

SPSS Statistics Command 19.2:
Creating an Index Using Compute Variable

Step 1: Create New Index

Transform → **Compute Variable** → Type name of new index in "Target Variable:" field → Type/paste 0 in "Numeric Expression:" field → **OK**

Step 2: Add Points to Index

Transform → **Compute Variable** → Keep new index name in "Target Variable:" field → Begin assigning points by adding appropriate points to new variable (e.g., IND = #) → **If** → **Include if case satisfies condition** → Type/paste condition under which points should be added to new index (e.g., if respondent was liberal: POLREC = 1) → **Continue** → **OK** → **OK**

Repeat this step as many times as necessary until you have completed adding points to index.

Step 3: Handle Missing Data

Transform → **Compute Variable** → Keep new index name in "Target Variable:" field

Type **$SYSMIS** in "Numeric Expression:" field → **If** → **Include if case satisfies condition** → Type "**NMISS** (variable names that compose index) > 0" → **Continue** → **OK** → **OK**

Step 4: Define New Index

Click **Variable View** tab → Set decimals, width, labels, etc.

Run Frequencies to Check IND

Having created such an index, it is always good practice to check the frequencies. Run Frequencies on IND, and you should get this:

ind Ideological Traditionalism (index)

		Frequency	Percent	Valid Percent	Cumulative Percent
Valid	0 Traditional	338	16.5	19.2	19.2
	1	474	23.2	27.0	46.2
	2	510	25.0	29.0	75.2
	3	280	13.7	15.9	91.2
	4 Nontraditional	155	7.6	8.8	100.0
	Total	1757	86.0	100.0	
Missing	System	287	14.0		
Total		2044	100.0		

Does IND Predict Attitudes Toward Abortion?

After all that work, let's finally see how well our index predicts abortion attitudes. To do this, run the following table.

abort Simple Abortion Index * ind Ideological Traditionalism (index) Crosstabulation

| | | | ind Ideological Traditionalism (index) | | | | | |
			0 Traditional	1	2	3	4 Nontraditional	Total
abort Simple Abortion Index	0 Yes/Approve	Count	27	91	136	102	67	423
		% within ind Ideological Traditionalism (index)	13.4%	34.2%	46.4%	60.0%	74.4%	41.5%
	1 Conditional Support	Count	58	111	107	44	18	338
		% within ind Ideological Traditionalism (index)	28.9%	41.7%	36.5%	25.9%	20.0%	33.1%
	2 No/Disapprove	Count	116	64	50	24	5	259
		% within ind Ideological Traditionalism (index)	57.7%	24.1%	17.1%	14.1%	5.6%	25.4%
Total		Count	201	266	293	170	90	1020
		% within ind Ideological Traditionalism (index)	100.0%	100.0%	100.0%	100.0%	100.0%	100.0%

As you can see, the index provides a very strong prediction of support for the unconditional right to an abortion: from 13% to 74% for an epsilon of 61 percentage points. Let's see if we can improve on our ability to predict.

Sexual Attitudes and Abortion

Earlier, we discovered that attitudes about various forms of sexual behavior were also related to abortion attitudes. As you'll recall, people were asked whether they felt that premarital sex and homosexuality were "always wrong," "almost always wrong," "sometimes wrong," or "not wrong at all." In addition, respondents were asked whether they had attended an X-rated movie during the past year.

Each of these items was related to abortion attitudes, with those most permissive in sexual matters also being more permissive about abortion.

Demonstration 19.5: Recode PREMARSX and HOMOSEX

Because we want to pursue this phenomenon, why don't you recode HOMOSEX and PREMARSX into *dichotomies* (only two values) of "Always or almost always wrong" versus "Only sometimes or never wrong." That will make it easier to conduct the following analysis.

Just as when we recoded ATTEND and RELIG, we want to recode HOMOSEX and PREMARSX "into different variables":[3] HOMOSEX2 and PREMARSX2. To review the steps involved in recoding into the same variable, see SPSS Statistics Command 19.1. In addition, the following hints may be useful to get you started.

Keep in mind that because both HOMOSEX and PREMARSX have the same values and labels, you can create the dichotomies for these two sexual attitude variables with a single command, then one other command to create new value labels.

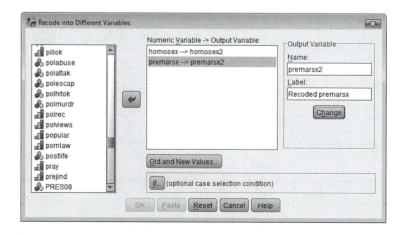

---

[3]For HOMOSEX and PREMARSX, the values 0 and 5 to 9 should be defined as "missing."

Transform → Recode → Into Different Variables.

Old Values for Both HOMOSEX and PREMARSX		New Values	Labels
1 to 2	→	1	Wrong
3 to 4	→	2	Okay

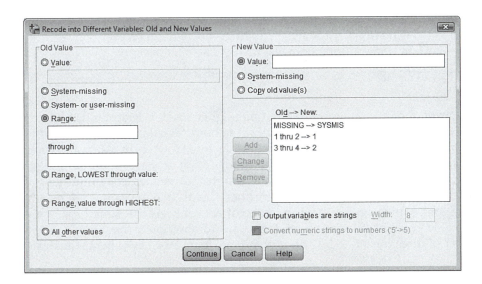

Demonstration 19.6: The Relationship Between Sexual Permissiveness and IND

Now let's see whether the sexual behavior attitudes are related to our "political-religious" index that predicts abortion attitudes so powerfully. Why don't you run each of those tables now. Here's a summary of what you should find.

Index of Ideological Traditionalism: Percentage Who Think PREMARSX or HOMOSEX Is Okay

Percentage Who Are Permissive About . . .	0 Traditional	1	2	3	4 Nontraditional
Premarital sex	35	65	79	93	96
Homosexuality	18	41	52	76	87

Even though there is a great difference in the overall level of permissiveness on these issues, we can see that the political–religious index is clearly related to each one. People are generally more permissive of premarital sex than homosexuality at all levels of ideological traditionalism. The gap is largest among those who are classified as traditional (or close to traditional) and smallest among those identified as nontraditional (4 on the scale).

Why don't you also go ahead and check to see if the index predicts whether people have taken in an X-rated movie during the past year (XMOVIE), whether people have had casual sex on a date (PIKUPSEX), or whether married individuals have ever cheated on their spouses (EVSTRAY). Once you have completed your analysis, compare your findings to those in Writing Box 19.1.

Writing Box 19.1

We see by inspection that the index of ideological traditionalism strongly predicts whether people have attended an X-rated movie: progressing steadily from 13% of those scored 0 (traditional) on the index to 37% among those scored 4 (nontraditional). The nontraditional respondents are far more likely to have seen an X-rated movie. Almost a third of traditional respondents had casual sex on a date (is that traditional behavior?), while just less than half of nontraditional respondents engaged in casual sex on a date; clearly, there seems to be a difference here, as well. There does not, however, seem to be a noteworthy difference among traditional respondents, nontraditional respondents, and those in between on the question of which married respondents had ever strayed from their spouses. The percentages range from 8% up to 15%, then back to 14%.

If you want to continue your analysis, you might consider determining whether the previously observed relationship between attitudes about sexual behavior and attitudes about abortion are a product of religious and political factors. You might also consider expanding the index to include other items on sexual attitudes.

The suggestions above are just a few of the many avenues you could take if you are interested in pursuing this line of inquiry further. Clearly, we have gone a long way toward accounting for people's opinions on the issue of abortion. At the same time, we've moved quickly to give you a broad view. If attitudes toward abortion interest you, you could go in any number of directions to follow up with a more focused and deliberate analysis.

Conclusion

You've had an opportunity now to see how social scientists might set out to understand people's attitudes toward abortion. We know that this is a topic about which you hear a great deal in the popular media, and it may be an issue that concerns you and about which you have strong opinions. The analyses above should give you some insight into the sources of opinions on this topic.

We hope this chapter has also expanded your understanding of the possibilities for multivariate analysis. Whereas bivariate analysis allows for some simple explanations of human thoughts and behaviors, multivariate analysis permits more sophisticated investigations and discoveries.

Main Points

- You can use your multivariate analysis skills to delve more deeply into the sources of opinion on controversial issues such as abortion.
- The primary goal in this chapter is to try to predict attitudes toward abortion.
- To that end, we went through a number of items that earlier were found to have an impact on abortion attitudes: religious affiliation, measures of religiosity, political philosophy, and sexual attitudes.
- We constructed a composite measure or index, called IND, to predict support for a woman's right to have an abortion.
- IND was made up of the following component variables: ATTEND, POLREC, and RELIG.
- After working through the demonstrations in this chapter, you should have a better understanding of how a social scientist might try to begin to account for opinion on an issue such as abortion.

■ With the univariate, bivariate, and multivariate analysis skills you have learned, you should now be able to design a more careful and deliberate study dealing with this or another issue/topic that interests and excites you.

Key Terms

Dichotomies   Compute Variable command

SPSS Statistics Commands Introduced in This Chapter

19.1   Recoding Into Different Variables

19.2   Creating an Index Using Compute Variable

Review Questions

Answer T (true) or F (false) for Questions 1 through 5.

1. In the 2010 General Social Survey (GSS), frequency of attendance at religious services was related to abortion attitudes, but religious affiliation was not.

2. According to the 2010 GSS, opposition to abortion tends to decrease with decreasing measures of religiosity.

3. According to the 2010 GSS, religious affiliation and church attendance are related to each other, but neither is related to abortion attitudes.

4. In the 2010 GSS, political philosophy was related to abortion attitudes, but party identification was not.

5. In the 2010 GSS, those who were somewhat permissive about premarital and homosexual sex also tended to oppose abortion.

6. List the commands for instructing SPSS Statistics to recode an original variable (Into Different Variables).

7. If you save your data set after recoding an original variable, what should you remember to do?

8. List two findings from our multivariate analysis of the relationship between abortion, religious affiliation, and church attendance.

9. Name the three items we used to construct our index of ideological traditionalism, IND.

10. Scores on IND ranged from _____ to _____.

11. Does a score of 4 on IND indicate support for or opposition to abortion?

12. How well did IND predict abortion attitudes?

13. How might you expand the index IND to include one (or more) of the variables dealing with sexual attitudes?

SPSS STATISTICS LAB EXERCISE 19.1

NAME ______________________________

CLASS ______________________________

INSTRUCTOR ______________________________

DATE ______________________________

To complete the following exercises, you need to load the data file AdventuresPLUS.SAV.

In this exercise, we are going to create a simple index combining attitudes toward teen sex (PILLOK, SEXEDUC, and TEENSEX). The goal is to construct an index/composite measure called IND2 that can be used to predict attitudes toward sex and family roles. To create IND2, simply follow the steps listed below. After we set up the index initially, you will be given a chance to expand it and explore its capabilities on your own.

1. Define the following values for each item listed below as "missing":

   a. PILLOK—0, 8, 9
   b. SEXEDUC—0, 3 to 9 (Hint: Use "Range plus one discrete missing value.")
   c. TEENSEX—0, 5 to 9 (Hint: Use "Range plus one discrete missing value.")
   d. FEPRESCH—0, 8, 9
   e. FECHLD—0, 8, 9

2. Recode PILLOK, TEENSEX, FEPRESCH, and FECHLD into dichotomies as shown below. Use **Transform → Recode → Into Different Variables** to recode the original variables. As noted earlier, if you save your data set after this exercise, be sure to use "Save As" and give the data set a new name so you are able to get back to the original, unrecoded data.

   a. PILLOK

Old Values		*New Values*	*Labels*
1 to 2	→	1	Okay/Agree
3 to 4	→	2	Wrong/Disagree

   b. TEENSEX

Old Values		*New Values*	*Labels*
1 to 2	→	2	Wrong
3 to 4	→	1	Okay

   c. FEPRESCH

Old Values		*New Values*	*Labels*
1 to 2	→	1	Agree
3 to 4	→	2	Disagree

d. FECHLD

Old Values		New Values	Labels
1 to 2	→	1	Agree
3 to 4	→	2	Disagree

3. Create a simple index called IND1 built on the items PILLOK, SEXEDUC, and TEENSEX (which we just recoded). There are a number of ways to create IND1. Feel free to design your own method or follow the general steps below. You may want to refer to our discussion in the chapter (and SPSS Statistics Command 19.1) if you need to refresh your memory.

Step 1: Create IND2.

Step 2: Assign points for IND2 based on scores for three component items (PILLOK, SEXEDUC, TEENSEX) as follows:

- If Wrong/Disagree (2) on PILLOK, give one point on IND2.
- If Oppose (2) on SEXEDUC, give one point on IND2.
- If Wrong (2) on TEENSEX, give one point on IND2.

Our goal is to create an index (IND2) with scores ranging from 0 (least traditional attitudes toward sex and family roles—or, if you like, least conservative/most liberal views) to 3 (most traditional attitudes toward sex and family roles—most conservative/least liberal views).

Step 3: Missing data.

Step 4: Define IND2.

When you are done, you will have created an index (IND2) that captures predispositions to hold traditional attitudes toward sex and family roles. Again, scores on the index range from 0 (least traditional) to 3 (most traditional).

4. Run Frequencies for IND2, and then use your output to fill in the blanks below.

IND2   *Valid Percentage*

0 Least traditional   _____

1   _____

2   _____

3 Most traditional   _____

5. See how well IND2 predicts attitudes toward sex and family roles. Run Crosstabs with column percentages and chi-square, specifying FEPRESCH and FECHLD as row variables and IND2 as column variable. Then use your output to fill in the blanks below:

		IND2			
		0 Least Traditional	1	2	3 Most Traditional
FEPRESCH	Agree				
FECHLD	Agree				

6. Does the index (IND2) provide a strong prediction of support for attitudes toward sex and family roles as measured by FEPRESCH and FECHLD? Discuss the findings of your analysis in detail below (keep in mind that FEPRESCH and FECHLD ask different questions).

7. You are now on your own! Expand IND2 to include another item of your choosing from your data file AdventuresPLUS.SAV (keep in mind that you may need to recode your item(s) before adding them to IND2). When you have expanded the index, run Frequencies, and then see how well your new expanded index predicts attitudes toward sex and family roles. Print both your frequencies and table(s), and then discuss your findings in detail below.

8. Continue your analysis by checking to see if your new expanded index predicts FEFAM and FEHIRE. Print your table(s), and discuss your findings in detail below (keep in mind that you may need to recode FEFAM and FEHIRE).

Chapter 20 Suggestions for Further Multivariate Analyses

In each of the previous analytic chapters, we have tried to leave a large number of "loose ends" for you to pursue on your own. If you are using this book in connection with a college class, you may want to follow up on some of those leads in the form of a term paper or a class project.

This chapter will build on some of those suggestions and offer some additional possibilities for your analyses. We begin with some suggestions regarding ideal family size. We will then move to discuss other possibilities involving issues such as child training, the Protestant ethic, and prejudice. You should realize, however, that even within these fairly limited data sets, there are a number of analytic possibilities we have not considered.

Ideal Family Size and Abortion

Before leaving the topic of abortion altogether, we'd like to suggest another explanatory variable you might consider: ideal family size. We've seen previously that those who favor small families are more supportive of abortion than are those who favor large families. Now you are able to pursue this matter further using the data on your AdventuresPLUS.SAV file. Here's the basic relationship with which you might start. Notice that we're using the recoded variable CHLDNUM that we created earlier.

abort Simple Abortion Index * chldnum Recoded chldidel Crosstabulation

| | | | chldnum Recoded chldidel | | |
			1 0 to 2	2 3 or more	Total
abort Simple Abortion Index	0 Yes/Approve	Count	154	101	255
		% within chldnum Recoded chldidel	48.3%	34.4%	41.6%
	1 Conditional Support	Count	99	99	198
		% within chldnum Recoded chldidel	31.0%	33.7%	32.3%
	2 No/Disapprove	Count	66	94	160
		% within chldnum Recoded chldidel	20.7%	32.0%	26.1%
Total		Count	319	294	613
		% within chldnum Recoded chldidel	100.0%	100.0%	100.0%

Symmetric Measures

		Value	Asymp. Std. Error[a]	Approx. T[b]	Approx. Sig.
Ordinal by Ordinal Gamma		.254	.063	3.923	.000
N of Valid Cases		613			

a. Not assuming the null hypothesis.
b. Using the asymptotic standard error assuming the null hypothesis.

Why don't you see whether the index of ideological traditionalism (IND) is related to the ideal family sizes that people reported. Then compare your findings with Writing Box 20.1.

Writing Box 20.1

There is an overall relationship between scores on the abortion opinion index and family-size preferences: Those who scored high on the index are more likely to prefer small families. The gamma calculation ($\gamma = .254$, and statistically significant) indicates that it is a relationship worth exploring further.

What happens, however, when you control for attitudes on abortion? Does this relationship stand up? Among those supporting abortion, does the index of traditional values have an impact? As a somewhat more focused approach, you might review two articles that appear on the *Adventures in Social Research* student study website (http://www.sagepub.com/babbie8estudy/): Renzi's "Ideal Family Size as an Intervening Variable Between Religion and Attitudes Towards Abortion" (1975) and D'Antonio and Stack's "Religion, Ideal Family Size, and Abortion: Extending Renzi's Hypothesis" (1980). While somewhat dated, these articles can be used as a starting point and in conjunction with more recent studies in this area. Remember, you can access an extensive bibliography using the National Opinion Research Center site we recommended in Chapter 16.

Once you have accessed and read some of these articles, see if you can replicate portions of the published analyses. Perhaps you can identify additional variables that should be taken into account.

Child Training

In Chapter 16, we began to look into some of the factors that might affect people's views of the qualities most important to develop in children. As a reminder, the qualities were as follows:

- OBEY—to obey
- POPULAR—to be well liked or popular
- THNKSELF—to think for himself or herself
- WORKHARD—to work hard
- HELPOTH—to help others when they need help

We suggested that you start your analysis with some basic demographic variables such as age, sex, and race. We encouraged you to consider relationships you might logically expect, such as more-educated respondents placing higher value on children's thinking for themselves.

Now you are able to explore all the possibilities in greater depth. To stay with the example of education and independence of thought in children, you could see if certain variables shed more light on that relationship. You might check POLVIEWS, for example. Before you do, ask yourself what you expect to find. You might also check the effect of gender while you're at it.

Earlier, we suggested that you look at the impact of the sexual attitude items. Those who are permissive regarding violations of the established norms for sexual conduct often feel that they are thinking for themselves in those regards. How do such people feel about encouraging that trait in young people? With your multivariate skills, you can examine that matter with some sophistication.

Perhaps you will find it useful to combine some of the "quality" items into an index. On the face of it, OBEY and THNKSELF appear to value opposite traits. Are they negatively related to each other? Would they fit into an index of "conformity–independence" perhaps? If that seems a fruitful index to create, to which other variables does it relate?

The Protestant Ethic

Perhaps one of the most famous of all books in the social sciences is Max Weber's *The Protestant Ethic and the Spirit of Capitalism* (1905/1958). In it, Weber traces the religious roots (in Calvinism) of the ethic of hard work and thrift, which he concluded forged the beginnings of capitalism. This ethic is something you might like to explore.

A central notion of the Protestant ethic is the idea of hard work, the belief that individuals are responsible for their own economic well-being. People's wealth is taken as a sign of how hardworking and diligent they are. In its most extreme form, the Protestant ethic sees poverty as the result of slothfulness and laziness and wealth as the product of persistence and hard work.

The GETAHEAD item in your General Social Survey (GSS) data set provides a measure of at least one dimension of the Protestant ethic that you might like to explore. To measure the degree to which people took responsibility for their own economic well-being, respondents were read this statement: "Some people say that people get ahead by their own hard work; others say that lucky breaks or help from other people are more important." Following the statement, they were asked, "Which do you think is most important?" Why don't you begin by examining how GETAHEAD is related to other variables. It seems logical that GETAHEAD would be related to WORKHARD.

Are people who are in disadvantaged groups or who have low incomes more apt to believe that their fates are governed by luck rather than hard work? Or does the belief in luck for success provide a rationalization for people who have little power to control the events that affect their lives? Because these variables are not the ratio measures that Pearson's *r* assumes, you can't take the results literally, but perhaps they can give you a first glimpse. Do not use this technique as a substitution for other, statistically appropriate examinations.

If you follow these suggestions, you'll need to be prepared for some surprises—even disappointments—so it may take some hard work (not inappropriate) for you to create a measure of commitment to the Protestant ethic that you like. Then you can begin the multivariate search for the causes of this point of view. Be sure to check out the Protestants, but don't expect an easy answer.

Capital Punishment, Gender, and Race

In your analysis of the Protestant ethic, you may have looked at the place of racial items such as MARBLK or NATRACE. If not, you may want to do so now. Logically, the belief that people get ahead by hard work alone would seem to rule out the limits imposed by prejudice and discrimination. You can explore the relationships among some of the work ethic and prejudice items to see if this is the case.

The variables on our AdventuresPLUS.SAV file allow us to explore other aspects of race and racial prejudice. You probably recall that the file contains the variable CAPPUN, which measures support and opposition to the death penalty; by utilizing some of the bivariate and multivariate skills we have acquired, we can explore this issue and its relationship to other variables such as gender and race in more depth.

According to Amnesty International, "over half the countries in the world have now abolished the death penalty." While 72 countries and territories, including the United States, retain capital punishment, the number that actually practices it is much smaller. In 2005, for instance, 94% of all executions took place in just four countries around the world: China, Iran, Saudi Arabia, and the United States.

In the United States today, the death penalty is still practiced in 37 states and provided for under U.S. federal civilian and military law. In all, 43 prisoners were put to death in the United States during 2011, raising the total number of executions since 1977 to more than 1,200. As of May 2011, more than 3,200 inmates in the United States remained on death row awaiting execution.[1] Often, executions are delayed by inmates' appeals or court challenges to death penalty laws. This can be explained in part not only by the nature of the U.S. judicial system but by the fact that capital punishment remains an enormously controversial issue. While attitudes toward the death penalty in the United States have fluctuated over time, currently, about two thirds of the population favor capital punishment and about one third would like to see it abolished.

Our data let us explore some of the factors influencing attitudes toward the death penalty in the United States.

Demonstration 20.1: CAPPUN by SEX

Some argue that gender influences attitudes toward the death penalty—specifically, that women are taught to be more nurturing and, especially when life without parole is an alternative to execution, are more likely to favor abolition of the death penalty. With the help of SPSS Statistics, we can quickly test this theory.

Before running Crosstabs, be sure to designate the appropriate values as "missing" for CAPPUN and SEX. Then run Crosstabs with percentages. Your results should match the table below.

cappun FAVOR OR OPPOSE DEATH PENALTY FOR MURDER * sex RESPONDENTS SEX Crosstabulation

| | | | sex RESPONDENTS SEX | | |
			1 MALE	2 FEMALE	Total
cappun FAVOR OR OPPOSE DEATH PENALTY FOR MURDER	1 FAVOR	Count	614	683	1297
		% within sex RESPONDENTS SEX	72.3%	63.7%	67.5%
	2 OPPOSE	Count	235	389	624
		% within sex RESPONDENTS SEX	27.7%	36.3%	32.5%
Total		Count	849	1072	1921
		% within sex RESPONDENTS SEX	100.0%	100.0%	100.0%

The table shows a relationship between these variables. Women are more likely than men to oppose the death penalty by more than 8 percentage points. That's a notable difference but

---

[1]For additional facts and figures surrounding this issue, you may want to check out Amnesty International's webpage at http://web.amnesty.org/pages/deathpenalty-facts-eng. Alternatively, you may want to access the Bureau of Justice Statistics' "Capital Punishment Statistics" page at http://bjs .ojp.usdoj.gov/index.cfm?ty=tp&tid=18. The information at http://www.deathpenaltyinfo.org/ is also illuminating. Remember, you can also find a host of other information on this topic by using your favorite search engine or visiting your library.

not one to get too excited about. Perhaps other variables in our data set, such as RACE, will allow us to explore attitudes toward the death penalty further.

Demonstration 20.2: CAPPUN by SEX, Controlling for RACE

In our country, because of long-standing patterns of economic, residential, and educational discrimination, Blacks have had very different community experiences with violence and the criminal justice system. Although they are a minority, Blacks make up the majority of our prison population. Violence is commonly experienced in impoverished Black urban neighborhoods.

As we discussed in earlier chapters, the introduction of a control variable will sometimes bring about changes in the nature of a simple two-variable relationship, such as CAPPUN and SEX. Relationships can become stronger when a control variable is applied, and under different circumstances, a relationship can become weaker or disappear. Sometimes the addition of a control variable will have no effect at all.

Why don't you see what happens to the relationship between CAPPUN and SEX when we control for RACE. As usual, before you run your table, be sure to designate the appropriate values as "missing" for RACE.

As the table shows, for Whites, we get a very similar (although opposition was a little lower) point spread to that in the earlier table. As you can see, 31% of the women and 23% of the men oppose the death penalty.

cappun FAVOR OR OPPOSE DEATH PENALTY FOR MURDER * sex RESPONDENTS SEX * race RACE OF RESPONDENT Crosstabulation

race RACE OF RESPONDENT					sex RESPONDENTS SEX		Total
					1 MALE	2 FEMALE	
1 WHITE	cappun FAVOR OR OPPOSE DEATH PENALTY FOR MURDER	1 FAVOR	Count		513	557	1070
			% within sex RESPONDENTS SEX		76.8%	69.4%	72.7%
		2 OPPOSE	Count		155	246	401
			% within sex RESPONDENTS SEX		23.2%	30.6%	27.3%
	Total		Count		668	803	1471
			% within sex RESPONDENTS SEX		100.0%	100.0%	100.0%
2 BLACK	cappun FAVOR OR OPPOSE DEATH PENALTY FOR MURDER	1 FAVOR	Count		58	73	131
			% within sex RESPONDENTS SEX		56.3%	41.2%	46.8%
		2 OPPOSE	Count		45	104	149
			% within sex RESPONDENTS SEX		43.7%	58.8%	53.2%
	Total		Count		103	177	280
			% within sex RESPONDENTS SEX		100.0%	100.0%	100.0%
3 OTHER	cappun FAVOR OR OPPOSE DEATH PENALTY FOR MURDER	1 FAVOR	Count		43	53	96
			% within sex RESPONDENTS SEX		55.1%	57.6%	56.5%
		2 OPPOSE	Count		35	39	74
			% within sex RESPONDENTS SEX		44.9%	42.4%	43.5%
	Total		Count		78	92	170
			% within sex RESPONDENTS SEX		100.0%	100.0%	100.0%
Total	cappun FAVOR OR OPPOSE DEATH PENALTY FOR MURDER	1 FAVOR	Count		614	683	1297
			% within sex RESPONDENTS SEX		72.3%	63.7%	67.5%
		2 OPPOSE	Count		235	389	624
			% within sex RESPONDENTS SEX		27.7%	36.3%	32.5%
	Total		Count		849	1072	1921
			% within sex RESPONDENTS SEX		100.0%	100.0%	100.0%

When we look at Blacks, we see that women remain more likely to oppose the death penalty for murder. However, the opposition is much higher among Blacks than Whites for both men and women.

Equally interesting are the total percentage differences between Whites and Blacks opposing the death penalty. Of the Blacks, 53% oppose, while only 27% of the Whites oppose—a 26-point spread. In the "Other" racial category, men are actually more likely to oppose utilizing the death penalty for murder than are women. This illustrates a *conditional relationship* dependent on race.

The introduction of RACE as a control variable raises important questions about why men and women of different racial groups are so different in their attitudes toward the death penalty. In the background lies an even bigger question of why such large differences exist between Blacks' and Whites' attitudes. These are the types of questions for which social researchers use the bivariate and multivariate skills we discussed in the past several chapters to explore in more detail. We leave the exciting possibilities for additional exploration in your able, bivariate and multivariate hands.

Conclusion

The National Rifle Association and other proponents of extensive gun distribution (see GUNLAW) are fond of saying, "Guns don't kill people; people kill people." Well, in a more pacific spirit, we'd like to suggest that SPSS Statistics doesn't analyze data; analysts analyze data. And the good news is that you are now a bona fide, certifiable data analyst.

We've taken you through bivariate and multivariate analysis to demonstrate the importance of ascertaining the sometimes-hidden meanings lying behind the responses people give to survey questions. That's a big part of the *adventure* of social research. (We wanted to call this book *Earl, Fred, Billy, and Jeanne's Excellent Adventure*, but you know publishers.) We've given you about all the guidance and assistance we planned when we started this adventure. Remember, in Chapter 1 we said you'll just add yourself to the mix; well, it's show time. You're on your own now, although you should have a support network behind you. If you read this book in connection with a college course, you have your instructor.

The final section of this book expands the horizons of social research even further. It suggests a number of ways you might reach out beyond the GSS data set we have provided for your introductory experience with the adventure of social research.

Main Points

- This chapter offers suggestions for further multivariate analyses.
- If you are using this book in connection with a college course, these suggestions may be useful in helping you design a term paper or class project.
- Even within the limited data sets that accompany this text, there are numerous analytic possibilities we have not explored or considered.
- The suggestions offered here are built on the questions and issues addressed in your data file AdventuresPLUS.SAV.
- The suggestions cover issues such as ideal family size, child training, the Protestant ethic, and prejudice.
- These suggestions are just a small portion of the analytic possibilities stemming from the data contained in your AdventuresPLUS.SAV file.

Key Terms

Conditional relationship

SPSS Statistics Commands Introduced in This Chapter

No new commands were introduced in this chapter.

Review Questions

1. Use the National Opinion Research Council website discussed at the end of Chapter 16 ("Additional Resources") to access at least one article or study that pertains to each of the following topics. Read the articles and see if you can identify a variable related to each of the following topics that was NOT discussed in this chapter. List the variable and explain its relationship to the topic.

   a. Ideal family size
   b. Abortion
   c. Child training
   d. Protestant ethic
   e. Capital punishment

2. Which of the studies that you accessed in response to Question 1 would you be interested in attempting to replicate and why?

3. Discuss how you might apply the techniques and procedures we covered in Chapters 17 through 20 on multivariate analysis to examine the following topics/issues:

   a. Ideal family size
   b. Abortion
   c. Child training
   d. Protestant ethic
   e. Capital punishment

NAME _____

CLASS _____

INSTRUCTOR _____

DATE _____

In Lab Exercise 16.1, we asked you to use bivariate techniques to examine one of the topics from the AdventuresPLUS.SAV file: gender roles, law enforcement, environment, mass media (use and confidence), national government spending priorities, teen sex, affirmative action and equalization.

In this exercise, you will be given an opportunity to expand your analysis of this issue by applying some of the multivariate techniques discussed in Chapters 17 through 19.

1. List the general topic/issue you examined in Lab Exercise 16.1.

2. List the variables (including any recoded variables or indexes) in your AdventuresPLUS .SAV file that pertain to this issue.

3. In Lab Exercise 16.1, you developed two hypotheses. List Hypotheses 1 and 2 below, identifying the dependent and independent variables in each.

   a. Hypothesis 1:

      Independent variable: _____

      Dependent variable: _____

   b. Hypothesis 2:

      Independent variable: _____

      Dependent variable: _____

4. Describe the strength and significance of the relationship between the variables in Hypothesis 1. In particular, you should note whether the relationship is worth exploring further.

5. Describe the strength and significance of the relationship between the variables in Hypothesis 2. In particular, you should note whether the relationship is worth exploring further.

6. Based on your responses to Questions 4 and 5, choose one hypothesis you would like to explore further (list abbreviated variable names). If neither hypothesis is worth exploring further, develop a new hypothesis and run through the exercises in SPSS Statistics Lab Exercise 16.1 again before continuing with this exercise. Once you have identified a hypothesis worth exploring further, fill in the blanks below.

   Independent variable: _____

   Dependent variable: _____

7. List a variable from your AdventuresPLUS.SAV file that you think may be related to the independent and dependent variables listed in response to Question 6, and explain your reasoning.

   Third variable A: _____

8. List another variable from your AdventuresPLUS.SAV file that you think may be related to the independent and dependent variables listed in response to Question 6, and explain your reasoning.

   Third variable B: _____

9. From the list of techniques reviewed in previous chapters, circle the procedures appropriate to examine the relationship between your independent variable, dependent variable, and third variable A (Question 7). (Hint: You should circle at least one technique in each group.)

   a. Group 1:

   ■ Running Crosstabs with a control variable

   b. Group 2:
   ■ Recoding to create a dummy variable
   ■ Multiple regression
   ■ Recoding into different variables
   ■ Creating an index using Compute

   c. Group 3:
   ■ Measure of association (lambda, gamma, correlation matrix with Pearson's r, r^2, regression, scatterplot with regression line)
   ■ Tests of significance (chi-square, t test, ANOVA)

10. From the list of techniques reviewed in previous chapters, circle the procedures appropriate to examine the relationship between your independent variable, dependent variable, and third variable B (Question 8). (Hint: You should circle at least one technique in each group.)

    a. Group 1:

    ■ Running Crosstabs with a control variable

    b. Group 2:
    ■ Recoding to create a dummy variable
    ■ Multiple regression
    ■ Recoding into different variables
    ■ Creating an index using Compute

c. Group 3:

- Measure of association (lambda, gamma, correlation matrix with Pearson's r, r^2, regression, scatterplot with regression line)
- Tests of significance (chi-square, t test, ANOVA)

11. Use the techniques you identified as appropriate in response to Question 9 to examine the relationship between your independent variable, dependent variable, and third variable A. Once you have conducted your analysis, do the following:

- Print and attach your output to this sheet.
- Summarize your findings in prose.

12. Use the techniques you identified as appropriate in response to Question 10 to examine the relationship between your independent variable, dependent variable, and third variable B. Once you have conducted your analysis, do the following:

- Print and attach your output to this sheet.
- Summarize your findings in prose.

Part V   The Adventure Continues

Chapter 21 Designing and Executing Your Own Survey

Chapter 22 Further Opportunities for Social Research

In the concluding chapters, we want to explore several different ideas that may support your continued investigations into the nature of human beings and the societies they create.

It struck us that you might be interested in conducting your own survey—perhaps as a class project. Consequently, in Chapter 21, we begin by taking a step back to look at the process of social research and the research proposal. We then focus specifically on survey research. In particular, we delve into designing and administering a survey, defining and entering data in SPSS Statistics, and writing a research report. In Chapter 22, we suggest other avenues for pursuing your social investigations. Among other things, we talk about the unabridged General Social Survey (GSS), other data sources you might explore, and other computer programs you might find useful.

We hope that by the time you finish these chapters, you will have fully realized the two purposes that lie behind our writing this book: to help you learn (1) the logic of social research and (2) how to pursue that logic through the use of SPSS Statistics. In addition, we hope you will have experienced some of the excitement and challenges that make social research such a marvelous adventure.

Chapter 21   Designing and Executing Your Own Survey

The General Social Survey (GSS) data sets provided with this book are of special interest to researchers because they offer a window on American public opinion. We thought you might be interested in learning about the thoughts and actions of people across the country.

At the same time, it occurred to us that many of you may be interested in conducting your own survey, investigating topics, issues, and populations not addressed by the GSS or other data sets. Conducting your own survey is a good idea because data analysis in a classroom setting is sometimes more meaningful if the data are more personal. Perhaps you would like to analyze the opinions and behaviors of your own class (if it's a large one), the opinions of students elsewhere in your school, or even the attitudes of members of your community.

If you are interested in conducting a survey, this chapter and some of the material contained on the *Adventures in Social Research* student study website (http://www.sagepub .com/babbie8estudy/) are designed to help you get started developing and administering the survey, defining and entering the data in SPSS Statistics, and writing a research report (you already know how to analyze the data!). This is our answer (albeit brief) to those students and readers who ask, "How can I do this myself?" While we cannot hope to provide you with an in-depth understanding of the intricacies of survey research in such a short time, we do give you a brief overview of the steps involved in designing and implementing your own survey. When possible, we will also point you toward a few other resources that you may find useful as you launch your own original study.

The Social Research Process and Proposal

Before we focus on designing and executing your own survey, we want to take a step back and place survey research and data analysis, which have been the central focus of this book, in the context of the social research process as a whole.

Our primary focus in this book has been a secondary analysis of a portion of a nationwide survey, the 2010 GSS. As you are aware, survey research is just one of many methods of data collection used by social researchers. Other methods include—but are not limited to—experimentation, participant observation, focus groups, intensive interviews, and content analysis.

Not only has our focus been limited to just one method of data collection, but it has also been limited to one step (albeit crucial) in the social research process—data analysis. We have focused on the use of one of the most popular computer software programs, SPSS Statistics, to analyze survey data. As we discuss in the next chapter, there are other software packages you might want to consider using as well.

Because our focus has been on the statistical analysis of survey data using SPSS Statistics, our discussions of the practice and process of social research have been necessarily limited to data analysis. Before conducting your own study, you may find it helpful to review the entire social research process and get some tips on writing a research proposal (which is often the first step in the research process) and, most important, to locate texts that will give you more-detailed descriptions of parts of the research process not covered here.

The *Adventures in Social Research* student study website (http://www.sagepub.com/babbie8estudy/) includes an excerpt from Russell K. Schutt's comprehensive research methods text, *Investigating the Social World* (2006). The excerpt focuses on the major steps involved in designing a research project and gives some guidelines for writing a research proposal. As Schutt notes, the research proposal is a crucial step in the social research process because it goes "a long way toward shaping the final research report and will make it easier to progress at later research stages." Schutt's text is just one of many research methods texts that you may find useful.

Others include Earl Babbie's *The Practice of Social Research* (2010), Kenneth Bailey's *Methods of Social Research* (1994), Thomas Sullivan's *Methods of Social Research* (2001), Chava Frankfort-Nachmias and David Nachmias's *Research Methods in the Social Sciences* (2000), Bridget Somekh and Cathy Lewin's *Research Methods in the Social Sciences* (2005), Alan Bryman's *Social Research Methods* (2004), and W. Lawrence Neuman's *Social Research Methods: Qualitative and Quantitative Approaches* (2006). While this list is by no means exhaustive, it should be useful in beginning your search for a text that will provide you with a comprehensive overview of social research.

Designing and Executing Your Own Survey

Like all methods of data collection, survey research has its own peculiar advantages and disadvantages, benefits and liabilities. Nevertheless, a carefully constructed and executed survey can yield results that are considered valid, reliable, and generalizable. The 2010 GSS is a good example of a carefully designed and implemented survey, yielding data from which findings can be generalized to all English-speaking, non-institutionalized American adults 18 years of age and older.

To design and conduct a survey that meets scientific standards, you need to pay particular attention to the steps required for survey research: research design, sampling, questionnaire construction, data collection, data processing, data analysis, and report writing.

To help you design and implement your own survey, we have included an excerpt from a guide called "SPSS Survey Tips" on the *Adventures in Social Research* website (http://www.sagepub.com/babbie8estudy/). This guide discusses some of the key considerations in planning, developing, and executing a survey. You should think of these as useful tips rather than hard-and-fast rules. Keep in mind it is unlikely that all social scientists or pollsters would agree with every recommendation. Moreover, in the "real world," the process of developing and implementing a survey is seldom as neat as the sequential ordering of the tips may suggest.

Despite these caveats, we think that you will find "SPSS Survey Tips" a useful and concise overview of the steps involved in developing and implementing your own survey. In addition, you may also want to consult one of the many texts devoted to the principles and techniques of survey research. In particular, we draw your attention to Pamela Alreck and Robert Settle's *Survey Research Handbook* (2004), Earl Babbie's *Survey Research Methods* (1990), Floyd J. Fowler, Jr.'s *Survey Research Methods* (2002), and Peter Nardi's *Doing Survey Research* (2005).

Sample Questionnaire

To help you collect your own data, we have included in Appendix B of this book a sample questionnaire that asks some of the same questions found in the subset of GSS items included in your Adventures.SAV file. We've also provided the tools you may need to make any modifications you might like in the interest of local relevance. If you wish, you can make copies of the questionnaire and use it to collect your own data.[1] If you are more ambitious, you might choose to design a totally different questionnaire that deals with whatever variables interest you. There is no need for you to be limited to the GSS variables we've analyzed in this book or suggested in the sample questionnaire. You have learned a technology that is much more broadly usable than that. To get ideas about the kinds of things sociologists and other social scientists study, you might go to the library and thumb through the research categories in Sociological Abstracts.

Whatever survey you choose to conduct, the following section tells you how to get your data into a form that SPSS Statistics will accept for analysis.

Getting Ready for Data Analysis Using SPSS Statistics

Once you have designed and administered your questionnaire (or collected your data), you are ready to begin analyzing your data. Getting data ready for analysis with SPSS Statistics is really a three-step process:

Step 1: Defining your data

Step 2: Editing and coding your data

Step 3: Entering your raw data

Step 1: Defining Your Data

The first step, often called *data definition*, involves giving each item from your questionnaire a variable name, setting the data type, describing the variable, indicating values and labels, designating missing values, setting the item's level of measurement, and designating a place to store this information.

This process can begin even before you have completed collecting the data if you are certain that the questionnaire itself is not going to be altered or revised.

Demonstration 21.1: Example 1—Defining ID

To walk you through this process, we are going to use the sample questionnaire provided in Appendix B of this book. You may want to take a moment to review the questionnaire before we begin defining the first variable, ID.

Go ahead and launch SPSS Statistics. If you get the "What would you like to do now?" dialog box, simply click **Cancel**. Now you should be looking at the Data Editor. Since we will define our data in the Variable View screen, make sure to click on the tab with that name if it is not already displayed.

As you know from previous chapters, Variable View contains descriptions of the variables contained in a data file. In Variable View, each row describes a variable, while the variable attributes (name, type, width, decimals, etc.) are contained in the columns. Unlike your earlier experience with Variable View, now you are creating a variable definition rather than simply modifying one that has been supplied by someone else.

We will define the first variable ID (the case identification number) from our sample questionnaire. Since each row in Variable View represents a variable, looking at our sample

---

[1] We have also included an empty data file, LOCAL.SAV, on the *Adventures in Social Research* website. As we will discuss later in the chapter, this file is already defined and can be used to enter data from a sample population.

questionnaire, you can see that the variables we'll be putting in the numbered rows start with ID, then move to CHLDIDEL, OBEY, POPULAR, and so on until we come to the last variable, SEX. For each variable, we will start by typing its name in the first column and then work our way across the columns from left to right ("Name" to "Measure").

Variable Names

To begin, double-click on the first cell in the upper left corner (the one that corresponds with Row No. 1 and the "Name" column). Now type the abbreviated variable name, ID.[2]

Now go ahead and hit **Enter**. Once you do that, you will notice that the cell that corresponds with Row No. 1 and the second column ("Type") becomes the active cell. Moreover, SPSS Statistics automatically gives you the default variable attributes for ID.

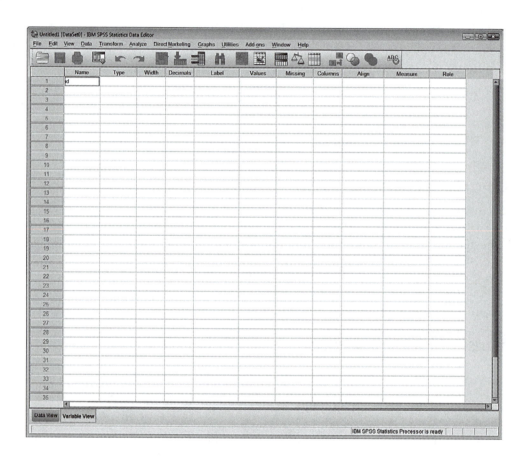

Type

There are several types of data, including "comma," "dot," "scientific notation," and others. To get a sense of the various types of data, double-click on the small gray box with the three dots (ellipses) in the right side of the Row No. 1 "Type" cell.

---

[2]If you are designing your own survey and planning to use SPSS Statistics to analyze the data, there are several "rules" to keep in mind when constructing abbreviated variable names. Among the most important rules to keep in mind are these: Each name must begin with a letter (letters, digits, periods, or the symbols @, #, _, or $ can follow), variable names should generally not exceed eight characters in length, variable names cannot end with a period, and each variable name on a file must be unique.

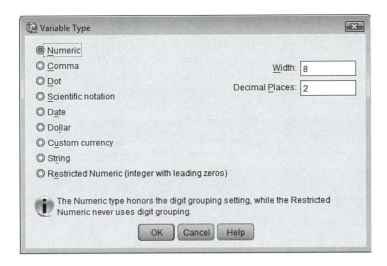

SPSS Statistics has probably chosen "Numeric" by default. If not, click the **Numeric** button and then select **OK**. If the "Numeric" type has already been selected, simply click **Cancel** or the **X** (Close button) to exit. All the data we have used in this text and certainly in the sample questionnaire are numeric.

Decimal

For a moment, skip over the "Width" column and click on the right side of the cell corresponding with Row No. 1 and the "Decimal" column. When you do, you will notice that a set of up and down arrows appears, allowing you to either increase or decrease the decimal places. Since our variable, ID, does not require any decimal places, click the down arrow until it reaches 0.

Width

Now go ahead and click on the right side of the cell corresponding with Row No. 1 and the "Width" column. Once again, you will see up and down arrows allowing you to specify the width or the number of characters for the column. While this column is not crucial for our purposes, you can go ahead and decrease the width to 3 or more if you wish.

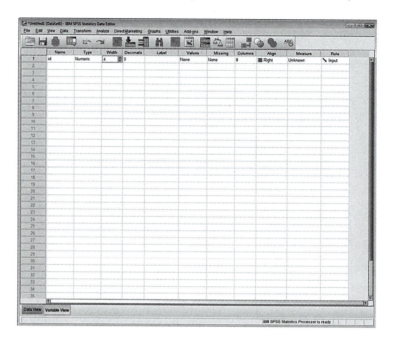

Label

Now go ahead and double-click in the cell corresponding with Row No. 1 and the "Label" column. This cell allows you to insert the variable label or a brief description of your variable. Variable labels can be about 255 characters long if necessary.

In this case, once you have double-clicked on the cell, go ahead and insert a variable label for ID, such as "Respondent's Identification Number." When you do that, you will see that the width of the column expands somewhat.

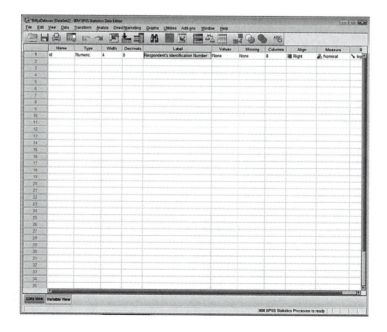

Once you are done, simply select **Enter** or use your cursor to click on the next cell, "Values."

Values

Once again, double-click on the gray box with the ellipses on the right side of the cell to open the Value Labels box. The Value Labels box allows you to assign values and labels for your variable. For instance, if we were defining the variable SEX, we would want to enter the values 1 and 2 for "male" and "female," respectively.

The numeral 1 would be placed in the "Value:" box and "male" in the "Value Label:" box. Then, simply click **Add**. The numeral 2 would then be placed in the "Value:" box and "female" in the "Value Label:" box. Then, again, simply click **Add** and then **OK**.

In the case of ID, however, we do not need to assign values and labels, because each respondent has a unique identification number. When we go through the next variable, CHLDIDEL, we will give you an opportunity to practice filling in the values and labels.

Click on the **Cancel** or **X** button to close the Value Labels box.

Missing

There are no missing values for ID, so we do not have to open the Missing Values dialog box in this case. However, if we wanted to define missing values (as we will in the case of CHLDIDEL), all we would have to do is double-click on the gray box with ellipses in the right side of the cell.

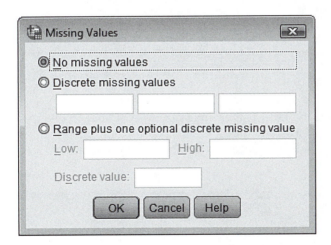

Then, in the Missing Values box, we would define the missing values as we have done previously by choosing "No missing values" (in the case of ID and the default setting in SPSS Statistics), "Discrete missing values," "Range of missing values," or "Range plus one discrete missing value."

If you opened the Missing Values box, you can click **Cancel** or **X** to exit.

Columns and Align

The following two columns ("Columns" and "Align") allow us to define the width of the columns in the Data Editor and specify the alignment of data values. By default, SPSS Statistics sets the column width to 8 and aligns the data on the right. If you wish to change either of these settings, you can double-click on the right-hand side of the cell and use the arrows. Since we don't need to change either the column width or alignment, we can skip to the last column, "Measure."

Measure

The last step in the process of defining your variable is indicating the level of measurement. When you click on the right side of the cell that corresponds with Row No. 1 and the "Measure" column, you will see a down arrow. By clicking on the down arrow, you can specify the level of measurement as "Scale" (interval/ratio), "Ordinal," or "Nominal."

In the case of ID, the level of measurement is nominal, each case having its own category. Remember that an ID number of 20 is not twice as "ID" as an ID number of 10, and no ID number is "more" ID than any other.

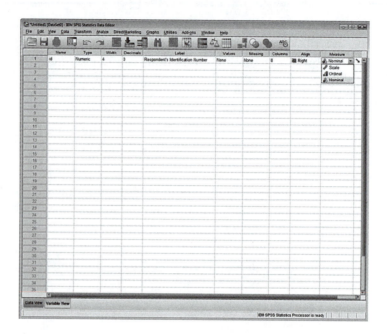

After you change the level of measurement to "Nominal," you are done defining the first variable. If you like, click on the **Data View** tab at the bottom left side of the screen. You will now see that ID is magically placed in the first column at the top left-hand corner of your Data Editor.

To return to the Variable View screen, either double-click on the variable name (ID)at the top of the column or click the **Variable View** tab.

Demonstration 21.2: Example 2— Defining CHLDIDEL

Now that you have successfully defined the first variable from our sample questionnaire, why don't we go ahead and practice by defining the second item, CHLDIDEL, together as well. This time, we will list the commands in an easy-to-read format. You should already be in the Variable View screen. We will place CHLDIDEL in Row No. 2. Each of the following commands refers to variable attributes in Row No. 2.

- Double-click on the cell corresponding with Row No. 2 and the "Name" column.
- Type **CHLDIDEL.**
- Click **Enter**.
- By default, SPSS Statistics will set the "Type" to numeric, so there is no need to change that setting.
- Click on the right side of the cell corresponding with Row No. 2 and the "Decimal" column.
- Use the down arrow to set the decimal to 0.
- Click on the right side of the cell corresponding with Row No. 2 and the "Width" column.
- Use the down arrow to change the width to 2.
- Double-click the cell corresponding with Row No. 2 and the "Label" column.
- Type a descriptive variable label, such as "Respondent's ideal number of children."
- Click **Enter** or use cursor to move to next column, "Values."
- Double-click in the gray box to open Value Labels dialog box.
- Type –1 in the "Value:" box, type IAP (inapplicable) in the "Value Labels:" box, and then click **Add**.

You will notice that "–1 = IAP" is now specified in the box below. If you wanted to change or revise the statement for any reason, simply highlight "–1 = IAP." Once you do that, you will notice that you are able to select **Remove** to delete the statement.

- Now go ahead and enter the other values and labels for CHLDIDEL: seven plus DK and NA.

Keep in mind that the sample questionnaire contains questions taken from your Adventures.SAV file. Consequently, you can refer to that file or the codebook (Appendix A) to verify the values and labels for the variables in the sample questionnaire (Appendix B).

- Click **Enter** or use your cursor to move to the next column.
- Click in the gray box on the right side of the "Missing" column to open the Missing Values dialog box.
- Select **Discrete missing values** and type the values –1 and 9.
- Click **OK**.
- By default, SPSS Statistics has probably already selected "Scale" as the level of measurement. If not, use the down arrows to set the level of measurement to "Scale."
- Click on the **Data View** tab, and you will see CHLDIDEL in the second column in your Data Editor window.

You have now successfully defined the first two items in our sample questionnaire, ID and CHLDIDEL. As you can probably see, while it is not a difficult process, it can be fairly time-consuming to define all the variables from a questionnaire, even one as small as our sample questionnaire.

SPSS Statistics Command 21.1: Defining a Variable

Make sure the Variable View screen is the active window

→ Double-click on the cell corresponding with the appropriate row and the "Name" column → Type abbreviated variable name → **Enter**

→ Move through the cells corresponding with that row and each applicable column from left ("Name") to right ("Measure")

→ Click or double-click on the right side of the cell to access either the arrows or the appropriate dialog box → For our purposes, the most important columns are "Name," "Decimals," "Labels," "Values Missing," and "Measure"

→ Once you are done, click on the **Data View** tab → Your new variable will be situated in a column at the top of the screen

Copying Variables With Shared Attributes: Abortion Variables

One way to cut down on the time-consuming process of defining variables is to copy and paste variables that share similar attributes or characteristics.

The seven abortion variables in our sample questionnaire, for instance, all use the numeric code 1 to represent "Yes," 2 to represent "No," 0 to represent "IAP," 8 to represent "DK," and 9 to represent "NA." Moreover, in each case, 0, 8, and 9 are defined as missing values. In cases such as these, it is not necessary to define each variable individually; instead, you can simply copy a variable. We will use the abortion variables to illustrate how this works.

Demonstration 21.3: Copying a Variable

Now that you are comfortable defining variables with SPSS Statistics, go ahead and define ABDEFECT, the first abortion variable in our sample questionnaire. Place ABDEFECT in Row No. 3 directly below CHLDIDEL.[3] If you need to review the steps for defining a variable, refer to SPSS Statistics Command 21.1.

Once you have defined ABDEFECT, you can easily define the other six abortion variables (ABNOMORE, ABHLTH, ABPOOR, ABRAPE, ABSINGLE, and ABANY) by copying and pasting. To do this, click on the row number (3) to the left of the cell containing ABDEFECT. When you do that, you will notice that the row is highlighted.

---

[3]We are placing ABDEFECT in Row No. 3 for illustrative purposes only. If you were actually to set up this data file, Row No. 3 would be occupied by OBEY, Row No. 4 by POPULAR, Row No. 5 by THNKSELF, and so on. The proper order is displayed on the LOCAL.SAV file on the *Adventures in Social Research* website.

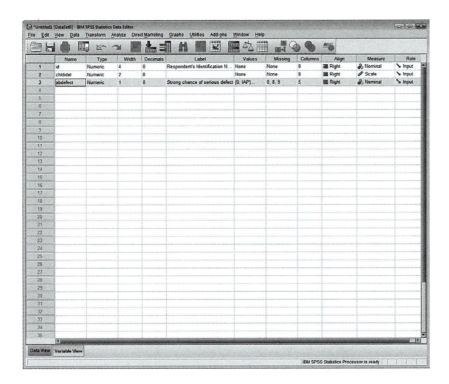

Now choose **Edit** → **Copy** from the drop-down menu.[4]

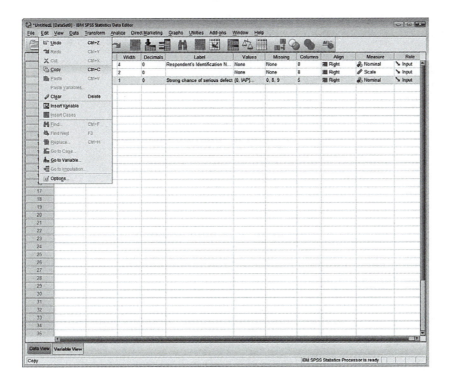

---

[4]Alternatively, you can right-click with your mouse and choose **Copy** (or <ctrl> + click on a Macintosh computer with a one-button mouse).

Then go ahead and select the row (most likely Row No. 4) where you would like to place the next variable (ABNOMORE). Now all you have to do is select **Edit → Paste Variables.**[5] By selecting "Paste Variables" instead of the simple Paste command, you'll get the following box, which allows you to add all the like variables at one time without having to repeat the same pasting over and over.

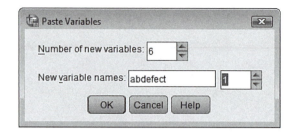

Select the number of variables you would like to create with the same specifications as ABDEFECT, and then click **OK.** Voilà!

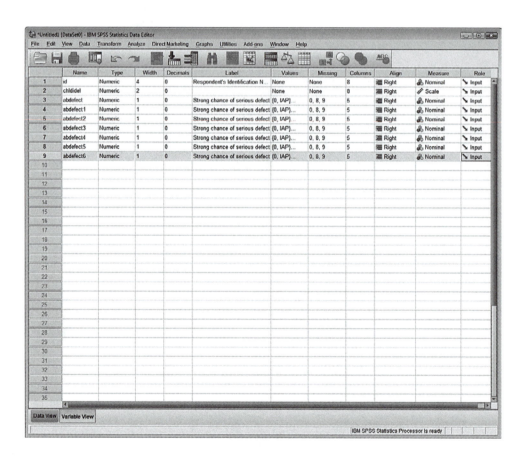

The only thing you need to do after that is insert a correct variable name and label for each.

_____

[5]You can also right-click with your mouse and select **Paste**, though we recommend you use the Paste Variables command.

SPSS Statistics Command 21.2: Copying a Variable

In Variable View → Click on left side of row to be copied

→ **Edit** → **Copy** → *OR* right-click with mouse → **Copy** (*OR* for Macintosh one-button mouse users, <**ctrl**> + click → **Copy**)

→ Click on left side of row you're copying to → **Edit** → **Paste Variables**

   → Select the number of variables to produce from the specifications of the copied variable

   → Click **OK**

Demonstration 21.4: Saving Your New File

Before you end this (or any other) data definition session, remember to save the new file with a name that will remind you of its contents.

To save a new data file, make the Data Editor the active window. Then, from the menu, choose **File** → **Save As** to open the Save Data As dialog box. Now click on the down arrow to the right of the "Save in:" field to choose the appropriate drive.

Then, in the "File name:" field, type a descriptive file name and click **Save.**

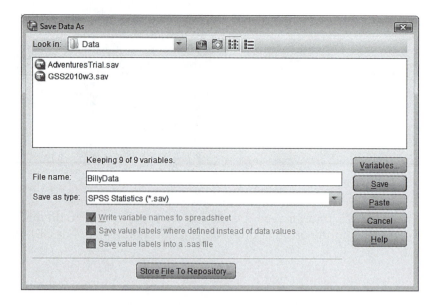

SPSS Statistics Command 21.3: Saving a New Data File

Data Editor active → **File** → **Save As**

   → Use down arrow to right of "Save in:" field to select appropriate drive

   → Type file name in "File name:" field → **Save**

LOCAL.SAV

Now that you have mastered the commands for defining a variable and saving your new data file, we are going to let you in on a little secret. To save you the time and trouble of having to define all the variables in the sample questionnaire, we have included a file named LOCAL.SAV on the *Adventures in Social Research* website. It contains the data definition for this particular questionnaire.

In short, because we know you are getting toward the end of the book and are probably anxious to begin your own adventure, we wanted to save you the time and work of having to define all the variables included in the sample questionnaire; so we did it for you (think of it as a gift of sorts)! For now, however, just keep the existence of LOCAL.SAV in the back of your mind. We'll show you how to use it after we discuss the next step (Step 2) in the process of getting your data ready for analysis with SPSS Statistics—editing and coding your data.

Step 2: Editing and Coding Your Data

The second step in the process of getting your data ready for analysis with SPSS Statistics is editing and *coding* your data. People do not always follow instructions when filling out a questionnaire. Verbal or written responses have to be transformed into a numeric code for processing with SPSS Statistics. For ease of entry, questionnaires should be edited for proper completion and coding before you attempt to key or enter them into an SPSS Statistics file.

To the extent that we could, we designed the sample questionnaire to be self-coding. By having a number next to each response for the *closed-ended questions*, the interviewer assigns a code as the respondent answers the question. In some cases, we have had to use *open-ended questions*, either because there would be too many responses to print on the questionnaire or because we couldn't anticipate all the possible responses.

Essentially, four steps are involved in editing and coding your data: making sure each questionnaire has a unique ID number, coding open-ended questions, making sure the codes are easy to read, and editing each questionnaire.

Unique ID Number

Each questionnaire should have a unique number in the "ID" field. We do this not because we want to identify individuals but because errors that show up later are frequently made in the coding process. For instance, we coded SEX as 1 for "male" and 2 for "female," but in our analysis, we find a respondent with SEX coded 7.

What we need to do is find the record with the erroneous code 7. To do this, we look up the ID number, go back to the original questionnaire, find out what code SEX should have been, and fix it.

Coding Open-Ended Questions

Next in the editing process, we have to code the open-ended questions. All the people coding questions should be following the same written instructions for coding. For instance, at the end of Appendix B of this book, we include a list of occupations and their socioeconomic statuses for coding SEI, the socioeconomic index. Other coding schemes might not be as elaborate. For instance, in a medical study, patients might be asked about the illnesses that brought them to the hospital.

Coding might be as simple as classifying the illnesses as acute or chronic.

Ensuring That Codes Are Easy to Read

Finally, the codes need to be written so they are easy to read. We have designed our questionnaire to be *edge-coded*. If you look in the right margin, you will see that we put a space for each variable's code. We included these numbers because some statistics software requires that variables be placed in specific columns across each record. Happily, that is not a requirement of SPSS Statistics for Windows or Macintosh.

We'll just use the numbered blanks as a convenient place to write our codes.

Editing Questionnaires

Whether you are going to edit the sample questionnaire we have been using as an example throughout this section or another questionnaire you designed on your own, you should get a copy of the codes used to define the file with which you are working (e.g., a copy of the work you did in Step 1 when you defined your data). For this example, we will access the codes used to define our LOCAL.SAV file.

Demonstration 21.5: Accessing File Information for Coding and Editing

When you open the LOCAL.SAV data file that is available on the *Adventures in Social Research* website, you will notice that the data cells are empty. Don't panic; they are supposed to be empty. This is the empty, but defined, data file LOCAL.SAV we mentioned earlier.

To display the codes for your use in coding your data on SPSS Statistics, you can either check the Variable View portion of the Data Editor or click **Utilities → Variables**.

Here is an example of the coding information you will find for the variable MARITAL, responses to Question 4 on the questionnaire: "Are you—married, widowed, divorced, separated, or have you never been married?"

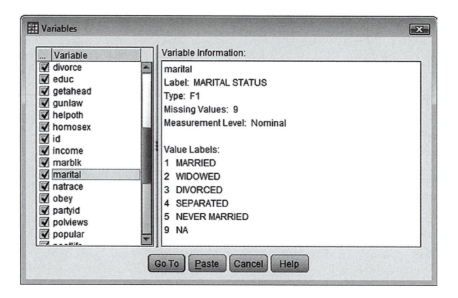

For people who are married, code 1 is used. For people who are widowed, 2 is used; divorced, 3; and so on. Notice that code 9, "No answer," has an NA next to it. This means that 9 has been designated a missing code. All missing codes are thrown out of the analysis. If respondents fail to answer a question or if they give two responses for one question, they should be assigned the missing value code for that variable.

Now that you have the information, you can go through each questionnaire, one at a time, to make sure the coding is correct, that there is only one code per question, and so on.

SPSS Statistics Command 21.4:
Accessing File Information for Coding and Editing

Utilities → Variables

→ From the list on the left, select the variable for which you want coding information.

Step 3: Entering Your Raw Data

Once you have edited and coded each of your questionnaires, you can move to the final step in the process: entering your raw data. Once again, we are going to use our sample questionnaire as an example.

Because the LOCAL.SAV file is on the *Adventures in Social Research* website, in order to enter data on it, you need to save the file to a local disk or drive (or a network drive, if you are connected to such). Go ahead and do that now. Once you are done, you can open the file and we will run through the process of entering data.

After opening your file, you should be looking at an empty data matrix in the Data View portion of the Data Editor. This is where we will enter our raw data. At this point, the matrix should be empty except for the variable names across the top and record numbers down the left side. You will notice that the order of the variables across the record is the same as the order of the variables on the questionnaire. We placed them in that order to make data entry easier and less error-prone.

Demonstration 21.6: Moving Through Data View

You can easily move from cell to cell in the data matrix. Pressing just the **Tab** key moves the *active cell* to the right, and pressing the **Shift** and **Tab** keys moves the active cell to the left. You can tell the active cell by its thick black border.

Pressing the **Enter** key moves the active cell down to the next row (record or respondent). The directional arrows on your keyboard will also move the active cell, one cell at a time. The mouse can be used to make a cell active just by pointing and clicking. Long-distance moves can be made by pressing <**Ctrl**> and <**Home**> (OR <**APPLE**> and <**Home**> for Macintosh computers) to move to the leftmost cell on the first case or <**Ctrl**> and <**End**> (OR <**APPLE**> and <**Home**> for Macintosh computers) to move to the rightmost cell on the last case.

SPSS Statistics Command 21.5:
Some Tips for Moving Through Data View

Tab: moves active cell to right

Shift + Tab: moves active cell to left

Enter: moves active cell down to next row

Directional arrows: move active cell, one cell at a time

<**Ctrl**> + <**Home**> on Windows computers OR <**APPLE**> + <**Home**> on Macintosh computers: moves active cell to leftmost cell, first case

<**Ctrl**> + <**End**> on Windows computers OR <**APPLE**> + <**End**> on Macintosh computers: moves active cell to rightmost cell, last case

Demonstration 21.7: Entering Data

To enter data, simply select the appropriate cell in the Data Editor, making it the active cell. When you do that, you will notice that the record number and the variable name appear in the rectangle at the top left-hand corner of the Data Editor, directly below the toolbar. You can now enter the appropriate value.

When you enter your data, you will notice that they first appear in the *Cell Editor* at the top of the Data Editor window, just below the toolbar. Simply press **Enter** or **Tab** or move to another cell to record your data.

SPSS Statistics Command 21.6:
Entering Numeric Data[6]

In Data View → Click appropriate cell → Type value → **Enter** OR move to another cell to record data

---

[6]We specify "Numeric" because the data we work with in the text are numeric. More important, to enter anything other than numeric data in SPSS Statistics, you must first specify the variable type. Consequently, the commands for entering nonnumeric data are somewhat different from those discussed here.

Demonstration 21.8: Revising or Deleting Data

Data may be changed at any time just by moving to the appropriate cell, keying in new values, and pressing **Enter** or moving to another cell.

If a particular case turns into a disaster, you can get rid of the entire case by clicking on the record number at the extreme left of a record and pressing the **Delete** key.

SPSS Statistics Command 21.7: Deleting an Entire Case

Click record number at far left side → **Delete**

Demonstration 21.9: Saving Your Data File

After you are done entering data, or if you want to stop entering data and continue at a later time, make sure you save your data. By clicking **File** and **Save**, you will save your data under the name used at the beginning of the session (for instance, LOCAL.SAV or whatever other name you used when you saved the LOCAL.SAV file on your local or network disk or drive).

If you wish to save it under another name, click on **File** and **Save As**. Be sure to note to which drive or other device you save.

Writing a Research Report

After you design and execute your survey and enter and analyze the data, finally it is time to report your results. To help you through this crucial aspect of the research process, we have included an excerpt on the *Adventures in Social Research* website (http://www.sagepub.com/babbie8estudy/) that, among other things, presents guidelines for reporting analyses you may find useful.

In addition, many research methods texts (including some of those we referenced earlier, such as Schutt, Alreck and Settle, Babbie, and others) devote sections to writing research reports.

You may also want to reference one of the many books devoted to writing, particularly those designed with social scientists in mind, such as Howard Becker's *Writing for Social Scientists* (1986) or Lee Cuba's *A Short Guide to Writing About Social Science* (2001).

Conclusion

Although a fair amount of work is involved in doing your own survey, entering and analyzing your own data, and then reporting your results, there can also be a special reward or excitement about coming to an understanding of the opinions and behaviors of a group of people with whom you are directly familiar. The benefits of conducting your own study are even more evident if your survey focuses on topics and issues that are meaningful, relevant, or otherwise important to you or to those in your community.

In fact, the main advantage of designing and executing your own original or ***primary research*** is that you can focus specifically on those issues, topics, populations, and variables that interest you. Since you are in charge of the design, implementation, analysis, and reporting of your study, you do not have to depend on the work of other social scientists who may have approached the issue, topic, or problem differently. The downside, of course, is the fact that the time and resources involved in conducting your own study can often be tremendous.

Nevertheless, we think you will find that even though both primary and secondary research have various advantages and disadvantages, the rewards and excitement that come from both types make the time, energy, and commitment involved worth it.

Main Points

■ This chapter is designed to help you construct and administer your own survey, enter the data in SPSS Statistics, and write a research report.

■ The *Adventures in Social Research* website (http://www.sagepub.com/babbie8estudy/) includes an excerpt designed to give you a sense of the steps involved in the social research process and some hints on writing a research proposal (often the first step in the research process).

■ Various considerations are involved in constructing and implementing a survey that is reliable, valid, and generalizable.

■ "SPSS Survey Tips" gives you an overview of how to design and execute your own survey.

■ Appendix B of this book contains a sample questionnaire with questions drawn from the 2010 General Social Survey.

■ There are three main steps involved in getting ready for data analysis using SPSS Statistics:

- Step 1: Defining your data
- Step 2: Editing and coding your data
- Step 3: Entering your raw data

■ The LOCAL.SAV file includes the data definition for the sample questionnaire.

■ The excerpt on the *Adventures in Social Research* website (http://www.sagepub.com/babbie8estudy/) discusses one of the most important and difficult aspects of the social research process: writing a research report.

Key Terms

Data definition	Active cell
Open-ended questions	Closed-ended questions
Cell Editor	Coding
Edge-coded	Primary research

SPSS Statistics Commands Introduced in This Chapter

21.1 Defining a Variable

21.2 Copying a Variable

21.3 Saving a New Data File

21.4 Accessing File Information for Coding and Editing

21.5 Some Tips for Moving Through Data View

21.6 Entering Numeric Data

21.7 Deleting an Entire Case

Review Questions

1. Name two methods social scientists use to collect data (besides survey research).

2. Name three decisions involved in designing a social research project.

3. Why is it important to write a research proposal?

4. Briefly describe at least four of the tips included in "SPSS Survey Tips."

5. List the three primary steps involved in getting ready for data analysis using SPSS Statistics.

6. To what does "data definition" refer?

7. Name two guidelines or rules for choosing a variable name that SPSS Statistics will accept.

8. What is a shortcut that can be used to define variables that share similar attributes?

9. What does the LOCAL.SAV file on the *Adventures in Social Research* website contain?

10. List two things you need to do when editing and coding your data.

11. List two ways you can move through the data matrix.

12. When you key in data, where does it appear before it is recorded (i.e., before you press **Enter** or move to another cell)?

13. Name two considerations involved in writing a research report.

14. Name two guidelines for reporting analyses.

NAME _____

CLASS _____

INSTRUCTOR _____

DATE _____

Our goal in this exercise is to allow you to practice defining, editing, coding, and entering data that you collect from a small number of respondents.

1. Make 15 copies of the sample questionnaire available in Appendix B and also on the *Adventures in Social Research* website (as an alternative, make 15 copies of a questionnaire you designed on your own). Then administer the questionnaire to 15 respondents (since this is practice only, it can be friends, family, colleagues, etc.).

2. After collecting the data, go through the three steps involved in getting data ready for analysis using SPSS Statistics:

   a. Define your data.

   b. Edit and code your data.

   c. Enter your raw data.

3. After you have entered your data, choose five variables, run Frequencies, and attach your output to this sheet.

4. Access the SPSS Statistics Tutorial **(Help → Tutorial).** Then work through the portion labeled "Using the Data Editor."

Chapter 22 **Further Opportunities for Social Research**

Well, we've come to the end of our introduction to SPSS Statistics. In this final chapter, we want to suggest ways in which you can expand beyond the scope of this book. In particular, we are going to focus briefly on the unabridged General Social Survey (GSS), other data sets, and other computer programs you may find useful as you continue your social research adventure.

The Unabridged GSS

The GSS data set that accompanies this book, Adventures.SAV, includes 150 variables. At the same time, you should realize that the data available through the GSS are vastly more extensive. In 2010, researchers asked 2,044 respondents hundreds of questions. The cumulative data file that merges all the General Social Surveys conducted between 1972 and 2010 contains more than 5,400 variables and 50,000 cases.

Since the first GSS was conducted in 1972, data have been collected on thousands of variables of interest to researchers around the country and, indeed, around the world. As we noted earlier, the National Opinion Research Center (NORC) maintains an Internet site called GSSDIRS (General Social Survey Data and Information Retrieval System). You can access the GSSDIRS at http://www.norc.org/GSS+Website/.

The GSSDIRS site houses an interactive cumulative index to the GSS, indexing variables by subject as well as abbreviated variable name. In addition, it contains a search engine that locates abstracts of research reports produced from GSS data, an electronic edition of the complete GSS codebook, an extensive GSS bibliography, and a data extraction system that permits downloading GSS data in a format that can be used with any version of SPSS Statistics.

Now that you are familiar with the use of GSS data through SPSS Statistics, you may be able to locate other GSS data sets through your school. You can get additional information about accessing GSS data from the following source:

The Roper Center for Public Opinion Research

www.ropercenter.uconn.edu → **Quick Links** → **General Social Surveys**

OR

www.ropercenter.uconn.edu/data_access/data/datasets/general_social_survey
.html

An important reference book for your use with the data is the *GSS 1972–2010 Cumulative Codebook* (2011), published by James A. Davis, Tom W. Smith, and Peter V. Marsden. Available both in book and computer form, the *Cumulative Codebook* may be purchased from the Roper Center. Or, if you wish, use the electronic version mentioned above. It is available at the *Adventures in Social Research* website (http://www.sagepub.com/babbie8estudy/).

The 2011 edition of the *Cumulative Codebook* (Davis et al.) contains information on all aspects of the GSS from 1972 to 2010, including exact questions, response codes, frequency counts, and so on. The table of contents for the latest edition, which is shown below, will give you an overview of the type of information included in this valuable resource.

Cumulative Codebook **Table of Contents**

Introduction

New Developments

Abbreviations and Data Identification Numbers

Index to Data Set

Codes for the 1972 to 2010 Surveys (Includes variable names, exact question text, response categories and codes, and unweighted marginal frequencies for questions in selected years)

General Social Survey Topical Modules

International Social Survey Program (ISSP) Modules

Appendices

Here is a sample of some of the information available to you in the main body of the *Cumulative Codebook*. This selection shows how coding has been accomplished for two of the abortion items, ABDEFECT and ABNOMORE, included in your Adventures.SAV file.

210. Please tell me whether or not you think it should be possible for a pregnant woman to obtain a legal abortion if . . . READ EACH STATEMENT, AND CIRCLE ONE CODE FOR EACH.

In addition to the advantage of working with many more variables than we've included in the present file, the larger GSS data set permits you to conduct longitudinal research—to analyze changes in opinions and behaviors over the years.

The initial experience of the GSS data set in connection with this book has put you in touch with a powerful research resource that we hope you'll be able to use more extensively in the future.

Other Data Sets

As if the full GSS weren't enough, thousands of other data sets exist that are appropriate for analysis with SPSS Statistics.

To begin, there is a global network of data archives, or data libraries, which operate somewhat like book libraries. Instead of lending books, these archives lend or sell sets of data that have been collected previously. NORC, which administers the GSS, is one such archive. Other major archives include the Roper Center for Public Opinion Research and ICPSR at University of Michigan.

210. Please tell me whether or not you think it should be possible for a pregnant woman to obtain a legal abortion if. . . READ EACH STATEMENT, AND CIRCLE ONE CODE FOR EACH.

A. If there is a strong chance of serious defect in the baby?

[VAR: ABDEFECT]

Response	Punch	Year																All
		1972-82	1982B	1983-87	1987B	1988-91	1993-96	1998	2000	2002	2004	2006	2008	2010				
Yes	1	10980	236	4625	236	3056	3939	1415	1397	703	624	1425	952	921				30509
No	2	2195	92	1210	94	707	870	385	377	193	232	495	321	304				7475
Don't know	8	419	25	187	21	141	171	79	83	24	33	68	61	32				1344
No answer	9	32	1	50	2	16	14	3	4	4	9	15	18	24				192
Not applicable	BK	0	0	1470	0	1987	2508	950	956	1841	1914	2507	671	763				15567

B. If she is married and does not want any more children?

[VAR: ABNOMORE]

Response	Punch	Year																All
		1972-82	1982B	1983-87	1987B	1988-91	1993-96	1998	2000	2002	2004	2006	2008	2010				
Yes	1	5920	111	2387	125	1640	2263	758	720	400	360	818	579	587				16668
No	2	7133	216	3452	206	2129	2506	1033	1050	493	502	1107	716	636				21179
Don't know	8	542	26	181	20	132	207	86	85	29	30	65	40	36				1479
No answer	9	31	1	52	2	19	18	5	6	2	6	13	17	22				194
Not applicable	BK	0	0	1470	0	1987	2508	950	956	1841	1914	2507	671	763				15567

A great deal of the data housed in data archives today is accessible via the World Wide Web. We have already given you the web addresses for NORC and the Roper Center. Other academic research centers, some of which include similar archives or links to archives, are as follows:

- Survey Research Center, at Princeton University (http://www.princeton.edu/~psrc/)
- Inter-University Consortium for Political and Social Research, at the University of Michigan (http://www.icpsr.umich.edu/icpsrweb/ICPSR/)
- Survey Research Center, at the University of Michigan (http://www.src.isr.umich.edu/)
- Marist College Institute for Public Opinion, at Marist College (http://maristpoll.marist.edu/)

For a more complete list of links, visit the Roper Center at http://www.ropercenter.uconn.edu/research/other_data_archives.html.

The data sets available for *secondary analysis* at these and other archives include, for example, studies conducted by university faculty researchers that may have been financed by federal research grants. Thus, the results of studies that may have cost hundreds of thousands of dollars to conduct can be yours now for the nominal cost of copying and shipping the data.

The Roper Center, for instance, is a repository not only for all data on American public opinion but for data collected in other countries as well. For instance, the Center maintains a Latin American Data Bank that acquires, processes, and archives public opinion surveys conducted by the survey research community in Latin America. Other archives maintain similar data banks for opinion collected by researchers around the world. To begin, you may want to access the following site for a list of some of the many international links (lower portion of the page, under "International"): http://www.ropercenter.uconn.edu/research/other_data_archives.html.

You probably also realize that, like many governments, the U.S. government conducts a great many surveys that produce data suitable for analysis with SPSS Statistics. The U.S. Census is a chief example, and it is possible for researchers to obtain and analyze data collected in the decennial censuses. You can visit the U.S. Census Bureau online at http://www.census.gov/.

Other examples include the following:

- Bureau of Justice Statistics (http://www.ojp.usdoj.gov/bjs)
- Bureau of Labor Statistics (http://www.bls.gov/data/)

For access to an almost full range of statistics produced by more than 100 U.S. federal agencies, you may want to log on to FedStats: One Stop Shopping for Federal Statistics (http://www.fedstats.gov/).

It is worth recalling that when Émile Durkheim set about his major analysis of suicide in Europe, he was forced to work with printed government reports of suicide rates in various countries and regions. Moreover, his analysis needed to be done by hand. With the advent of computers for mass data storage and analytic programs such as SPSS Statistics, the possibilities for secondary analysis have been revolutionized. You live at an enviable time in that regard.

This is, of course, just the tip of the iceberg. These limited recommendations don't even begin to scratch the surface of the numerous sites available to help you locate and access statistical data on the World Wide Web. While you should make use of the web to locate and, in some cases, access data, please bear in mind that you also have to be cautious and wary when conducting online searches. This is particularly true in light of the fact that anyone can post information online, so the accuracy of the data or other information you locate may not necessarily be conducive to sound social research. Also bear in mind that different search engines are likely to identify different sites, even when provided with the same search terms.

Other Computer Programs

We have organized this book around the use of SPSS Statistics because this program is so widely used by social scientists. Quite frankly, we spent a fair amount of time considering other programs as the focus for the book, but we ultimately decided that it would be most useful to the greatest number of students if we used SPSS Statistics. And by the way, the popularity of SPSS Statistics is due largely to its excellence as a tool for social research.

At the same time, we want you to know that there are several other excellent analytic programs available. We'll mention some of these here in the event that you may have access to them. You should realize that the commands and procedures used in other programs will differ from those of SPSS Statistics, but the logic of data analysis that we've presented in this book applies across the various programs available.

The program package that is probably the next most commonly used by social scientists is *SAS (Statistical Analysis System)*. Like SPSS Statistics, it is an omnibus package of techniques that goes far beyond those we've introduced you to in this book. Should it be useful to you, SAS functions within a broader set of programs for accounting, management, and other activities. You can learn more about SAS by accessing the SAS home page at http://www.sas.com/.

SYSTAT is another widely used package of programs designed for social science research. Like SAS and SPSS Statistics, SYSTAT is also an omnibus package with graphing, statistical, and other analytical capabilities much more advanced than those we have discussed here. SYSTAT was developed by Dr. Leland Wilkinson while a professor at the University of Illinois, Chicago. For a complete overview of this software, its general features, and system requirements, see the SYSTAT home page at http://www.systat.com.

A few other programs you may find useful for social science data analysis include Minitab (http://www.minitab.com/), P-STAT (http://www.pstat.com/), and STATA (http://www.stata.com/).

Conclusion

We hope that you've had some fun as you've worked through this book and that you've learned some important research techniques and some facts about the American public. We hope you have discovered that social research, although very important—especially considering the range of social problems we confront today—is also a fascinating enterprise.

Like the investigative detective, the social researcher must possess large amounts of curiosity and ingenuity. Sit beside a social researcher at work, and you'll hear him or her muttering things such as, "Wait a minute. If that's the case, then I'd expect to find . . ."; "Hey! That probably explains why . . ."; "OMG! That means. . . ." Maybe you've caught yourself practicing these social scientific incantations. If so, congratulations and welcome. If not, you have a delightful adventure waiting for you just around the corner. We'll see you there.

Main Points

- The unabridged 2010 General Social Survey (GSS) contains 2,044 cases and 790 variables.
- The Adventures.SAV file includes all 2,044 cases and 150 of the variables from the 2010 GSS.
- Since the first GSS in 1972, more than 50,000 respondents have been asked more than 5,400 different questions.
- There are many ways to access the unabridged GSS.
- The National Opinion Research Center website, the GSSDIRS, is a very useful resource.
- The *Cumulative Codebook* is a valuable resource for researchers using the GSS.
- The GSS is just one of many data sets available to social researchers.

- Data sets of interest to social scientists (of which there are many) are often housed in one of the data archives, and in some cases, can be accessed via the World Wide Web.
- The federal government is an important source of statistical data of interest to social scientists.
- We focused this text on SPSS Statistics because it is so widely used by social researchers. It is easy for beginners to learn and also has a great expanse of capabilities and power for the advanced user; however, it is not the only statistical package available.
- In addition to SPSS Statistics, other excellent software programs often used by social researchers include (but are not limited to) SAS, SYSTAT, and STATA.

Key Terms

Secondary analysis
SYSTAT
SAS (Statistical Analysis System)

SPSS Statistics Commands Introduced in This Chapter

No new commands were introduced in this chapter.

Review Questions

1. How many variables are contained on your Adventures.SAV file?

2. How many cases are contained on your Adventures.SAV file?

3. How many variables are contained in the entire 2010 General Social Survey (GSS)?

4. How many respondents took part in the GSS in 2010?

5. What is the GSSDIRS, and what type of information does it contain?

6. How can you access the GSS data?

7. What type of information does the *Cumulative Codebook* contain?

8. What are data archives?

9. Name two data archives.

10. Name two specific data sets relevant to your particular discipline.

11. Besides SPSS Statistics, what statistical package is most commonly used by social scientists?

APPENDICES

APPENDIX A: THE CODEBOOK

This codebook contains the following information for each of the items contained on your Adventures.SAV file: abbreviated variable name, the full wording used in the interview (item wording), values, labels, and level of measurement. The Adventures.SAV data file, formatted for use with SPSS Statistics, can be downloaded from the *Adventures in Social Research* student study website at http://www.sagepub.com/babbie8estudy/.

Please note that we have excluded those values and labels normally designated as "missing" (NA—No answer, DK—Don't know, IAP—Inapplicable [formerly NAP—Not applicable], Can't choose). You can use SPSS Statistics to identify these values/labels in a number of different ways. You can, for instance, click on the **Variable View** tab and then locate the information in the Values column. Alternatively, you can select **Utilities → Variables.**

Adventures.SAV

As noted in Chapter 3, Adventures.SAV is used in the demonstrations in the body of the chapters and is also to be used to complete the lab exercises at the end of each chapter (starting with Chapter 4). This subsample contains all 2,044 cases and 150 of the 790 variables drawn from the 2010 General Social Survey. The items are arranged according to the following subject categories:

Abortion
Children
Family
Religion
Social–political opinions

Sexual attitudes and behaviors
Gender roles
Health
Police
Mass media
National government spending priorities
Teen sex
Affirmative action and equalization
Educators and education
Voting and politics
Science and the environment
Respondent's parents (education and occupational prestige)
The Internet
Social life
Demographics

Abortion

Please tell me whether or not *you* think it should be possible for a pregnant woman to obtain a legal abortion if . . . .

ABANY   The woman wants it for any reason.
         [Nominal]
         1 Yes
         2 No

ABDEFECT There is a strong chance of serious defect in the baby.
          [Nominal]
          1 Yes
          2 No

ABHLTH   The woman's own health is seriously endangered by the pregnancy.
         [Nominal]
         1 Yes
         2 No

ABNOMORE She is married and does not want any more children.
[Nominal]
1 Yes
2 No

ABPOOR The family has a very low income and cannot afford any more children.
[Nominal]
1 Yes
2 No

ABRAPE She became pregnant as a result of rape.
[Nominal]
1 Yes
2 No

ABSINGLE She is not married and does not want to marry the man.
[Nominal]
1 Yes
2 No

Children

CHLDIDEL What do you think is the ideal number of children for a family to have?
[Interval/ratio (I/R)][1]
0 None
1 One
2 Two
3 Three
4 Four
5 Five
6 Six
7 Seven or more
8 As many as you want

If you had to choose, which thing on this list would you pick as the most important for a child to learn to prepare him or her for life: to obey, to be well liked or popular, to think for himself or herself, to work hard, to help others when they need help? Which comes next in importance? Which comes third? Which comes fourth?

OBEY To obey
[Ordinal]
1 Most important
2 Second important
3 Third important
4 Fourth important
5 Least important

POPULAR To be well liked or popular
[Ordinal]
1 Most important
2 Second important
3 Third important
4 Fourth important
5 Least important

THNKSELF To think for himself or herself
[Ordinal]
1 Most important
2 Second important
3 Third important
4 Fourth important
5 Least important

WORKHARD To work hard
[Ordinal]
1 Most important
2 Second important
3 Third important
4 Fourth important
5 Least important

HELPOTH To help others when they need help
[Ordinal]
1 Most important
2 Second important

[1]SPSS Statistics refers to interval and ratio data as "scale," as opposed to distinguishing between the two. In order to simplify reference, we will use "I/R" to

3 Third important
4 Fourth important
5 Least important

Family

MARITAL
Are you currently married, widowed, divorced, separated, or have you never been married?
[Nominal]
1 Married
2 Widowed
3 Divorced
4 Separated
5 Never married

DIVORCE
[If currently married or widowed] Have you ever been divorced or legally separated?
[Nominal]
1 Yes
2 No

DIVLAW
Should divorce in this country be easier or more difficult to obtain than it is now?
[Nominal]
1 Easier
2 More difficult
3 Stay as is

HAPMAR
Taking things all together, how would you describe your marriage? Would you say that your marriage is very happy, pretty happy, or not too happy?
[Ordinal]
1 Very happy
2 Pretty happy
3 Not too happy

CHILDS
How many children have you ever had? Please count all that were born alive at any time (including any you had from a previous marriage).
[I/R*]
O None
1 One
2 Two
3 Three
4 Four
5 Five
6 Six
7 Seven
8 Eight or more

SIBS
How many brothers and sisters did you have? Please count those born alive but no longer living, as well as those alive now. Also include stepbrothers and stepsisters, and children adopted by your parents.
[I/R]
Responses in actual number (i.e., 00, 01, 02, etc.)

Religion

ATTEND
How often do you attend religious services?
[Ordinal]
0 Never
1 Less than once a year
2 Once a year
3 Several times a year
4 Once a month
5 Two to three times a month
6 Nearly every week
7 Every week
8 More than once a week

POSTLIFE Do you believe there is a life after death?
[Nominal]
1 Yes
2 No
8 Undecided

PRAY About how often do you pray?
[Ordinal]
1 Several times a day
2 Once a day
3 Several times a week
4 Once a week
5 Less than once a week
6 Never

RELIG What is your religious preference? Is it Protestant, Catholic, Jewish, some other religion, or no religion?
[Nominal]
1 Protestant
2 Catholic
3 Jewish
4 None
5 Other (specify)
6 Buddhism
7 Hinduism
8 Other Eastern
9 Moslem/Islam
10 Orthodox Christian
11 Christian
12 Native American
13 Inter-nondenominational

RELIG16 In what religion were you raised?
[Nominal]
1 Protestant
2 Catholic
3 Jewish
4 None
5 Other
6 Buddhism
7 Hinduism
8 Other Eastern
9 Moslem/Islam
10 Orthodox-Christian
11 Christian
12 Native American
13 Inter-nondenomenational

GOD Which statement comes closest to expressing what you believe about God?
[Nominal]
1 Don't believe
2 Don't know, no way to find out
3 Higher power
4 Believe sometimes
5 Believe with doubts
6 No doubts

Social–Political Opinions

CAPPUN Do you favor or oppose the death penalty for persons convicted of murder?
[Nominal]
1 Favor
2 Oppose

GETAHEAD

Some people say that people get ahead by their own hard work; others say that lucky breaks or help from other people are more important. Which do you think is most important?

[Ordinal]
1 Hard work
2 Both equally
3 Luck or help
4 Other

GUNLAW

Would you favor or oppose a law that would require a person to obtain a police permit before he or she could buy a gun?

[Nominal]
1 Favor
2 Oppose

GRASS

Do you think the use of marijuana should be made legal or not?

[Nominal]
1 Should
2 Should not

MARHOMO

Homosexual couples should have the right to marry one another.

[Ordinal]
1 Strongly agree
2 Agree
3 Neither agree or disagree
4 Disagree
5 Strongly disagree

LETDIE1

When a person has a disease that cannot be cured, do you think doctors should be allowed by law to end the patient's life by some painless means if the patient and his family request it?

[Nominal]
1 Yes
2 No

PORNLAW

Which of these statements comes closest to your feelings about pornography laws?

[Ordinal]
1 There should be laws against the distribution of pornography whatever the age.
2 There should be laws against the distribution of pornography to persons under 18.
3 There should be no laws forbidding the distribution of pornography.

SPANKING

Do you strongly agree, agree, disagree, or strongly disagree that it is sometimes necessary to discipline a child with a good, hard spanking?

[Ordinal]
1 Strongly agree
2 Agree
3 Disagree
4 Strongly disagree

TRDUNION

Workers need strong trade unions to protect their interests.

[Ordinal]
1 Strongly agree
2 Agree
3 Disagree
4 Strongly disagree

Note: For the following six questions, interviewers were instructed to use the term Black or African American, depending on the customary usage in their area.

NATRACE

We are faced with many problems in this country, none of which can be solved easily or inexpensively. I'm going to name some of these problems, and for each one, I'd like you to tell me whether you think we're spending too much money on it, too little money, or about the right amount. Are we spending too much money, too little money, or about the right amount on improving the conditions of Blacks?
[Ordinal]
1 Too little
2 About right
3 Too much

Sexual Attitudes and Behaviors

HOMOSEX

What about sexual relations between two adults of the same sex—do you think it is always wrong, almost always wrong, wrong only sometimes, or not wrong at all?
[Ordinal]
1 Always wrong
2 Almost always wrong
3 Sometimes wrong
4 Not wrong at all
5 Other

XMOVIE

Have you seen an X-rated movie in the past year?
[Nominal]
1 Yes
2 No

PREMARSX

There's been a lot of discussion about the way morals and attitudes about sex are changing in this country. If a man and woman have sexual relations before marriage, do you think it is always wrong, almost always wrong, wrong only sometimes, or not wrong at all?
[Ordinal]
1 Always wrong

RACDIF1

On the average, (Blacks/African Americans) have worse jobs, income, and housing than White people. Do you think these differences are . . .

Mainly due to discrimination?
[Nominal]
1 Yes
2 No

RACDIF2

Because most (Blacks/African Americans) have less in-born ability to learn?
[Nominal]
1 Yes
2 No

RACDIF3

Because most (Blacks/African Americans) don't have the chance for education that it takes to rise out of poverty?
[Nominal]
1 Yes
2 No

RACDIF4

Because most (Blacks/African Americans) just don't have the motivation or will power to pull themselves up out of poverty?
[Nominal]
1 Yes
2 No

MARBLK

How about having a close relative or family member marry a Black person? Would you be very in favor of its happening, somewhat in favor, neither in favor nor opposed to its happening, somewhat opposed, or very opposed to its happening?
[Ordinal]
1 Strongly favor
2 Favor
3 Neither favor nor oppose
4 Oppose

3 Sometimes wrong
4 Not wrong at all
5 Other

EVPAIDSX Thinking about the time since your 18th birthday, have you ever had sex with a person you paid or who paid you for sex?
[Nominal]
1 Yes
2 No

EVSTRAY Have you ever had sex with someone other than your husband or wife while you were married?
[Nominal]
1 Yes
2 No
3 Never married

FRNDSEX [Have you had sexual partner(s) in the past 12 months who were . . .] a close personal friend?
[Nominal]
1 Yes
2 No

NUMMEN Now thinking about the time since your 18th birthday (including the past 12 months) how many male partners have you had sex with?
[I/R]
Actual number
989 More than 989

NUMWOMEN Now thinking about the time since your 18th birthday (including the past 12 months) how many female partners have you had sex with?
[I/R]
Actual number
989 More than 989

PAIDSEX [Have you had sexual partner(s) in the past 12 months who were . . .] a person you paid or were paid for sex?

[Nominal]
1 Yes
2 No

PARTNERS How many sex partners have you had in the past 12 months?
[Ordinal]
0 No partners
1 1 partner
2 2 partners
3 3 partners
4 4 partners
5 5–10 partners
6 11–20 partners
7 21–100 partners
8 More than 100 partners

PARTNRS5 Now think about the past 5 years—the time since February/March [2005], and including the past 12 months—how many sex partners have you had in that 5-year period?
[Ordinal]
0 No partners
1 1 partner
2 2 partners
3 3 partners
4 4 partners
5 5–10 partners
6 11–20 partners
7 21–100 partners
8 More than 100 partners

PIKUPSEX [Have you had sexual partner(s) in the past 12 months who were . . .] a casual date or pick up?
[Nominal]
1 Yes
2 No

Gender Roles

Now I'm going to read several statements. As I read each one, please tell me whether you strongly agree, agree, disagree, or strongly disagree with it. For example, here is the statement.

FEPOL Tell me if you agree or disagree with this statement: Most men are better suited emotionally for politics than are most women.
[Nominal]
1 Agree
2 Disagree

FECHLD A working mother can establish just as warm and secure a relationship with her children as a mother who doesn't work.
[Ordinal]
1 Strongly agree
2 Agree
3 Disagree
4 Strongly disagree

FEPRESCH A preschool child is likely to suffer if his or her mother works.
[Ordinal]
1 Strongly agree
2 Agree
3 Disagree
4 Strongly disagree

FEFAM It is much better for everyone involved if the man is the achiever outside the home and the woman takes care of the home and family.
[Ordinal]
1 Strongly agree
2 Agree
3 Disagree
4 Strongly disagree

SEXFREQ How often did you have sex during the past 12 months?
[Ordinal]
0 Not at all
1 Once or twice
2 About once a month
3 Two or three times a month
4 About once a week
5 Two or three times a week
6 More than three times a week

SEXORNT Which of the following best describes you?
[Ordinal]
1 Gay, lesbian, or homosexual
2 Bisexual
3 Heterosexual or straight

SEXSEX Have your sex partners in the past 12 months been . . . .
[Ordinal]
1 Exclusively male
2 Both male and female
3 Exclusively female

SEXSEX5 Have your sex partners in the past 5 years been . . . .
[Ordinal]
1 Exclusively male
2 Both male and female
3 Exclusively female

XMARSEX What is your opinion about a married person having sexual relations with someone other than the marriage partner—is it always wrong, almost always wrong, wrong only sometimes, or not wrong at all?
[Ordinal]
1 Always wrong
2 Almost always wrong
3 Wrong only sometimes
4 Not wrong at all

Now I'm going to read several statements. As I read each one, please tell me whether you strongly agree, agree, neither agree nor disagree, disagree, or strongly disagree.

FEHIRE   Because of past discrimination, employers should make special efforts to hire and promote qualified women.

[Ordinal]
Strongly agree
Agree
Neither agree nor disagree
Disagree
Strongly disagree

Health (Emotional and Physical)

HEALTH   Would you say your own health, in general, is excellent, good, fair, or poor?

[Ordinal]
Excellent
Good
Fair
Poor

HAPPY   Taken all together, how would you say things are these days—would you say that you are very happy, pretty happy, or not too happy?

[Ordinal]
1 Very happy
2 Pretty happy
3 Not too happy

SATFIN   We are interested in how people are getting along financially these days. So far as you and your family are concerned, would you say that you are pretty well satisfied with your present financial situation, more or less satisfied, or not satisfied at all?

[Ordinal]
1 Pretty well satisfied
2 More or less satisfied
3 Not satisfied at all

SATJOB   On the whole, how satisfied are you with the work you do—would you say you are very satisfied, moderately satisfied, a little dissatisfied, or very dissatisfied?

[Ordinal]
1 Very satisfied
2 Moderately satisfied
3 A little dissatisfied
4 Very dissatisfied

Police

POLHITOK   Are there any situations you can imagine in which you would approve of a policeman striking an adult male citizen?

[Nominal]
1 Yes
2 No
3 Not sure

If Yes or Not Sure: Would you approve of a policeman striking a citizen who . . .

POLABUSE   Had said vulgar and obscene things to the policeman?

[Nominal]
1 Yes
2 No

POLMURDR   Was being questioned as a suspect in a murder case?

[Nominal]
1 Yes
2 No

POLESCAP Was attempting to escape from custody?
[Nominal]
1 Yes
2 No

POLATTAK Was attacking the policeman with his fists?
[Nominal]
1 Yes
2 No

Mass Media

NEWS How often do you read the newspaper—every day, a few times a week, once a week, less than once a week, or never?
[Ordinal]
1 Every day
2 Few times a week
3 Once a week
4 Less than once a week
5 Never

TVHOURS On the average day, about how many hours do you personally watch television?
[I/R]
Responses in actual hours

I am going to name some institutions in this country. As far as the people running these institutions are concerned, would you say you have a great deal of confidence, only some confidence, or hardly any confidence at all in them?

CONPRESS Press?
[Ordinal]
1 A great deal
2 Only some
3 Hardly any

CONTV TV?
[Ordinal]
1 A great deal
2 Only some
3 Hardly any

National Government Spending Priorities

We are faced with many problems in this country, none of which can be solved easily or inexpensively. I'm going to name some of these problems, and for each one, I'd like you to tell me whether you think we're spending too much money on it, too little money, or about the right amount.

Are we spending too much money, too little money, or about the right amount on . . .

NATHEAL Improving and protecting the nation's health?
[Ordinal]
1 Too little
2 About right
3 Too much

NATCITY Solving the problems of the big cities?
[Ordinal]
1 Too little
2 About right
3 Too much

NATCRIME Halting the rising crime rate?
[Ordinal]
1 Too little
2 About right
3 Too much

NATDRUG Dealing with drug addiction?
[Ordinal]
1 Too little
2 About right
3 Too much

NATEDUC Improving the nation's education system?
[Ordinal]
1 Too little
2 About right
3 Too much

NATRACE Improving the conditions of Blacks?
[Ordinal]
1 Too little

2 About right
3 Too much

NATFARE Welfare?
[Ordinal]
1 Too little
2 About right
3 Too much

NATCHLD Assistance for children?
[Ordinal]
1 Too little
2 About right
3 Too much

NATPARK Parks and recreation?
[Ordinal]
1 Too little
2 About right
3 Too much

Teen Sex

PILLOK Do you strongly agree, agree, disagree, or strongly disagree that methods of birth control should be available to teenagers between the ages of 14 and 16 if their parents do not approve?
[Ordinal]
1 Strongly agree
2 Agree
3 Disagree
4 Strongly disagree

SEXEDUC Would you be for or against sex education in the public schools?
[Nominal]
1 For
2 Against
3 Depends

TEENSEX What if they are in their early teens, say, 14 to 16 years old? In that case, do you think sex relations before marriage are always wrong, almost always wrong, wrong only sometimes, or not wrong at all?
[Ordinal]
1 Always wrong
2 Almost always wrong
3 Wrong only sometimes
4 Not wrong at all

Affirmative Action and Equalization

AFFRMACT Some people say that because of past discrimination, Blacks should be given preference in hiring and promotion. Others say that such preference in hiring and promotion of Blacks is wrong because it discriminates against Whites. What about your opinion? Are you for or against preferential hiring and promotion of Blacks? IF FAVOR: Do you favor preference in hiring and promotion strongly or not strongly? IF OPPOSE: Do you oppose preference in hiring and promotion strongly or not strongly?
[Ordinal]
1 Strongly support preference
2 Support preference
3 Oppose preference
4 Strongly oppose preference

WRKWAYUP Do you agree strongly, agree somewhat, neither agree nor disagree, disagree somewhat, or disagree strongly with the following statement?
Irish, Italians, Jewish and many other minorities overcame prejudice and worked their way up. Blacks should do the same without special favors.
[Ordinal]
1 Agree strongly
2 Agree somewhat
3 Neither agree nor disagree
4 Disagree somewhat
5 Disagree strongly

RACWORK Are the people who work where you work all Whites, mostly Whites, about half and half, mostly Blacks, or all Blacks?
[Ordinal]
1 All Whites

Educators and Education

CONEDUC   As far as the people running [education] institutions are concerned, would you say you have a great deal of confidence, only some confidence, or hardly any confidence at all in them?
[Ordinal]
1   A great deal
2   Only some
3   Hardly any

COLATH   What about somebody who is against all churches and religion . . . should such a person be allowed to teach in a college or university, or not?
[Nominal]
4   Yes, allowed to teach
5   Not allowed

COLCOM   What about a man who admits he is a Communist . . . should such a person be allowed to teach in a college or university, or not?
[Nominal]
4   Yes, allowed to teach
5   Not allowed

COLHOMO   What about a man who admits that he is a homosexual . . . should such a person be allowed to teach in a college or university, or not?
[Nominal]
4 Yes, allowed to teach
5 Not allowed

COLMSLM   What about a Muslim clergyman . . . should such a person be allowed to teach in a college or university, or not?
[Nominal]
4   Yes, allowed to teach
5   Not allowed

2   Mostly Whites
3   Half Whites, half Blacks
4   Mostly Blacks
5   All Blacks
6   Work alone

DISCAFF   What do you think the chances are these days that a White person won't get a job or promotion while an equally or less qualified Black person gets one instead? Is this very likely, somewhat likely, or not very likely to happen these days?
[Ordinal]
1   Very likely
2   Somewhat likely
3   Not very likely

EQWLTH   Some people think that the government in Washington ought to reduce the income differences between the rich and the poor, perhaps by raising the taxes of wealthy families or by giving income assistance to the poor. Others think that the government should not concern itself with reducing the income difference between the rich and the poor. Here is a card with a scale from 1 to 7. Think of a score of 1 as meaning that the government ought to reduce the income differences between rich and poor and a score of 7 as meaning that the government should not concern itself with reducing income differences. What score between 1 and 7 comes closest to the way you feel?
[Ordinal]
1   Government reduce differences
2
3
4
5
6
7   No government action

COLRAC
What about a person who believes that Blacks are genetically inferior . . . should such a person be allowed to teach in a college or university, or not?
[Nominal]
4 Yes, allowed to teach
5 Not allowed

Voting and Politics

VOTE08
In 2008, you remember that Obama ran for president on the Democratic ticket against McCain for the Republicans. Do you remember for sure whether or not you voted in that election?
[Nominal]
1 Voted
2 Did not vote
3 Ineligible

PRES08
[If voted] Did you vote for Obama or McCain?
[Nominal]
1 Obama
2 McCain
3 Other candidate
4 Didn't vote for president

POLVIEWS
We hear a lot of talk these days about liberals and conservatives. I'm going to show you a 7-point scale on which the political views that people might hold are arranged from extremely liberal—point 1—to extremely conservative—point 7. Where would you place yourself on this scale?
[Ordinal]
1 Extremely liberal
2 Liberal
3 Slightly liberal
4 Moderate
5 Slightly conservative
6 Conservative
7 Extremely conservative

PARTYID
Generally speaking, do you usually think of yourself as a Republican, Democrat, Independent, or what?
[Ordinal]
0 Strong Democrat
1 Not strong Democrat
2 Independent, near Democrat
3 Independent
4 Independent, near Republican
5 Not strong Republican
6 Strong Republican
7 Other party

Science and the Environment

BIGBANG
The universe began with a huge explosion. (Is that true or false?)
[Nominal]
1 True
2 False

EVOLVED
Human beings, as we know them today, developed from earlier species of animals. (Is that true or false?)
[Nominal]
1 True
2 False

GRNDEMO
In the past 5 years, have you taken part in a protest or demonstration about an environmental issue?
[Nominal]
1 Yes, I have
2 No, I have not

GRNECON
How much do you agree or disagree with the following statement: *We worry too much about the future of the environment and not enough about prices and jobs today?*

GRNEXAGG

How much do you agree or disagree with the following statement: *Many of the claims about environmental threats are exaggerated?*

[Ordinal]
1 Strongly agree
2 Agree
3 Neither agree nor disagree
4 Disagree
5 Strongly disagree

GRNINTL

How much do you agree or disagree with the following statement: *For environmental problems, there should be international agreements that American and other countries should be made to follow?*

[Ordinal]
1 Strongly agree
2 Agree
3 Neither agree nor disagree
4 Disagree
5 Strongly disagree

GRNMONEY

In the past 5 years, have you ever given money to an environmental group?

[Nominal]
1 Yes, I have
2 No, I have not

GRNPRICE

How willing would you be to pay much higher prices in order to protect the environment?

[Ordinal]
1 Very willing
2 Fairly willing
3 Neither willing nor unwilling
4 Not very willing
5 Not at all willing

GRNPROG

How much do you agree or disagree with the following statement: *People worry too much about human progress harming the environment?*

[Ordinal]
1 Strongly agree
2 Agree
3 Neither agree nor disagree
4 Disagree
5 Strongly disagree

GRNSOL

How willing would you be to accept cuts in your standard of living in order to protect the environment?

[Ordinal]
1 Very willing
2 Fairly willing
3 Neither willing nor unwilling
4 Not very willing
5 Not at all willing

GRNTAXES

How willing would you be to pay much higher taxes in order to protect the environment?

[Ordinal]
1 Very willing
2 Fairly willing
3 Neither willing nor unwilling
4 Not very willing
5 Not at all willing

GRWTHARM How much do you agree or disagree with the following statement: *Economic growth always harms the environment?*
[Ordinal]
1 Strongly agree
2 Agree
3 Neither agree nor disagree
4 Disagree
5 Strongly disagree

GRWTHELP How much do you agree or disagree with the following statement: *In order to protect the environment, America needs economic growth?*
[Ordinal]
1 Strongly agree
2 Agree
3 Neither agree nor disagree
4 Disagree
5 Strongly disagree

RECYCLE How often do you make a special effort to *sort glass or cans or plastic or papers and so on* for recycling?
[Ordinal* → set #5 to missing]
1 Always
2 Often
3 Sometimes
4 Never
5 Recycling not available where I live

Respondent's Parents

MADEG Respondent's mother's highest degree
[Ordinal]
1 Less than high school
2 High school
3 Associate/junior college
4 Bachelor's
5 Graduate

MAEDUC Respondent's mother's years of education
[I/R]
Responses coded in actual years.

MAPRES80 Prestige of respondent's mother's occupation
[I/R]
Actual prestige score #

MASEI Hodge-Siegel-Rossi socioeconomic ratings for respondent's mother's occupation[2]
[I/R]
Actual SEI #

PADEG Respondent's father's highest degree
[Ordinal]
1 Less than high school
2 High school
3 Associate/junior college
4 Bachelor's
5 Graduate

PAEDUC Respondent's father's years of education
[I/R]
Responses coded in actual years

PAPRES80 Prestige of respondent's father's occupation
[I/R]
Actual prestige score #

PASEI Hodge-Siegel-Rossi socioeconomic ratings for respondent's father's occupation[2]
[I/R]
Actual SEI #

The Internet

INTRHOME Do you have access to the Internet in your home?
[Nominal]
1 Yes
2 No

---

[2]Information about the Hodge-Siegel-Rossi occupational prestige score is available in the *General Social Surveys 1972–2010 Cumulative Codebook,* Appendix F, by Davis, Smith, and Marsden (2011). See also Nakao and Treas (1994) "Updating Occupational Prestige and Socioeconomic Scores: How the New Measures Measure Up," in *Sociological Methodology,* for an in-depth discussion of the updated occupational prestige and socioeconomic scores.

Demographic, Classifying, and Identifying

ID
Respondent identification number [assigned by the interviewer]
[Nominal]

AGE
Date of birth?
[I/R]
Recoded into actual age in years

AGEKDBRN
How old were you when your first child was born?
[I/R]
Actual age at time first child was born

CLASS
If you were asked to use one of the four names for your social class, which would you say you belong to: the lower class, the working class, the middle class, or the upper class?
[Ordinal]
1 Lower class
2 Working class
3 Middle class
4 Upper class
5 No class

DWELOWN
(Do you/does your family) own your (home/apartment), pay rent, or what?
[Nominal]
1 Own or is buying
2 Pays rent
3 Other

DEGREE
Respondent's [highest] degree
[Ordinal]
1 Less than high school
2 High school
3 Associate/junior college
4 Bachelor's
5 Graduate

USEWWW
Other than for e-mail, do you ever use the Internet or World Wide Web?
[Nominal]
1 Yes
2 No

WWWHR
Not counting e-mail, about how many minutes or hours per week do you use the web? (Include time you spend visiting regular websites and time spent using interactive Internet services such as chat rooms, Usenet groups, discussion forums, bulletin boards, and the like.)
[I/R]
Actual # of hours

Social Life

SOCBAR
How often do you do the following: *Go to a bar or tavern?*
[Ordinal]
1 Almost every day
2 Once or twice a week
3 Several times a month
4 About once a month
5 Several times a year
6 About once a year
7 Never

SOCFREND
How often do you do the following: *Spend a social evening with friends who live outside the neighborhood?*
[Ordinal]
1 Almost every day
2 Once or twice a week
3 Several times a month
4 About once a month
5 Several times a year
6 About once a year
7 Never

EDUC

Years of education were coded from answers to six items respondents were asked regarding their education. Codes 0 through 12 indicate elementary and high school grades completed, codes 13 through 16 indicate years of college completed, and codes 17 through 20 indicate years of postgraduate education completed.
[I/R]
Responses coded in actual years

0   No formal education

1–12   Elementary and high school grades completed (1 = 1st grade, 2 = 2nd grade, 3 = 3rd grade, 4 = 4th grade, etc.)

13–16   Years of college completed (i.e., 13 = 1 year, 14 = 2 years, 15 = 3 years, 16 = 4 years)

17–20   Years of postgraduate education completed (i.e., 17 = 1 year, 18 = 2 years, 19 = 3 years, 20 = 4 years)

HOMPOP

Household size
[I/R]
Actual # in household

INCOME06

In which of these groups did your total family income, from all sources, fall last year before taxes?
[Ordinal]

1   UNDER $1,000
2    $1,000 to $2,999
3    $3,000 to $3,999
4    $4,000 to $4,999
5    $5,000 to $5,999
6    $6,000 to $6,999
7    $7,000 to $7,999
8    $8,000 to $9,999
9    $10,000 to $12,499
10   $12,500 to $14,999
11   $15,000 to $17,499
12   $17,500 to $19,999

13   $20,000 to $22,499
14   $22,500 to $24,999
15   $25,000 to $29,999
16   $30,000 to $34,999
17   $35,000 to $39,999
18   $40,000 to $49,999
19   $50,000 to $59,999
20   $60,000 to $74,999
21   $75,000 to $89,999
22   $90,000 to $109,999
23   $110,000 to $129,999
24   $130,000 to $149,999
25   $150,000 or over

MAJORCOL

In what field was [your highest] degree?
[Nominal]

1   Accounting/bookkeeping
2   Advertising
3   Agriculture
4   Allied health
5   Anthropology
6   Architecture
7   Art
8   Biology
9   Business administration
10* *(the number 10 was intentionally skipped in this list)*
11   Chemistry
12   Communications/speech
13   Communication disorders
14   Computer science
15   Dentistry
16   Education
17   Economics
18   Engineering
19   English
20   Finance

423

PRESTG80

Prestige of respondent's occupation

[I/R]

Actual # representing occupational prestige score (1980 codes)

RACE

What race do you consider yourself?

[Nominal]

1 White
2 Black
3 Other

RINCOM06

In which of these groups did your earnings from [occupation from a previous question] fall for last year? That is, before taxes or other deductions?

[Ordinal]

1 UNDER $1,000
2 $1,000 to $2,999
3 $3,000 to $3,999
4 $4,000 to $4,999
5 $5,000 to $5,999
6 $6,000 to $6,999
7 $7,000 to $7,999
8 $8,000 to $9,999
9 $10,000 to $12,499
10 $12,500 to $14,999
11 $15,000 to $17,499
12 $17,500 to $19,999
13 $20,000 to $22,499
14 $22,500 to $24,999
15 $25,000 to $29,999
16 $30,000 to $34,999
17 $35,000 to $39,999
18 $40,000 to $49,999
19 $50,000 to $59,999
20 $60,000 to $74,999
21 $75,000 to $89,999
22 $90,000 to $109,999
23 $110,000 to $129,999
24 $130,000 to $149,999
25 $150,000 or ...

21 Foreign language
22 Forestry
23 Geography
24 Geology
25 History
26 Home economics
27 Industry and technology
28 Journalism
29 Law
30 Law enforcement
31 Library science
32 Marketing
33 Mathematics
34 Medicine
35 Music
36 Nursing
37 Optometry
38 Pharmacy
39 Philosophy
40 Physical education
41 Physics
42 Psychology
43 Political science
44 Sociology
45 Special education
46 Theater arts
47 Theology
48 Veterinary medicine
49 Liberal arts
50 Other
51 General sciences
52 Social work
53 General studies
54 Other vocational
55 Health

OWNGUN

Do you happen to have in your home (or garage) any guns or revolvers?

[Nominal]

1 Yes

WRKSTAT

Last week were you working full-time, part-time, going to school, keeping house, or what?
[Nominal]
1 Working full-time
2 Working part-time
3 With a job, but not at work because of temporary illness, vacation, strike
4 Unemployed, laid off, looking for work
5 Retired
6 In school
7 Keeping house
8 Other

ZODIAC

Astrological sign of respondent
[Nominal]
1 Aries
2 Taurus
3 Gemini
4 Cancer
5 Leo
6 Virgo
7 Libra
8 Scorpio
9 Sagittarius
10 Capricorn
11 Aquarius
12 Pisces

WTSSnr

Weight of variables[3]

SEI

Hodge-Siegel-Rossi socioeconomic ratings for respondent's occupation[2]
[I/R]
Actual SEI #

SEX

Coded by the interviewers, based on observation
[Nominal]
1 Male
2 Female

REGION

Region of interview
[Nominal]
1 New England (ME, VT, NH, MA, CT, RI)
2 Middle Atlantic (NY, NJ, PA)
3 East North Central (WI, IL, IN, MI, OH)
4 West North Central (MN, IA, MO, ND, SD, NB, KS)
5 South Atlantic (DE, MD, WV, VA, NC, SC, GA, FL, DC)
6 East South Central (KY, TN, AL, MS)
7 West South Central (AR, OK, LA, TX)
8 Mountain (MT, ID, WY, NV, UT, CO, AR, NM)
9 Pacific (WA, OR, CA, AK, HI)

UNION

Do you (or your [SPOUSE]) belong to a labor union? (Who?)
[Nominal]
1 Yes, respondent belongs
2 Yes, spouse belongs
3 Yes, both belong
4 No, neither belong

---

[3]For information on how the variables were weighted, see *General Social Surveys 1972–2010 Cumulative Codebook*, Appendix A.

APPENDIX B: QUESTIONNAIRE FOR CLASS SURVEY

Questionnaire for Class Survey

ID _____ _____ _____ (CODE LEADING ZEROS)

CHLDIDEL 1. What do you think is the ideal number of children ____ ____ ____
   for a family to have?   1 2 3
   _____ (CODE LEADING ZERO
   IF < 10)

   2. If you had to choose, which thing on this list would you ____ ____
   say is the most important for a child to learn to prepare 4 5
   him or her for life? (CODE 1)

   [INTERVIEWER: READ CHOICES BELOW]

OBEY   _____ A. To obey   _____
                    6

POPULAR _____ B. To be well liked or popular   _____
                    7

THNKSELF _____ C. To think for himself or herself _____
                    8

WORKHARD _____ D. To work hard   _____
                    9

HELPOTH _____ E. To help others   _____
                   10

   a. Which comes next in importance? (Code 2)
   b. Which comes third? (Code 3)
   c. Which comes fourth? (Code 4)
   d. Which comes fifth? (Code 5) (no response: Code 9)

SIBS   3. How many brothers and sisters did you have? Please ____ ____
   count those born alive, but no longer living, as well as 11 12
   those alive now. Also include stepbrothers and
   stepsisters, and children adopted by your parents.
   _____ (CODE NUMBER WITH LEADING ZERO)

MARITAL 4. Are you married, widowed, divorced, separated, or have _____
   you never been married?   13

   _____ 1. Married
   _____ 2. Widowed
   _____ 3. Divorced
   _____ 4. Separated
   _____ 5. Never married
   _____ 9. No answer

[INTERVIEWER: ASK QUESTION 5 ONLY IF MARRIED OR WIDOWED]

DIVORCE 5. Have you ever been divorced or legally separated? _____
                   14

   _____ 1. Yes
   _____ 2. No
   _____ 3. No answer or not applicable

PARTYID   6. Generally speaking, do you usually think of yourself as a   _____
              Republican, Democrat, Independent, or what?   15

              _____ 0. Strong Democrat
              _____ 1. Not very strong Democrat
              _____ 2. Independent, close to Democrat
              _____ 3. Independent, neither, or no response
              _____ 4. Independent, close to Republican
              _____ 5. Not very strong Republican
              _____ 6. Strong Republican
              _____ 7. Other party, refused to say
              _____ 9. No answer

POLVIEWS   7. We hear a lot of talk these days about liberals and conservatives.   _____
              I'm going to read you a set of seven categories of political views   16
              that people might hold, arranged from extremely liberal to
              extremely conservative. Which best describes you, or haven't you
              thought much about this?

              _____ 1. Extremely liberal
              _____ 2. Liberal
              _____ 3. Slightly liberal
              _____ 4. Moderate, middle-of-the-road
              _____ 5. Slightly conservative
              _____ 6. Conservative
              _____ 7. Extremely conservative
              _____ 9. No answer or not applicable

CAPPUN   8. Do you favor or oppose the death penalty for murder?   _____
                                                                                   17

              _____ 1. Favor
              _____ 2. Oppose
              _____ 9. Don't know, no answer

GUNLAW   9. Would you favor or oppose a law that would require a   _____
              person to obtain a police permit before he or she could   18
              buy a gun?

              _____ 1. Favor
              _____ 2. Oppose
              _____ 9. Don't know, no answer

GETAHEAD   10. Some people say that people get ahead by their own hard work;   _____
               others say that lucky breaks or help from other people is more   19
               important. Which do you think is most important?

              _____ 1. Hard work most important
              _____ 2. Hard work, luck equally important
              _____ 3. Luck most important
              _____ 9. Other, don't know, no answer

RACDIF4   11. On the average, African Americans have worse jobs, income and   _____
               housing than White people. Do you think this is because African   20
               Americans just don't have the motivation or will to pull
               themselves up out of poverty?

              _____ 1. Yes
              _____ 2. No

RACMAR   12. Do you think there should be laws against marriages between   _____
               (Negroes/Blacks/African Americans) and Whites?   21

              _____ 1. Yes
              _____ 2. No

_____ 8. Don't know
_____ 9. No answer, not applicable

RACPUSH   13. Some people have expressed the opinion that (Negroes/Blacks/ _____
            African Americans) shouldn't push themselves where they're   22
            not wanted. Which of these responses comes closest to how you,
            yourself, feel?

            _____ 1. Agree strongly
            _____ 2. Agree slightly
            _____ 3. Disagree slightly
            _____ 4. Disagree strongly
            _____ 8. No opinion
            _____ 9. No answer, not applicable

            14. Please tell me whether or not you think it should be possible for
            a pregnant woman to obtain a legal abortion if . . .

ABDEFECT   A. There is a strong chance of serious defect in the baby   _____
                                                                                23

            _____ 1. Yes
            _____ 2. No
            _____ 3. Don't know, no answer

ABNOMORE   B. She is married and does not want any more children   _____
                                                                                24

            _____ 1. Yes
            _____ 2. No
            _____ 3. Don't know, no answer

ABHLTH   C. The woman's own health is seriously endangered by the   _____
               pregnancy   25

            _____ 1. Yes
            _____ 2. No
            _____ 3. Don't know, no answer

ABPOOR   D. The family has a very low income and cannot   _____
               afford any more children   26

            _____ 1. Yes
            _____ 2. No
            _____ 3. Don't know, no answer

ABRAPE   E. She became pregnant as a result of rape   _____
                                                                                27

            _____ 1. Yes
            _____ 2. No
            _____ 3. Don't know, no answer

ABSINGLE   F. She is not married and does not want to marry the man   _____
                                                                                28

            _____ 1. Yes
            _____ 2. No
            _____ 3. Don't know, no answer

ABANY   G. The woman wants it for any reason   _____
                                                                                29

            _____ 1. Yes
            _____ 2. No
            _____ 3. Don't know, no answer

RELIG   15. What is your religious preference? Is it Protestant, Catholic, _____
                Jewish, some other religion, or no religion?   30

_____ 1. Protestant
_____ 2. Catholic
_____ 3. Jewish
_____ 4. None
_____ 5. Other
_____ 9. No answer, no response

ATTEND   16. How often do you attend religious services?   _____

31

_____ 0. Never
_____ 1. Less than once a year
_____ 2. About once or twice a year
_____ 3. Several times a year
_____ 4. About once a month
_____ 5. Two or three times a month
_____ 6. Nearly every week
_____ 7. Every week
_____ 8. Several times a week
_____ 9. Don't know, no answer

POSTLIFE   17. Do you believe there is a life after death?   _____

32

_____ 1. Yes
_____ 2. No
_____ 9. Undecided, no answer, not applicable

PRAY   18. About how often do you pray?   _____

33

_____ 1. Several times a day
_____ 2. Once a day
_____ 3. Several times a week
_____ 4. Once a week
_____ 5. Less than once a week
_____ 6. Never
_____ 9. Don't know, no answer, not applicable

PREMARSX   19. There has been a lot of discussion about the way morals and   _____
attitudes about sex are changing in this country. If a man and   34
woman have sex relations before marriage, do you think it is
always wrong, almost always wrong, wrong only sometimes, or
not wrong at all?

_____ 1. Always wrong
_____ 2. Almost always wrong
_____ 3. Wrong only sometimes
_____ 4. Not wrong at all
_____ 9. Don't know, no answer

HOMOSEX   20. What about sexual relations between two adults of the same   _____
sex? Do you think it is always wrong, almost always wrong,   35
wrong only sometimes, or not wrong at all?

_____ 1. Always wrong
_____ 2. Almost always wrong
_____ 3. Wrong only sometimes
_____ 4. Not wrong at all
_____ 9. Don't know, no answer

XMOVIE   21. Have you seen an X-rated movie in the past year?   _____

36

_____ 1. Yes
_____ 2. No
_____ 3. Don't know, no answer

EDUC   22. What is the highest grade in elementary school or high   ___ ___
school that you finished and got credit for?   37   38

[INTERVIEWER: IF RESPONDENT COMPLETED 12 GRADES, ASK QUESTION 23.]

23. Did you complete one or more years of college for credit not
including schooling such as business college or technical or
vocational school?

_____ 1. Yes
_____ 2. No

[INTERVIEWER: IF YES TO QUESTION 23, GO TO QUESTION 24;
ELSE GO TO QUESTION 25.]

24. How many years did you complete?

CODING INSTRUCTION: ADD YEARS FROM QUESTION 22 TO YEARS FROM
QUESTION 24 AND CODE WITH LEADING ZERO IF NECESSARY IN
COLUMNS 37 AND 38. USE CODE 99 FOR NO RESPONSE.

SEI   25. What is your occupation?   ___ ___
_____ (write in)   39   40

[INTERVIEWER: IF RESPONDENT HAS NOT BEEN EMPLOYED FULL-TIME,
RECORD HIS OR HER FATHER'S OCCUPATION.]

CODING INSTRUCTION: LOOK UP OCCUPATION ON HODGE, SIEGEL, AND
ROSSI PRESTIGE SCALE AND RECORD SCORE IN COLUMNS 39 AND 40. USE
CODE 99 FOR NO RESPONSE.

CLASS   26. If you were asked to use one of four names for   ___
your social class, which would you say you   41
belong in: the lower class, the working class, the
middle class, or the upper class?

_____ 1. Lower class
_____ 2. Working class
_____ 3. Middle class
_____ 4. Upper class
_____ 9. Don't know, no answer

AGE   27. In what year were you born?   ___ ___
42   43

CODING INSTRUCTION: SUBTRACT BIRTH YEAR FROM THIS YEAR TO
GET AGE IN YEARS. CODE IN COLUMNS 42 AND 43. CODE 99 FOR NO
RESPONSE.

INCOME98   28. In which of these groups did your total family income, from   ___ ___
all sources, fall last year, before taxes, that is? Just tell me to   44   45
stop when I say the category that describes your family.

_____ 1. Under $1,000
_____ 2. $1,000 to $2,999
_____ 3. $3,000 to $3,999
_____ 4. $4,000 to $4,999
_____ 5. $5,000 to $5,999
_____ 6. $6,000 to $6,999
_____ 7. $7,000 to $7,999

```
_____  8.  $8,000 to $9,999
_____  9.  $10,000 to $12,499
_____ 10.  $12,500 to $14,999
_____ 11.  $15,000 to $17,499
_____ 12.  $17,500 to $19,999
_____ 13.  $20,000 to $22,499
_____ 14.  $22,500 to $24,999
_____ 15.  $25,000 to $29,999
_____ 16.  $30,000 to $34,999
_____ 17.  $35,000 to $39,999
_____ 18.  $40,000 to $49,999
_____ 19.  $50,000 to $59,999
_____ 20.  $60,000 to $74,999
_____ 21.  $75,000 to $89,999
_____ 22.  $90,000 to $109,999
_____ 23.  $110,000 or over
_____ 99.  Refused, don't know, not applicable
```

RACE   30. CODING INSTRUCTION: CODE WITHOUT ASKING ONLY IF _____
            THERE IS NO DOUBT IN YOUR MIND.   46

```
_____ 1.  White
_____ 2.  Black
_____ 3.  Other
```

SEX   31. CODING INSTRUCTION: CODE RESPONDENT'S SEX.   _____
                                                           47

```
_____ 1.  Male
_____ 2.  Female
```

Occupational Title and Socioeconomic Prestige Scores

Below are occupational titles and socioeconomic index scores for use in coding Question 25, SEI, on the above questionnaire. To find a respondent's SEI score, look through the occupational titles until you find one that matches or is similar to the respondent's occupation. Write the number that appears to the right of the occupation in the space provided for it in the right margin of Question 25 on the questionnaire.

The occupational prestige scores were developed by Hodge, Siege, and Rossi at the University of Chicago. You can learn more about how the prestige scores were obtained by reading Appendices F and G in *General Social Surveys, 1972–2010: Cumulative Codebook* (Davis, Smith, & Marsden, 2011).

Occupational Title	SEI Score
PROFESSIONAL, TECHNICAL, AND KINDRED WORKERS	
Accountants	57
Architects	71
Computer specialists	
Computer programmers	51

Occupational Title	SEI Score
Computer systems analysts	51
Computer specialists, n.e.c.	51
Engineers	
Aeronautical astronautical engineers	71
Chemical engineers	67

Occupational Title	SEI Score
Civil engineers	68
Electrical and electronic engineers	69
Industrial engineers	54
Mechanical engineers	62
Metallurgical and materials engineers	56
Mining engineers	62
Petroleum engineers	67
Sales engineers	51
Engineers, n.e.c.	67
Farm management advisers	54
Foresters and conservationists	54
Home management advisers	54
Lawyers and judges	
Judges	76
Lawyers	76
Librarians, archivists, and curators	
Librarians	55
Archivists and curators	66
Mathematical specialists	
Actuaries	55
Mathematicians	65
Statisticians	55
Life and physical scientists	
Agricultural scientists	56
Atmospheric and space scientists	68
Biological scientists	68
Chemists	69
Geologists	67
Marine scientists	68
Physicists and astronomers	74
Life and physical scientists, n.e.c.	68

Occupational Title	SEI Score
Operations and systems researchers and analysts	51
Personnel and labor relation workers	56
Physicians, dentists, and related practitioners	
Chiropractors	60
Dentists	74
Optometrists	62
Pharmacists	61
Physicians, including osteopaths	82
Podiatrists	37
Veterinarians	60
Health practitioners, n.e.c.	51
Nurses, dietitians, and therapists	
Dietitians	52
Registered nurses	62
Therapists	37
Health technologists and technicians	
Clinical laboratory technologists and technicians	61
Dental hygienists	61
Health record technologists and technicians	61
Radiologic technologists and technicians	61
Therapy assistants	37
Health technologists and technicians, n.e.c.	47
Religious workers	
Clergymen	69
Religious workers, n.e.c.	56
Social scientists	
Economists	57
Political scientists	66
Psychologists	71

Occupational Title	SEI Score
Sociologists	66
Urban and regional planners	66
Social scientists, n.e.c.	66
Social and recreation workers	
Social workers	52
Recreation workers	49
Teachers, college and university	
College agriculture teachers	78
College atmospheric, earth, marine, and space teachers	78
College biology teachers	78
College chemistry teachers	78
College physics teachers	78
College engineering teachers	78
College mathematics teachers	78
College health specialists teachers	78
College psychology teachers	78
College business and commerce teachers	78
College economics teachers	78
College history teachers	78
College sociology teachers	78
College social science teachers, n.e.c.	78
College art, drama, and music teachers	78
College coaches and physical education teachers	78
College education teachers	78
College English teachers	78
College foreign language teachers	78
College home economics teachers	78
College law teachers	78
College theology teachers	78

Occupational Title	SEI Score
College trade, industrial, and technical teachers	78
Miscellaneous teachers, college and university	78
Teachers, college and university, subject not specified	78
Teachers, except college and university	
Adult education teachers	43
Elementary school teachers	60
Pre-kindergarten and kindergarten teachers	60
Secondary school teachers	63
Teachers, except college and university, n.e.c.	43
Engineering and science technicians	
Agriculture and biological technicians, except health	47
Chemical technicians	47
Draftsmen	56
Electrical and electronic engineering	
Technicians	47
Industrial engineering technicians	47
Mechanical engineering technicians	47
Mathematical technicians	47
Surveyors	53
Engineering and science technicians, n.e.c.	47
Technicians, except health, engineering, and science	
Airplane pilots	70
Air traffic controllers	43
Embalmers	52
Flight engineers	47
Radio operators	43

Occupational Title	SEI Score	Occupational Title	SEI Score
Tool programmers, numerical control	47	Health administrators	61
Technicians, n.e.c.	47	Construction inspectors, public administration	41
Vocational and educational counselors	51	Inspectors, except construction, public administration	41
Writers, artists, and entertainers		Managers and superintendents, building	38
Actors	55	Office managers, n.e.c.	50
Athletes and kindred workers	51	Officers, pilots, and pursers; ship	60
Authors	60	Officials and administrators; public administration, n.e.c.	61
Dancers	38	Officials of lodges, societies, and unions	58
Designers	58		
Editors and reporters	51	Postmasters and mail superintendents	58
Musicians and composers	46		
Painters and sculptors	56	Purchasing agents and buyers, n.e.c	48
Photographers	41		
Public relations men and publicity writers	57	Railroad conductors	41
Radio and television announcers	51	Restaurant, cafeteria, and bar managers	39
Writers, artists, and entertainers, n.e.c.	51	Sales managers and department heads, retail	50
Research workers, not specified	51	Sales managers, except retail trade	50
Professional, technical, and kindred workers—allocated	51	School administrators, college	61
MANAGERS AND ADMINISTRATORS, EXCEPT FARM		School administrators, elementary and secondary	60
Assessors, controllers, and treasurers, local public administration	61	Managers and administrators, n.e.c.	50
Bank officers and financial managers	72	Managers and administrators, except farm—allocated	50
Buyers and shippers, farm products	41	*SALES WORKERS*	
Buyers, wholesale and retail trade	50	Advertising agents and salesmen	42
Credit men	49	Auctioneers	32
		Demonstrators	28
Funeral directors	52	Hucksters and peddlers	18

Occupational Title	SEI Score
Insurance agents, brokers, and underwriters	47
Newsboys	15
Real estate agents and brokers	44
Stocks and bonds salesmen	51
Salesmen and sales clerks, n.e.c.	34
Sales representatives, manufacturing industries	49
Sales representatives, wholesale trade	40
Sales clerks, retail trade	29
Salesmen, retail trade	29
Salesmen of services and construction	34
Sales workers—allocated	34
CLERICAL AND KINDRED WORKERS	
Bank tellers	50
Billings clerks	45
Bookkeepers	48
Cashiers	31
Clerical assistants, social welfare	36
Clerical supervisors, n.e.c	36
Collectors, bill and account	26
Counter clerks, except food	36
Dispatchers and starters, vehicle	34
Enumerators and interviewers	36
Estimators and investigators, n.e.c.	36
Expediters and production controllers	36
File clerks	30
Insurance adjusters, examiners, and investigators	48

Occupational Title	SEI Score
Library attendants and assistants	41
Mail carriers, post office	42
Mailhandlers, except post office	36
Messengers and office boys	19
Meter readers, utilities	36
Office machine operators	
Bookkeeping and billing machine operators	45
Calculating machine operator	45
Computer and peripheral equipment operators	45
Duplicating machine operators	45
Keypunch operators	45
Tabulating machine operators	45
Office machine operators, n.e.c.	45
Payroll and timekeeping clerks	41
Postal clerks	43
Proofreaders	36
Real estate appraisers	43
Receptionists	39
Secretaries	
Secretaries, legal	46
Secretaries, medical	46
Secretaries, n.e.c.	46
Shipping and receiving clerks	29
Statistical clerks	36
Stenographers	43
Stock clerks and storekeepers	23
Teacher aides, except school monitors	36
Telegraph messengers	30
Telegraph operators	44

Occupational Title	SEI Score
Telephone operators	40
Ticket, station, and express agents	35
Typists	41
Weighers	36
Miscellaneous clerical workers	36
Not specified clerical workers	36
Clerical and kindred workers—allocated	36
CRAFTSMEN AND KINDRED WORKERS	
Automobile accessories installers	47
Bakers	34
Blacksmiths	36
Boilermakers	31
Bookbinders	31
Brickmasons and stonemasons	36
Brickmasons and stonemasons, apprentice	36
Bulldozer operators	33
Cabinetmakers	39
Carpenters	40
Carpenter apprentices	40
Carpet installers	47
Cement and concrete finishers	32
Compositors and typesetters	38
Printing trades apprentices, except pressmen	40
Cranemen, derrickmen, and hoistmen	39
Decorators and window dressers	37
Dental laboratory technicians	47
Electricians	49
Electrician apprentices	41

Occupational Title	SEI Score
Electric power linemen and cablemen	39
Electrotypers and stereotypers	38
Engravers, except photoengravers	41
Excavating, grading and road machine operators, except bulldozer	33
Floor layers, except tile setters	40
Foremen, n.e.c.	45
Forgemen and hammermen	36
Furniture and wood finishers	29
Furriers	35
Glaziers	26
Heat treaters, annealers, and temperers	36
Inspectors, scalers, and graders: log and lumber	31
Inspectors, n.e.c.	31
Jewelers and watchmakers	37
Job and die setters, metal	48
Locomotive engineers	51
Locomotive firemen	36
Machinists	48
Machinist apprentices	41
Mechanics and repairmen	
Air conditioning, heating, and refrigeration	37
Aircraft	48
Automobile body repairmen	37
Automobile mechanics	37
Automobile mechanic apprentices	37
Data processing machine repairmen	34
Farm implements	33

Occupational Title	SEI Score	Occupational Title	SEI Score
Heavy equipment mechanics, including diesel	33	Pressmen and plate printers, printing	40
Household appliance and accessory installers and mechanics	33	Pressmen apprentices	40
Loom fixers	30	Rollers and finishers, metal	36
Office machines	34	Roofers and slaters	31
Radio and television	35	Sheetmetal workers and tinsmiths	37
Railroad and car shop	37	Sheetmetal apprentices	37
Mechanic, except auto, apprentices	41	Shipfitters	36
Miscellaneous mechanics and repairmen	35	Shoe repairmen	33
Millers; grain, flour, and feed	25	Sign painters and letterers	30
Millwrights	40	Stationary engineers	35
Molders, metal	39	Stone cutters and stone carvers	33
Molder, apprentices	39	Structural metal craftsmen	36
Motion picture projectionists	34	Tailors	41
Opticians, and lens grinders and polishers	51	Telephone installers and repairmen	39
Painters, construction and maintenance	30	Telephone linemen and splicers	39
Painter apprentices	30	Tile setters	36
Paperhangers	24	Tool and die makers	42
Pattern and model makers, except paper	39	Tool and die maker apprentices	41
Photoengravers and lithographers	40	Upholsterers	30
Piano and organ tuners and repairmen	32	Specified craft apprentices, n.e.c.	41
Plasterers	33	Not specified apprentices	41
Plasterer apprentices	33	Craftsmen and kindred workers, n.e.c.	47
Plumber and pipe fitters	41	Former members of the Armed Forces	47
Plumber and pipe fitter apprentices	41	Craftsmen and kindred workers—allocate	47
Power station operators	39	Current members of the Armed Forces	47
		OPERATIVES, EXCEPT TRANSPORT	
		Asbestos and insulation workers	28

Occupational Title	SEI Score
Assemblers	27
Blasters and powdermen	32
Bottling and canning operatives	23
Chainmen, rod men, and axmen; surveying	39
Checkers, examiners, and inspectors; manufacturing	36
Clothing ironers and pressers	18
Cutting operatives, n.e.c.	26
Dressmakers and seamstresses, except factory	32
Drillers, earth	27
Dry wall installers and lathers	27
Dyers	25
Filers, polishers, sanders, and buffers	19
Furnacemen, smeltermen, and pourers	33
Garage workers and gas station attendants	22
Graders and sorters, manufacturing	33
Produce graders and packers, except factory and farm	19
Heaters, metal	33
Laundry and dry cleaning operatives, n.e.c.	18
Meat cutters and butchers, except manufacturing	32
Meat cutters and butchers, manufacturing	28
Meat wrappers, retail trade	19
Metal platers	29
Milliners	33
Mine operatives, n.e.c.	26
Mixing operatives	29

Occupational Title	SEI Score
Oilers and greasers, except auto	24
Packers and wrappers, n.e.c	19
Painters, manufactured articles	29
Photographic process workers	36
Precision machine operatives	
Drill press operatives	29
Grinding machine operatives	29
Lathe and milling machine operatives	29
Precision machine operatives, n.e.c	29
Punch and stamping press operatives	29
Riveters and fasteners	29
Sailors and deckhands	34
Sawyers	28
Sewers and stitchers	25
Shoemaking machine operatives	32
Solderers	29
Stationary firemen	33
Textile operatives	
Carding, lapping, and combing operatives	29
Knitters, loopers, and toppers	29
Spinners, twisters, and winders	25
Weavers	25
Textile operatives, n.e.c.	29
Welders and flame-cutters	40
Winding operatives, n.e.c.	29
Machine operatives, miscellaneous specified	32
Machine operatives, not specified	32

Occupational Title	SEI Score
Miscellaneous operatives	32
Not specified operatives	32
Operatives, except transport—allocated	32
TRANSPORT EQUIPMENT OPERATIVES	
Boatmen and canalmen	37
Bus drivers	32
Conductors and motormen, urban rail transit	28
Deliverymen and routemen	28
Fork lift and tow motor operatives	29
Motormen; mine, factory, logging camp, etc.	27
Parking attendants	22
Railroad brakemen	35
Railroad switchmen	33
Taxicab drivers and chauffeurs	22
Truck drivers	32
Transport equipment operatives—allocated	29
LABORERS, EXCEPT FARM	
Animal caretakers, except farm	29
Carpenters' helpers	23
Construction laborers, except carpenters' helpers	17
Fishermen and oystermen	30
Freight and material handlers	17
Garbage collectors	17
Gardeners and groundkeepers, except farm	23
Longshoremen and stevedores	24
Lumbermen, raftsmen, and woodchoppers	26
Stockhandlers	17
Teamsters	12

Occupational Title	SEI Score
Vehicle washers and equipment cleaners	17
Warehousemen, n.e.c.	20
Miscellaneous laborers	17
Not specified laborers	17
Laborers, except farm—allocated	17
FARMERS AND FARM MANAGERS	
Farmers (owners and tenants)	41
Farm managers	44
Farmers and farm managers—allocated	41
FARMERS LABORERS AND FARM FOREMEN	
Farm foremen	35
Farm laborers, wage workers	18
Farm laborers, unpaid family workers	18
Farm laborers, unpaid family workers	18
Farm service laborers, self-employed	27
Farm laborers, farm foremen, and kindred workers—allocated	19
SERVICE WORKERS, EXCEPT PRIVATE HOUSEHOLD	
Cleaning service workers	
Chambermaids and maids, except private household	14
Cleaners and charwomen	12
Janitors and sextons	16
Food service workers	
Bartenders	20
Busboys	22
Cooks, except private household	26

Occupational Title	SEI Score
Dishwashers	22
Food counters and fountain workers	15
Waiters	20
Food service workers, n.e.c, except private household	22
Health service workers	
Dental assistants	48
Health aides, except nursing	48
Health trainees	36
Midwives	23
Nursing aides, orderlies, and attendants	36
Practical nurses	42
Personal service workers	
Airline stewardesses	36
Attendants, recreation and amusement	15
Attendants, personal service, n.e.c.	14

Occupational Title	SEI Score
Baggage porters and bellhops	14
Barbers	38
Boarding and lodging housekeepers	22
Bootblacks	9
Child-care workers, except private households	25
Elevator operators	21
Hairdressers and cosmetologists	33
Personal service apprentices	14
Housekeepers, except private households	36
School monitors	22
Ushers, recreation and amusement	15
Welfare service aides	14
Protective service workers	
Crossing guards and bridge tenders	24
Firemen, fire protection	44

Note: n.e.c. = not elsewhere classified.

INDEX/GLOSSARY

statistics used by social scientists. Statistics are widely used to describe and summarize.

Determination, coefficient of, 252

Deterministic relationship, 262

Deterministic relationship A (bivariate) relationship between two variables where one variable can be perfectly predicted from the other variable with no errors

Dialog box, 37–38

Dialog box In SPSS Statistics, most menu selections open dialog boxes, such as the Frequencies dialog box or the Help Topics dialog box. These boxes are then used to select variables and other options for analysis.

Dichotomies Variables with two mutually exclusive and exhaustive categories or attributes, such as SEX (i.e., "male" and "female").

Dichotomous variables, 226

Dichotomous variables A variable that consists of exactly two mutually exclusive and exhaustive categories or attributes

Dimensions of political orientation, 10

Dimensions Different aspects of a concept. For example, the concept of religiosity might involve such dimensions as ritual participation (e.g., church attendance), private devotionalism (e.g., prayer), belief, ethical behavior, and so forth.

Direction of association, 168, 227–228
  Pearson's r and, 247

Direction of association If an association exists between variables, it may be positive or negative. See negative association and positive association.

Discrete variables, 21

Discrete variables Variables whose values are completely separate from one another, such as race or sex. These are variables with a limited number of distinct (i.e., discrete) values or categories that cannot be reduced or subdivided into smaller units or numbers. Discrete variables are distinguished from continuous variables.

Dispersion and central tendency, 65–66

Distributions, frequency, 56–61

DK (don't know), 28, 63

DK (don't know) A value label used in the General Social Survey and other surveys to distinguish, for example, when a respondent did not know the answer to a particular question

Doing Survey Research, 378

Doubling time, 145

Doubling time Term used by demographers to refer to the number of years it takes for the world's population to double in size.

Dummy variables, 320
  recoding SEX to create MALE, 320–321

Dummy variables Placeholder or indicator variables. Because categorical variables do not contain numeric information, they should not be used in the regression model. However, each value of a categorical variable can be represented in the regression model by an indicator or dummy variable that contains only the values 0 and 1.

EDCAT, 194
  PARTY by, 193, 332
  POLREC by, 192–193
  POLREC by MARITAL2 by, 335
  recoding EDUC to, 190–192

Edge-coded questionnaires, 391

Edge-coded A type of coding whereby one space is designated in the far right-hand margin of the questionnaire for each variable's code

Editing
  accessing file information for, 391–392
  questionnaires, 391

EDUC, 190–192
  correlation coefficients, 258–260
  multiple regression, 323–325
  PARTYID and, 332
  scatterplots, 268–269
  t tests, 286–287

Education and politics, 190

Elaboration, 316

Elaboration A multivariate technique used to examine the relationship between the dependent and independent variables while controlling for a third (control) variable

Ending SPSS Statistics sessions, 46–47

Epsilon (ε), 175

Epsilon (ε) A measure of association appropriate to percentage tables. It is the difference in percentages (calculated from the distribution of attributes on the dependent variable) between the extreme categories of the independent variable in a table. If 30% of the men agree to some statement and 50% of the women agree, epsilon is 20 percentage points.

Error, sampling, 276

Exclusive categories, 8

Exclusive The requirement that every observation fit into only one category

Exhaustive categories, 8

Exhaustive The requirement that categories of each variable be comprehensive enough that it is possible to categorize every observation

Expected cell frequencies, 278

Expected cell frequencies The values in the cells of a bivariate table that would exist if there were no relationship between the variables

Floor effect, 206

Floor effect A term used to refer to situations in which, for example, the overall percentage of respondents agreeing to something approaches zero

SPSS Statistics Student Version A scaled-down version of SPSS Statistics that, unlike the standard/professional version, is limited to 50 variables and 1,500 cases
SPSS (or PASW) Statistics Viewer A window in SPSS Statistics that contains the results of procedures or output. The SPSS Statistics Viewer is divided into two panes: Outline (the left pane) and Contents (the right pane).
Standard deviation A measure of dispersion appropriate to ratio variables; indicates the extent to which the values are clustered around the mean or spread away from it
Statistical significance A measure of the likelihood that an observed relationship between variables in a probability sample represents something that exists in the population rather than being due to sampling error
Strength of association Statistical measures of association such as lambda, gamma, and Pearson's r indicate the strength of the relationship or association between two variables. For example, values of lambda can vary from 0 to 1. The closer the value of lambda to 1, the stronger the relationship between the two variables, while the closer the value of lambda to 0, the weaker the association between the two variables.
Substantive significance A term used to refer to the importance of an association between variables that cannot be determined by empirical analysis alone but depends, instead, on practical and theoretical considerations
SYSTAT A statistical analysis software program

10-percentage-point rule A rule of thumb stating that a percentage-point difference of 10 or more signifies a meaningful relationship between variables, warranting further examination
Theory An interrelated set of general principles that describes the relationships among a number of variables
Title bar The top bar in the computer window. It includes the name of the software program (e.g., IBM SPSS Statistics Data Editor), as well as the name of the data file (e.g., "Untitled").
Toggling The process of switching back and forth between numeric values and labels in the SPSS Statistics Data Editor
Toolbar A row of icons that allows easy activation of program features with mouse clicks. Each SPSS Statistics window has one or more toolbars for easy access to common tasks.
Total (N) Complete sample size, number of respondents or cases, often referenced in a bivariate table

t test A class of significance tests used to test for significant differences between different kinds of means. In this book, the independent samples t test is used to test whether the observed difference between two means could have easily happened by chance, or if it is so unlikely that we believe the difference we see in our sample also exists in the population from which the sample was taken.

Two-way frequency distribution. See crosstabulation.

Univariate analysis The analysis of a single variable for the purpose of description; contrasted with bivariate and multivariate analysis

Valid percentage In SPSS Statistics, usually the third column in a frequency distribution. This column contains the relative percentage of times each value of a variable occurred in a given sample, excluding any missing data.

Validity The extent to which an empirical measure actually taps the quality intended

Value labels The descriptive value labels assigned to each value of a variable. For example, the value labels for the variable CAPPUN are "favor" and "oppose," while the values are "1" and "2," respectively.

Variable View This view in the SPSS Data Editor window provides a listing of the variables in a data file and characteristics about them (e.g., label, value label, missing data, level of measurement). On Windows computers, Variable View can be selected by the toggle in the lower left side of the Data Editor window. On Macintosh computers, Variable View can be selected using the toggle switch in the lower middle of the Data Editor window.

Variables Logical sets of attributes. For example, gender is a variable comprising the attributes male and female.

Variance A statistic that measures the variability of a distribution and is determined by dividing the sum of the squared deviations by 1 less than the number of cases (N—1); often close to the mean of the sum of squared deviations

Vertical (y) axis In a graph, such as a line or bar chart, the axis that runs up and down on the far left side and usually represents the dependent variable

z scores Useful for comparing variables that come from distributions with different means and standard deviations